U1069529

CForBio “中国森林生物多样性监测网络”丛书 马克平 主编

湖南八大公山森林动态样地
——树种及其分布格局

Hunan Badagongshan Forest Dynamics Plot: Tree Species and Their Distribution Patterns

卢志军 鲍大川 刘海波 徐耀粘 黄汉东 江明喜 谷志容 著

中国林业出版社
China Forestry Publishing House

图书在版编目（CIP）数据

湖南八大公山森林动态样地——树种及其分布格局 / 卢志军等著. —北京：中国林业出版社，2016.10
(中国森林生物多样性监测网络丛书)

ISBN 978-7-5038-8765-9

I. ①湖… II. ①卢… III. ①自然保护区－森林植物－介绍－湖南 IV. ①S759.992.64

中国版本图书馆CIP数据核字(2016)第262298号

内容简介

本书以我国武陵山区湖南八大公山25 hm^2森林动态样地胸径大于1 cm的232种木本植物为对象，对每个树种的主要分类特征进行简要描述，以彩色照片展示树干、小枝、叶、花、果等；辅以树种在25 hm^2样地内的空间分布图以及个体数量和径级结构。同时，对八大公山样地的地形、土壤、植被等进行简单介绍。裸子植物采用郑万钧（1978）分类系统，被子植物采用恩格勒（1964）分类系统。本书可作为山地常绿落叶阔叶混交林研究的参考资料；同时，亦可作为植物爱好者认识华中地区森林植物，了解中国常绿落叶阔叶混交林的野外手册。

中国林业出版社·科技出版分社

策划、责任编辑：于界芬
电话：010-83143542

出　版：中国林业出版社（100009　北京西城区德内大街刘海胡同7号）
网　址：http://lycb.forestry.gov.cn/
发　行：中国林业出版社
印　刷：固安县京平诚乾印刷有限公司
版　次：2017年10月第1版
印　次：2017年10月第1次
开　本：889mm × 1194mm　1 / 16
印　张：16
字　数：403千字
定　价：168.00元

序 言 1

在过去的几十年时间里，中国科学院和林业、农业等相关部门陆续建立了数百个生态系统定位研究站。其中，中国科学院组建的中国生态系统研究网络 (CERN) 拥有分布于全国包括农田、森林、草地、湿地、荒漠等生态系统类型的36个生态站。国家林业局建立的中国森林生态系统研究网络 (CFERN) 由29个生态站组成，基本覆盖了我国典型的地带性森林生态系统类型和最主要的次生林、人工林类型。

随着研究的发展，特别是近年来人们对生物多样性和全球变化研究的关注，国际上正在推动生态系统综合研究网络平台的建立。在全球水平上，全球生物多样性综合观测网络 (GEO BON) 是一个有代表性的研究网络。它试图把全球与生态系统和生物多样性长期定位研究相关的网络整合起来，通过综合研究，探讨生态系统与生物多样性维持与变化机制以及系统之间的相互作用机理，为生态系统可持续管理与生物多样性的保护提供科学依据和管理模式。

近年来，中国科学院生物多样性委员会组织建立了中国森林生物多样性监测网络（Chinese Forest Biodiversity Monitoring Network，以下简称CForBio）。中国是生物多样性特别丰富的少数国家之一，也是唯一一个具有从北部寒温带到南部热带完整气候带谱的国家。截止到2016年，中国森林生物多样性监测网络包括大型监测样地18个，成为继美国史密森研究院热带研究所建立的热带森林生物多样性监测网络（CTFS-Forest GEO）之后又一大型区域监测网络。由于CForBio横跨多个纬度梯度，对于揭示中国森林生物多样性形成和维持机制，以及森林生物多样性对全球变化的响应，科学利用和有效保护中国森林生物多样性资源具有重要意义。

目前，CForBio已经有很好的研究进展，各样地研究成果陆续在国际著名生态学刊物如*Ecology, Journal of Ecology, Oikos*等上发表，受到国内外同行的高度评价。但这些文章都是关于某一具体问题的研究总结，还无法让国内外同行全面了解CForBio各个样地整体情况。因此，出版这套以中英文形式介绍各大样地基本情况的“中国森林生物多样性监测网络”丛书是非常必要的。感谢马克平研究员组织相关专家编写这套丛书。我相信该丛书不仅是国内外同行深入了解CForBio各样地的参考书，同时也将为我国森林生物多样性监测和森林生态系统联网研究奠定重要的基础。

（孙鸿烈）

中国科学院前副院长

Foreword 1

In the past few decades, hundreds of Ecosystem Research Stations have been set up by the Chinese Academy of Sciences, State Forestry Administration, Ministry of Agriculture and other relative departments. Among them, 36 ecological research stations were established by Chinese Ecosystem Research Network (CERN), supported by the Chinese Academy of Sciences. The 36 research stations are scattered over the country representing diverse ecosystems, including farmland, forest, grassland, wetland, desert and others. Moreover, the Chinese National Ecological Research Network (CFERN), supported by the State Forestry Administration, consists of 29 research stations, covering typical zonal forest ecosystems and main secondary forests and plantations in China.

With the development of research, especially the growing concern over researches on biodiversity and global change in recent years, the establishment of ecosystem research network have been promoted under international supports. So the Group on Earth Observations Biodiversity Observation Network (GEO-BON) is representative across the world, and it attempts to integrate worldwide networks relating to long-term research on ecosystem and biodiversity. Based on the comprehensive studies, the maintenance and change mechanism of ecosystem and biodiversity and their interactions have been explored, which provide scientific basis and management mode for sustainable development of ecosystem and protection of biodiversity.

In recent years, Biodiversity Committee of the Chinese Academy of Sciences has organized and established Chinese Forest Biodiversity Monitoring Network (CForBio). Chinese is among the few countries with particularly rich biodiversity, and the only country with the whole climate spectrum from north cold temperate zone to south tropical. Till 2016, CForBio consists of 18 large scale monitoring plots, and it is the second large scale regional monitoring network after ForestGEO organized by Center for Tropical Forest Science (CTFS), the Smithsonian

Tropical Research Institute. Due to its covering several latitude gradient, CForBio is significant for revealing formation and maintenance mechanisms of Chinese forest biodiversity, responses of biodiversity to global change, scientific utilization and effective protection of Chinese forest biodiversity resources.

Encouraging progress has been made in this area since the network built, for lots of research findings have been published in the international peer reviewed ecological journals, such as *Ecology, Journal of Ecology* and *Oikos,* etc., which brought about positive response from colleagues in the field of plant ecology. However, the published papers mostly focus on research of specific problems; scientists and public still can't understand the whole situation of each plot in details. So it is really necessary to publish this series, which introduce basic information of permanent forest plots in both Chinese and English. I am grateful to Professor Keping Ma for organizing related specialists to prepare the series. And I believe that this series would be a valuable reference book for scientists and public to further understand CForBio, and it will also lay a foundation for the forest biodiversity monitoring and forest ecosystem research in China.

Honglie Sun

The former Vice-President for the Chinese Academy of Sciences

序 言 2

森林在维持世界气候与水文循环中起着根本性的作用。森林是极为丰富多样的动物、植物与微生物的家园，而人类正是依靠这些生物获取各种产品，包括食品与药物。尽管对人类福祉如此重要，森林仍然遭受着来自土地利用与全球气候环境变化的巨大威胁。在这种不断变化的情况下，为了更好地管理全球剩余的森林，迫切需要树种在生长、死亡与更新方面的详细信息。

中国森林生物多样性监测网络 (CForBio) 正在中国沿着纬度与环境梯度建立大尺度森林监测样地。通过这个重要的全国行动倡议与来自中国科学院及若干其他单位的研究者的努力，CForBio开始搜集关于中国森林的结构与动态的关键信息。现在CForBio与史密森研究院及哈佛大学阿诺德树木园的热带森林监测网络 (CTFS) 形成了合作伙伴。CTFS是个在24个热带或温带国家拥有长期大尺度森林动态研究样地的全球性网络。CForBio与CTFS合作的目标是通过合作研究，了解森林是如何运作的，它们是如何随着时间而改变的，以及如何重建或者恢复，以确保森林提供的环境服务能可持续或者增长。森林及其提供的服务的长期可持续性有赖于我们预测森林对全球变化，包括气候与土地利用变化的响应的能力，以及我们去理解与创建适当的森林服务市场的能力。通过拥有63个森林大样地的全球网络及大量项目的训练与能力建设，CForBio与CTFS的伙伴关系是发展这些预测工具的重要基础。这种伙伴关系也将促进为全球各地的当地社区、林业管理者与政策制定者在森林的保育与管理方面发展应用性的林业项目建议，发展与示范利用乡土物种进行森林重建的方法，以及从经济学角度评估森林在减缓气候变化、生物多样性保护和流域保护上的价值的方法。

我祝贺作者们创作了这部关于样地植物的优秀丛书。本丛书为将来的森林监测提供了基准信息，是涉及森林恢复、碳存储、动植物关系、遗传多样性、气候变化、局地与区域保育等研究内容的研究者、学生与森林管理者们有价值的参考资料。

S. J. 戴维斯
主任
史密森热带研究所全球森林监测网络

Foreword 2

Forests play an essential role in regulating of world's climatic and hydrological cycles. They are home to a vast array of animal, plant and microorganism species on which humans depend for many products, including food and medicines. Despite the importance of forests to human welfare they are under enormous threat from changes in land-use and global climatic conditions. In order to better manage the world's remaining forests under these changing conditions detailed information on the dynamics of growth, mortality and recruitment of tree species is urgently needed.

The Chinese Forest Biodiversity Monitoring Network (CForBio) that aims to establish large-scale forest monitoring plots across latitudinal and environmental gradients in China. Through this important national initiative, researchers from the Chinese Academy of Sciences and several other research institutions in China, CForBio has begun to gather key information on the structure and dynamics of China's forests. The CForBio initiative is now partnering with the Center for Tropical Forest Science (CTFS) of the Smithsonian Research Institute and the Arnold Arboretum of Harvard University. CTFS is a global program of long-term large-scale forest dynamics plots in 24 tropical and temperate countries/areas. The goal of the partnership between CForBio and CTFS is to work together to understand how forests work, how they are changing over time, and how they can be re-created or restored to ensure that the environmental services provided by forests are sustained or increased. The long-term sustainability of forests and the services they provide depend on our ability to predict forest responses to global changes, including changes in climate and land-use, and our ability to understand and create appropriate markets for forest services. The CForBio-CTFS partnership is ideally poised to develop these predictive tools through a global network of 63 large forest plots and an extensive program of training and capacity building. The partnership will also lead to the development of applied forestry programs that advise local communities, forest managers and policy makers around the world on conservation and management of forests, to develop and demonstrate methods of native species reforestation, and to economically value the roles that forests play in climate mitigation, biodiversity conservation, and watershed protection.

I congratulate the authors on the production of this excellent new series of stand books. In addition to providing a baseline for future forest monitoring, these books provide a valuable resource for researchers, students, and forest managers dealing with issues of forest restoration, carbon storage, plant-animal interactions, genetic diversity, climate change, and local and regional conservation issues.

Stuart Davies

Director

Center for Tropical Forest Science / Forest GEO

The Smithsonian Tropical Research Insitute

前　言

山地常绿落叶阔叶混交林在我国主要分布在中亚热带1000～2000 m的中山地带。该类森林常绿与落叶树种数量大致相当，或常绿树种占据优势。生境特征为温度低、湿度大、雨量多、日照少，往往常年云雾笼罩 (吴征镒，1980)。我国的科研工作者针对该类型植被开展了大量的研究工作，积累了相当丰富的研究资料。

八大公山样地是中国森林生物多样性监测网络在华中地区的重要节点，位于中国具有国际意义的生物多样性关键地区之一的川鄂湘黔边界山地地区，样地设置在八大公山国家级自然保护区。该保护区位于中国种子植物特有属川东－鄂西分布中心 (孑遗中心) 和中国特有植物的环形地带 (川东、鄂西南、湘西北和黔东北)，是水青冈属 (*Fagus*) 的分布中心和可能的起源中心。八大公山的亮叶水青冈 (*Fagus lucida*) 林及其植物区系，是华中植物区最有代表性的典型森林群落和区系，与欧洲、美洲，以至整个北温带的水青冈林及其区系有着共同的历史渊源。在常绿落叶阔叶混交林中保存着许多古老孑遗类群，具有重要的学术价值和保存意义，为亚热带山地森林生物多样性长期动态监测提供了理想的对象。在中国科学院生物多样性委员会的资助下，中国科学院武汉植物园(以下简称武汉植物园)与湖南八大公山国家级自然保护区管理处等单位合作，按CTFS建设规范于2010～2011年在保护区内以亮叶水青冈和多脉青冈 (*Cyclobalanopsis multinervis*) 为优势物种的常绿落叶阔叶混交林内，建设了一个25 hm^2的永久固定样地。

本书由中国科学院生物多样性委员会组织并资助出版，由武汉植物园科研人员撰写，详细描述了八大公山样地232种木本植物的物种特征、分布格局和径级结构，以彩色照片对物种的主要识别特征加以呈现。

华中地区植物资源收集和生物多样性保护是武汉植物园的核心任务之一，但在其成立近60年来，野外生物多样性的定位监测研究工作一直是空白。在川鄂湘黔边界山地开展长期生物多样性定位研究在武汉植物园的历史上是第一次。八大公山位于湘鄂两省交界山区，地处偏远，道路曲折，多雨泥泞，大型监测样地的建设工作非常艰难，但在各方面的大力支持和协助下，在武汉植物园植被生态学学科组全体师生的共同努力和坚持下，样地得以建设成功，常规监测与研究工作得以正常开展。借此机会，诚挚感谢下列为八大公山样地建设做出贡献的单位和个人：

样地建设在中国科学院植物研究所(以下简称植物研究所)马克平研究员的策划下得以启动，样地的选点工作得到了植物研究所米湘成博士的大力帮助，在建设过程中得到了植物研究所任海保博士的技术指导。

参加样地建设工作除了本书作者外，还有武汉植物园植被生态学学科组的乔秀娟、魏新增、孟红杰，以及在样地建设中出大力、流大汗，已经离开学科组的何东、郭屹立、王庆刚、陈虎、路俊盟、张奎汉、廖建雄、付为弟，还包括中南林业科技大学的喻勋林教授、中国科学院上海辰山植物园的严岳鸿研究员、湖北科技学院的蔡朝晖副教授、张规富副教授、刘卫东副教授、黄俊潮博士；湖南科技大学、中南林业科技大学、湖北科技学院、三峡职业技术学院130余名实习学生；武汉植物园张守君、温丁朝、申健勇和刘司浪。

本书照片主要由作者卢志军 (420张)、刘海波 (109张)、江明喜 (57张)、鲍大川 (38张)、徐耀粘 (23张) 和谷志容 (4张) 拍摄；部分由徐文秀 (5张)、路俊盟 (3张)、张奎汉 (1张) 拍摄；此外，陈彬博士 (19张)、徐文斌 (15张)、喻勋林教授 (2张)、覃必武 (2张) 和段来军 (2张) 提供了部分物种照片。树种名称和分类特征描述主要参考《中国植物志》、*Flora of China* 和《湖北植物志》，植物英文名称主要参考《拉汉英种子植物名称》。

书稿完成后承蒙马克平研究员和徐学红女士多次审校，中南林业科技大学喻勋林教授和吉首大学张代贵教授精心审阅，于界芬女士细心编辑。由于时间仓促，水平所限，错误疏漏之处，敬请读者不吝赐教。

江明喜

2017年1月 武汉

Preface

Mountain evergreen deciduous broad-leaved mixed forest is mainly distributed in mid-subtropical medium mountainous regions with elevation ranging from 1000 to 2000 m. Evergreen species richness is usually equivalent to that of deciduous, or evergreen is dominant. The habitats are characterized by low temperature, heavy humidity and rainfall, little sunlight, frequently with year round cloud and mist hanging over (Wu, 1980). There have been many Chinese researches revolving around this type forest and abundant data have been obtained.

Badagongshan plot is an important node of Chinese Forest Biodiversity Monitoring Network (CForBio) in central China, located in one of the global biodiversity hotspots—the Wuling Mountains region. The plot's locality is in Badagongshan National Nature Reserve which is located in Eastern Sichuan-Western Hubei seed plant endemic genera center and ring-shaped distributian zone of endemic species (Western Sichuan, Southwestern Hubei, Northwestern Hunan and Northeastern Guizhou), distribution center and possible origin center of *Fagus*. The *Fagus lucida* forests and their flora are representative forest community types and flora for central China floristic region, sharing possible common historical origins with *Fagus* forests and their flora in Europe, America, even the whole North Temperate Zone. Tertiary flora conserved here could be more intact with more living fossils and ancient relic groups, with significant academic and conservation values, providing an ideal platform for long-term monitoring on mid-subtropical mountain forest biodiversity and research on its maintenance. In 2010-2011, financially supported by Biodiversity Committee, Chinese Academy of Sciences (CAS), Wuhan Botanical Garden (CAS), in collaboration with Badagongshan Nature Reserve and other two institutions, established one 25 hm^2 permanent monitoring plot in a *F. lucida* and *Cyclobalanopsis. multinervis* dominating forest. The protocols followed CTFS (Center for Tropical Forest Science) (Condit, 1998).

This book is organized and financially supported by Biodiversity Committee (CAS), written by a research team of Wuhan Botanical Garden. The species characteristics, spatial distribution pattern and diameter class of the Badagongshan Plot were described here. Colorful pictures demonstrated main taxonomic characteristics of 232 woody plants in the 25 hm^2 plot.

Plant resource collection and biodiversity conservation are the core tasks of Wuhan Botanical Garden. However, there have been no solid progresses on *in-situ* monitoring of wild biodiversity since its foundation. It is also the first case for *in-situ* research on biodiveristy in the mountainous regions of Sichuan-Hubei-Hunan-Guizhou. Badagongshan is located in the boundary of Hunan and Hubei, and remoteness, winding roads and heavy rainfall made it a tough task to establish the Badagongshan Plot. Fortunately, we obtained a lot of assistance from many people and institutions to conduct monitoring and research who we want to thank here.

The establishment of Badagongshan Plot was initiated by Prof. Keping Ma; Dr. Xiangcheng Mi helped us select the plot locality; Dr. Haibao Ren provided technical supports for woody plant survey.

Besides the authors of this book, many persons contributed to the establishment of the Badagongshan Plot including: Xiujuan Qiao, Xinzeng Wei, Hongjie Meng; former members of our group who contributed a lot but left including Dong He, Yili Guo, Qinggang Wang, Hu Chen, Junmeng Lu, Kuihan Zhang, Jianxiong Liao, Weidi Fu; Prof. Xunlin Yu (Central South University of Forestry and Technology); Prof. Yuehong Yan (Shanghai Chenshan Botanical Garden, CAS; Associate Professors Chaohui Cai, Guifu Zhang and Weidong Liu, Dr. Junchao Huang (Hubei University of Science and Technology); 130 interns (Hunan University of Science and Technology, Central South University of Forestry and Technology, Hubei University of Science and Technology, HuBei Three Gorges Polytechnic); Shoujun Zhang, Dingchao Wen, Jianyong Shen and Silang Liu (Wuhan Botanical Garden, CAS).

The photos in this book are mainly taken by Zhijun Lu (420 photos), Haibo Liu (109 photos), Mingxi Jiang (57 photos), Dachuan Bao (38 photos), Yaozhan Xu (23 photos) and Zhirong Gu (4 photos); some by Wenxiu Xu (5 photos), Junmeng Lu (3 photos) and Kuihan Zhang (one photo); photos of several species are kindly contributed by Dr. Bin Chen (19 photos), Wenbin Xu (15 photos), Prof. Xunlin Yu (2 photos), Biwu Qin (2 photos) and Laijun Duan (2 photos). Taxonomic names and characteristics descriptions mainly followed *Flora of China* (Chinese, English versions) and *Flora Hubeiensis*; English names of plants followed *Dictionary of Seed-plants Names Latin-Chinese-English*.

We are grateful to Prof. Keping Ma and Ms. Xuehong Xu, who reviewed the manuscript several times; Prof. Xunlin Yu and Daigui Zhang for their helpful comments; Ms. Jiefen Yu for her careful editing. Limited by time and knowledge, there possibly are mistakes in this book, and the readers are kindly welcome to point them out.

Mingxi Jiang
January 2017, Wuhan

目　录

Contents

八大公山自然保护区
Introduction to Badagongshan Nature Reserve

I

1.1 发展简史

八大公山国家级自然保护区位于湖南省桑植县境内，属于森林生态系统类型自然保护区，1982年经湖南省人民政府批准建立，1986年经国务院批准为我国首批国家级自然保护区。

1.1 A Brief History

Badagongshan Nature Reserve is located in Sangzhi county, Hunan province, a forest ecosystem nature reserve. It was established in 1982, approved by the government of Hunan province; promoted to National Nature Reserve among the first batch, approved by the State Council of China.

1.2 地理位置和自然环境

八大公山自然保护区地理位置29°39′18″～29°49′48″ N, 109°41′45″～110°09′50″ E，地处武陵山脉的北端，澧水水系和清江水系分水岭南翼，中亚热带北缘。总面积20000 hm^2。保护区属侵蚀溶蚀山原，岭高谷深、坡陡顶平、岩溶发育、地形崎岖，海拔一般在1000 m以上，最高峰为斗篷山 (1890.4 m)。本区属于山地湿润季风气候。年均温11.5℃，1月均温0.1℃，7月均温22.8℃。年降雨量平均2105.4 mm，最高达2840.1 mm，堪比热带雨林。雨日176 d，全年相对湿度90%，属于十分湿润地区。

1.2 Location and Natural Environment

Badagongshan Nature Reserve is located on the north end of the Wuling Mountains (29°39′18″~29°49′48″ N, 109°41′45″~110°09′50″ E); south wing of the Qingjiang River and the Lishui River, and north rim of mid-subtropical zone; with an area of 20,000 hm^2. The topography is characterized by deep valleys, steep slopes and flat tops, averaging more than 1000 m with the highest peak 1890.4 m (Doupengshan). It is mountain humid monsoon climate. The mean temperature ranges from 0.1 ℃ in January to 22.8 ℃ in July with an annual mean of 11.5℃. Annual rainfall averages 2,105.4 mm, with a maximum of 2,840.1 mm, comparable to tropical rainforests. Rains occur on average 176 d/yr^{-1}. Annual relative humidity is 90%, very moist.

1.3 主要植被类型

八大公山自然保护区原生型天然林集中分布在海拔1000 m以上的山地，其中亮叶水青冈 (*Fagus lucida*) 及其混交林是该地区的优势群落，主要分布在海拔1200～1890 m，约占天然林总面积的60%，优势群落主要包括：亮叶水青冈纯林、亮叶水青冈和多脉青冈共优势种群落，亮叶水青冈与其他阔叶树种共优势种群落。优势树种主要包括亮叶水青冈、香桦 (*Betula insignis*)、多种鹅耳枥 (*Carpinus* spp.)、茅栗 (*Castanea seguinii*)、檫木 (*Sassafras tzumu*)、石灰花楸 (*Sorbus folgneri*)、珙桐 (*Davidia involucrata*)、头状四照花 (*Cornus capitata*)、槭 (*Acer* spp.)、椴 (*Tilia* spp.)和稠李 (*Padus* spp.) 等落叶种类；多脉青冈、西南山茶 (*Camellia pitardii*)、细叶青冈 (*C. myrsinifolia*)、黄丹木姜子 (*Litsea elongata*)、四川杜鹃 (*Rhododendron sutchuenense*) 和交让木 (*Daphniphyllum macropodum*) 等常绿种类。

1.3 Main Vegetation Types

Natural forests of Badagongshan Nature Reserve are mainly distributed above elevation 1000 m. *Fagus lucida* and its mixed forests are dominant vegetation, mainly 1200～1890 m, accounting for 60% of the natural forests including pure *F. lucida* forests, mixed forest of *F. lucida* and *Cyclobalanopsis multinervis*, or mixed forests with other broadleaved trees. The dominant deciduous trees include *F. lucida, Betula insignis, Carpinus* spp., *Castanea seguinii, Sassafras tzumu, Sorbus folgneri, Davidia involucrata, Cornus capitata, Acer* spp., *Tilia* spp. and *Padus* spp; the dominant evergreen trees include *C. multinervis, Camellia pitardii, C. myrsinifolia, Litsea elongata, Rhododendron sutchuenense and Daphniphyllum macropodum*.

八大公山山地常绿落叶阔叶混交林25 hm^{2}森林动态样地

The 25 hm^2 Badagongshan Mountain Evergreen and Deciduous Broadleaved Mixed Forest Dynamics Plot

2

2.1 样地建设和群落调查

八大公山样地属于典型的山地常绿落叶阔叶混交林，东西长500 m，南北长500 m。样地由中国科学院武汉植物园负责，与中南林业科技大学、湖南科技大学、八大公山国家级自然保护区合作建成，2009年12月选址，2010年3月开始建设，2011年10月完成第一次木本植物调查。样地建设技术规程按照CTFS (Center for Tropical Forest Science) 的方法 (Condit, 1998)。用全站仪将样地划分为625个20 m × 20 m的小样方，调查样方内所有胸径 (DBH) ≥1 cm的木本植物，内容包括物种名称、坐标、胸径，并挂牌标记。

2.1 Plot Establishment and Community Census

The Badagongshan Plot belongs to typical mountain evergreen and deciduous broadleaved mixed forest, 500 m (west-east) by 500 m (south-north). This plot is jointly established by Wuhan Botanical Garden CAS, Central South University of Forestry and Technology, Hunan University of Science and Technology, and Badagongshan Nature Reserve. Locality of the plot was selected in December of 2009, and the construction was launched in March of 2010, and the first woody plants census was accomplished in October of 2011. The protocols of this plot followed the standards of CTFS (Center for Tropical Forest Science) (Condit, 1998). The plot was divided into 625 20 m by 20 m subplots. All woody individuals with DBH ≥1 cm were surveyed including species name, coordinates, DBH and tagged.

2.2 地形和土壤

八大公山样地地形复杂多变，坡陡顶平。最高海拔1456 m，最低海拔1355 m。土壤类型为山地黄棕壤。pH范围3.39～6.32。

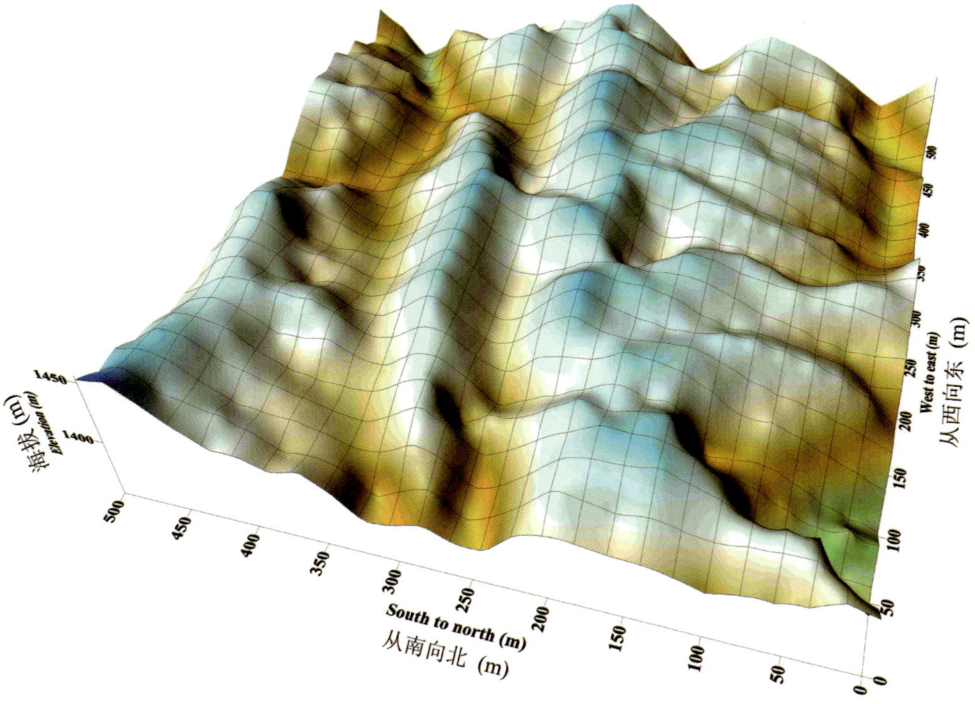

八大公山25 hm^{2}样地地形
Topography of the 25 hm^2 Badagongshan plot

2.2 Topography and Soil

The topography of Badagongshan Plot is very rugged and complicated with steep slopes and flat peaks, with elevation ranging from 1355 to 1456 m. The soil is mountain yellow brown type. pH value ranges from 3.39 to 6.32.

2.3 物种组成和群落结构

八大公山样地植物以泛热带分布科 (24.50%) 和北温带分布属 (24.56%) 占优势，包括9种珍稀濒危植物。样地内木本植物合计53科，114属，232种 (含种下分类单位)，186575株存活个体。落叶树139种，常绿树93种。壳斗科 (Fagaceae)、杜鹃花科 (Ericaceae)、樟科 (Lauraceae) 和山茶科 (Theaceae) 占据优势；个体数超过1000株的有38个物种 (常绿树种63.16%)；稀有种 (<25株) 103个 (占样地物种总数43.28%)。群落垂直结构可分为乔木层 (优势树种多脉青冈和亮叶水青冈)，亚乔木层 [优势种长蕊杜鹃 (*R. stamineum*) 和黄丹木姜子] 和灌木层 [优势种短柱柃 (*Eurya brevistyla*) 和薄叶山矾 (*Symplocos anomala*)]。个体数超过1000株的树种和稀有种主要来自灌木层。由于生长型限制，样地内树种平均胸径较小 (5.41 cm)，DBH<5 cm个体占优势 (68.40%)。主要优势树种的径级结构呈现倒J形。

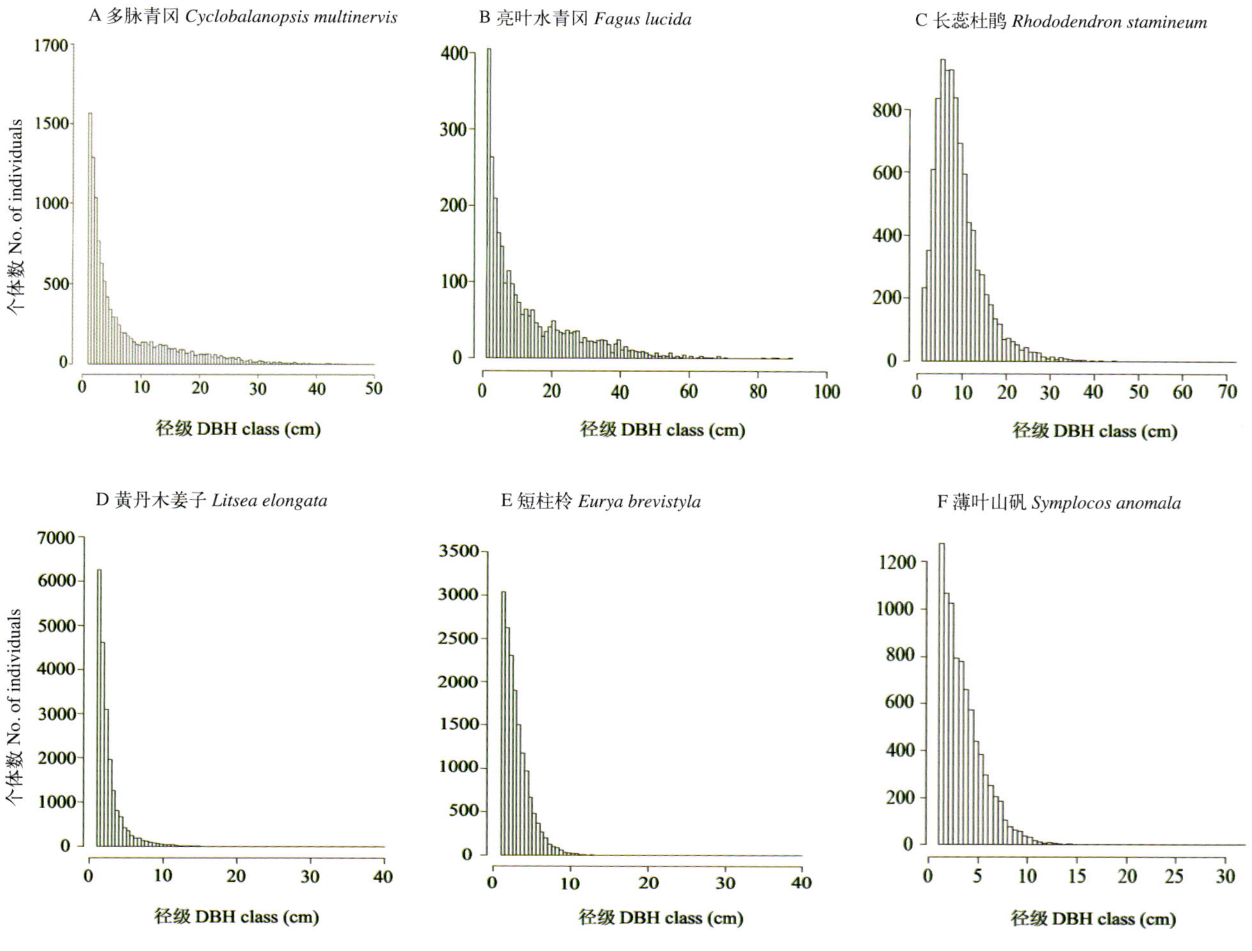

八大公山25 hm²样地内六种优势树种的径级结构
Size class of six dominant woody plants in the 25 hm² Badagongshan plot

2.3 Species Composition and Community Structure

The Badagongshan plot was dominated by pantropic distribution families (24.50%) and north temperate distribution genera (24.56%). There were nine rare and endangered woody plants. 186,575 individuals could be classified into 53 families, 114 genera and 232 species (93 evergreen and 139 deciduous). Dominant species were from the family of Fagaceae, Ericaceae, Lauraceae and Theaceae.There were 38 species with individuals>1000 (63.16% evergreen). 103 rare species (<25 individuals) accounted for 43.28% of all the woody plants. The strata could be divided as canopy layer dominated by *C. multinervis* and *F. lucida*, understory layer dominated by *R. stamineum* and *L. elongata*, shrub layer dominated by *Eurya brevistyla* and *Symplocos anomala*. Abundant (>1000 individuals) and rare species were mainly from shrub layer. Limited by growth form, the plot had a fairly small mean DBH (5.41 cm) and dominated by small sized individuals (68.40%). The size class of six dominant trees was reverse J.

树种及其分布格局
Tree Species and Their Distribution Patterns

3

树种多度、重要值、所在林层以及胸径分级说明 Notes to abundance, importance value (IV), stratum and DBH (diameter at breast height) class:

1. 树种多度只计主干；重要值=（相对多度+相对胸高断面积）× 100/2；胸高断面积统计包括主干、分枝和萌枝 Only main trunks were counted for abundance; IV=(relative abundance +relative basal area) × 100/2; trunks, branches and sprouts were all computed for basal area

2. 按照树种的高度和生长型，划分为3个林层：乔木层（≥15 m）包含50个物种（落叶37种，常绿13种）；亚乔木层（5～15 m）包含54个物种（落叶34种，常绿20种）；灌木层（<5 m）包含128个物种（落叶68种，常绿60种）According to height and growth form of tree species, the forest is divided into three strata: canopy stratum (≥15 m) consisting of 50 species (37 deciduous and 13 ever～green); under-story stratum (5-15 m) consisting of 54 species (34 deciduous and 20 ever-green); shrub stratum (<5 m) consisting of 128 species (68 deciduous and 60 ever-green)

3. 胸径分级标准（单位：厘米，上限排除法）DBH classification criteria (unit: cm, upper limit method)，乔木层 canopy stratum：[1.0～2.5), [2.5～5.0), [5.0～10.0), [10.0～25.0), [25.0～40.0), [40.0～100.0), [100.0, +∞)；亚乔木层 under-story stratum：[1.0～2.5), [2.5～5.0), [5.0～8.0), [8.0～11.0), [11.0～15.0), [15.0～20.0), [20.0, +∞)；灌木层 shrub stratum：[1.0～2.0), [2.0～3.0), [3.0～4.0), [4.0～5.0), [5.0～7.0), [7.0～10.0), [10.0,+∞)

4. 文字描述部分for plant description，fl.代表“开花”fl.for ‘flowering’，fr. 代表“结果”fr. for ‘fruiting’，yr代表“年”yr.for ‘year’

1 马尾松

mǎ wěi sōng | Masson Pine

Pinus massoniana，松属
松科 | Pinaceae

树干 Trunk
摄影：卢志军 Photo by: Zhijun Lu

代码（SpCode）= PINMAS
个体数（Individual number/25 hm^2）= 1
最大胸径（Max DBH）= 31.07 cm
所在林层（Stratum）= 乔木层 (Canopy)
重要值排序（Importance value rank）= 48/50

常绿乔木。树皮红褐色至灰褐色，不规则块裂。针叶2针一束，长12～20 cm，细柔，微扭曲。雄球花穗状，聚生于新枝下部；雌球花单生或2～4个聚生于新枝近顶端。球果卵圆形或圆锥状卵圆形，有短梗，下垂，成熟前绿色，熟时栗褐色。花期4～5月，球果翌年10～12月成熟。

Evergreen trees. Bark red-brown to gray brown, irregularly peeling in blocks. Needles 2 per bundle, 12-20 cm in length, gentle and slender, slightly twisted. Male cone in spikes, fasciculate on new branch bottom; female cone solitary or in 2-4, subterminal on new branch. Seed cones ovoid or conical-ovoid, shortly pedunculate, pendulous, green, chestnut brown at maturity. Fl. Apr.-May, seed cones maturity Oct.-Dec. of following yr.

小枝 Branchlets
摄影：陈彬 Photo by: Bin Chen

球果 Cone
摄影：卢志军 Photo by: Zhijun Lu

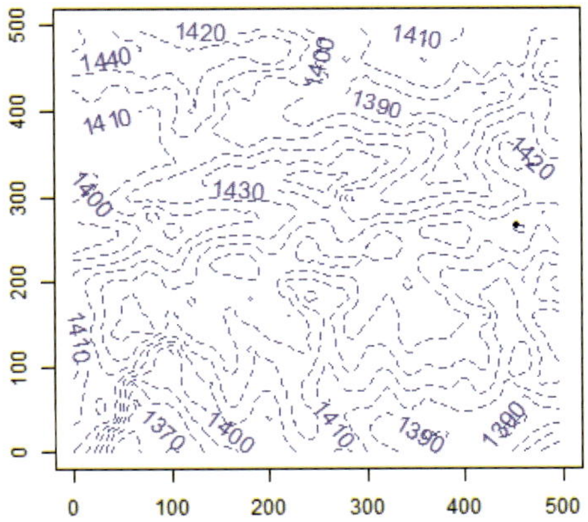

个体分布图 Distribution of individuals

径级分布表 DBH class

胸径区间 (Diameter class) (cm)	个体数 (No. of individuals in the plot)	比例 (Proportion) (%)
1～2.5	0	0.0
2.5～5	0	0.0
5～10	0	0.0
10～25	0	0.0
25～40	1	100.0
40～100	0	0.0
≥100	0	0.0

2 三尖杉 　　sān jiān shān | Fortune Plumyew

Cephalotaxus fortunei，三尖杉属
三尖杉科 | Cephalotaxus

代码（SpCode）= CEPFOR
个体数（Individual number/25 hm^2）= 13
最大胸径（Max DBH）= 9.77 cm
所在林层（Stratum）= 亚乔木层 (Understory)
重要值排序（Importance value rank）= 47/54

常绿乔木。树皮褐色或红褐色，片裂。叶排成两列，线状披针形，通常微弯，镰状，上部渐窄，先端有渐尖的长尖头，基部楔形或宽楔形，上面深绿色，中脉隆起，下面气孔带白色，绿色中脉不明显。雄球花聚生成头状。种子椭圆状卵圆形，外种皮成熟时紫色或红紫色，顶端有小尖头。花期4月，种子10月成熟。

Evergreen trees. Bark brown or dark reddish brown, peeling in strips. Leaves in two columns, linear-lanceolate, usually slightly bending, falcate, distally gradually attenuate, apex acuminate with a long tip, base cuneate or broadly cuneate, adaxially deep green, midvein prominent, abaxially stomata white, midvein green and indistinct. Pollen-cone capitula. Seeds ellipsoid-ovoid, aril purple or reddish purple when mature, apex shortly mucronate. Fl. Apr., seed maturity Oct..

树干 　Trunk
摄影：卢志军 　Photo by: Zhijun Lu

果枝 　Fruiting branch
摄影：鲍大川 　Photo by: Dachuan Bao

叶背 　Leaf abaxial surface
摄影：卢志军 　Photo by: Zhijun Lu

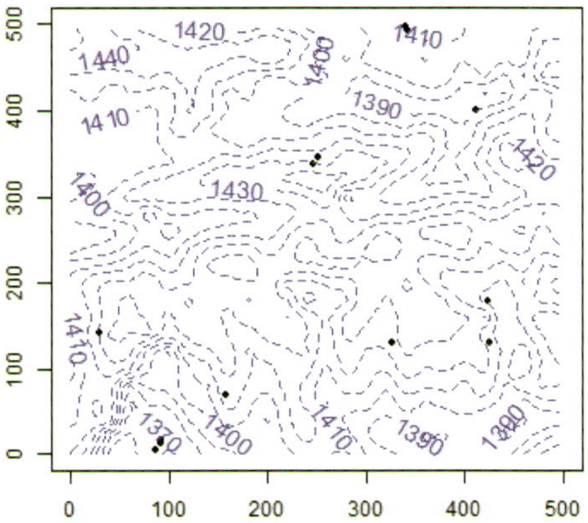

个体分布图 Distribution of individuals

径级分布表 DBH class

胸径区间 (Diameter class) (cm)	个体数 (No. of individuals in the plot)	比例 (Proportion) (%)
1～2.5	5	38.5
2.5～5	4	30.8
5～8	3	23.1
8～11	1	7.7
11～15	0	0.0
15～20	0	0.0
≥20	0	0.0

3 红豆杉 hóng dòu shān | Maíre Yew

Taxus wallichiana var. *chinensis*, 红豆杉属
红豆杉科 | Taxaceae

代码（SpCode）= TAXWAL
个体数（Individual number/25 hm^2）= 11
最大胸径（Max DBH）= 17.75 cm
所在林层（Stratum）= 亚乔木层 (Understory)
重要值排序（Importance value rank）= 44/54

树干 Trunk
摄影：卢志军 Photo by: Zhijun Lu

常绿乔木。树皮灰褐色或红褐色或暗褐色，条裂。叶线形，通常稍直或微呈镰状。雄球花淡黄色。种子卵圆形，较短，通常中上部较窄。花期4～5月，种子10月成熟。

Evergreen trees. Bark grayish brown, reddish brown or dark brown, peeling in strips. Leaves linear, usually slightly straight or slightly falcate. Male cone light yellow. Seeds ovoid, short, middle and upper usually narrower. Fl. Apr.-May, seed maturity Oct..

小枝 Branchlets
摄影：江明喜 Photo by: Mingxi Jiang

果枝 Fruiting branches
摄影：鲍大川 Photo by: Dachuan Bao

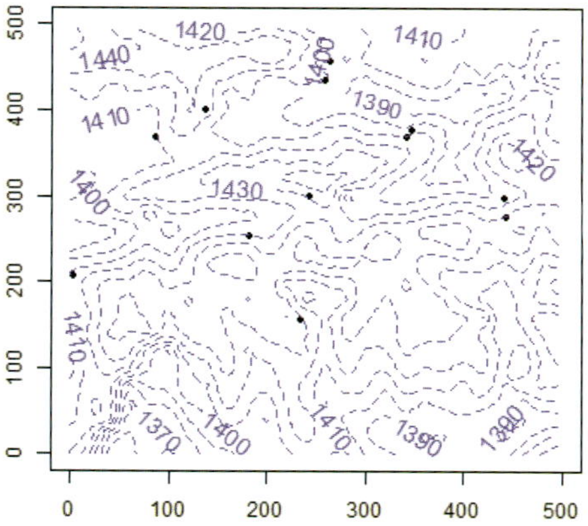

个体分布图 Distribution of individuals

径级分布表 DBH class

胸径区间 (Diameter class) (cm)	个体数 (No. of individuals in the plot)	比例 (Proportion) (%)
1～2.5	2	18.2
2.5～5	0	0.0
5～8	3	27.3
8～11	3	27.3
11～15	1	9.1
15～20	2	18.2
⩾20	0	0.0

4 化香树

huà xiāng shù | Dye Tree

Platycarya strobilacea，化香树属
胡桃科 | Juglandaceae

代码（SpCode）= PLASTR
个体数（Individual number/25 hm^2）= 5
最大胸径（Max DBH）= 24.58 cm
所在林层（Stratum）= 灌木层 (Shrub)
重要值排序（Importance value rank）= 81/128

落叶小乔木。树皮灰色。小枝黄褐色或栗褐色。奇数羽状复叶，小叶1～15。雌雄同株。两性花序2～10 cm；雄花序2～15 cm；雌花苞片直立或反折。果序卵状椭圆形至长椭圆状圆柱形，苞片革质，小坚果扁平。花期5～6月，果期9月。

Deciduous small trees. Bark brown. Branchlets yellowish brown or chestnut brown. Leaves imparipinnate, leaflets 1-15. Monoecious. Androgynous spike 2-10 cm; male spikes 2-15 cm; female flowers with bract straight or reflexed. Fruiting spike ovoid-ellipsoid or ellipsoid-cylindric, bracts leathery, nutlets flat. Fl. May–Jun., fr. Sept..

树干 Trunk
摄影：卢志军 Photo by: Zhijun Lu

果枝 Fruiting branches
摄影：陈彬 Photo by: Bin Chen

花序 Inflorescence
摄影：鲍大川 Photo by: Dachuan Bao

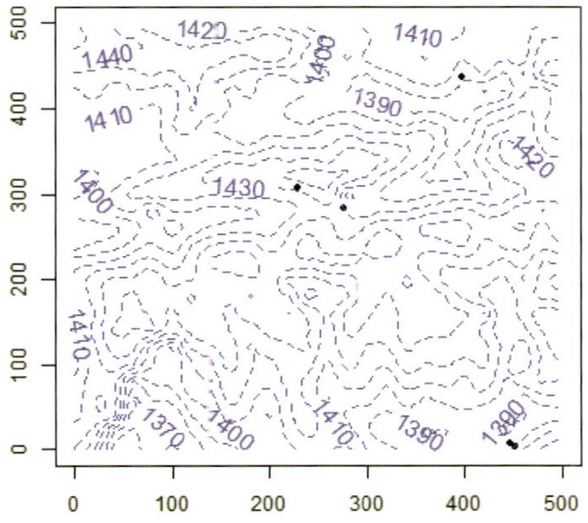

个体分布图 Distribution of individuals

径级分布表 DBH class

胸径区间 (Diameter class) (cm)	个体数 (No. of individuals in the plot)	比例 (Proportion) (%)
1～2	2	40.0
2～3	2	40.0
3～4	0	0.0
4～5	0	0.0
5～7	0	0.0
7～10	0	0.0
≥10	1	20.0

5 青钱柳

qīng qián liǔ | Round-wing-fruit Cyclocarya

Cyclocarya paliurus，青钱柳属
胡桃科 | Juglandaceae

代码（SpCode）= CYCPAL
个体数（Individual number/25 hm^2）= 3
最大胸径（Max DBH）= 25.92 cm
所在林层（Stratum）= 乔木层 (Canopy)
重要值排序（Importance value rank）= 38/50

落叶乔木。幼枝被茸毛或鳞片。奇数羽状复叶，小叶长圆状卵形至长圆状披针形。雄性柔荑花序7～18 cm，3条或2～4一束；雌性葇荑花序单独顶生。果序总状，果翅圆盘形。花期4～6月，果期7～11月。

树干 Trunk
摄影：卢志军 Photo by: Zhijun Lu

Deciduous trees. Young branchlets tomentose or palta. Imparipinnate leaves, blade oblong-elliptic ovate to oblong-elliptic lanceolate. Male inflorescence in catkins, 7-18 cm, 3 or 2-4 per cluster; female catkins solitary, terminal. Fruiting spikes raceme; fruit wings disc-shaped. Fl. Apr.–Jun., fr. Jul.–Nov..

花枝 Flowering branches
摄影：江明喜 Photo by: Mingxi Jiang

果枝 Fruiting branches
摄影：江明喜 Photo by: Mingxi Jiang

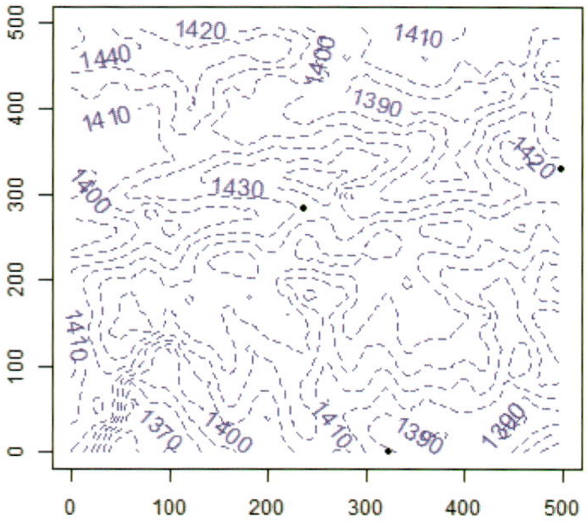

个体分布图 Distribution of individuals

径级分布表 DBH class

胸径区间 (Diameter class) (cm)	个体数 (No. of individuals in the plot)	比例 (Proportion) (%)
1～2.5	0	0.0
2.5～5	1	33.3
5～10	0	0.0
10～25	1	33.3
25～40	1	33.3
40～100	0	0.0
≥100	0	0.0

6 茅栗　máo lì | Seguin Chinkapin

Castanea seguinii，栗属
壳斗科 I Fagaceae

代码（SpCode）= CASSEG
个体数（Individual number/25 hm^2）= 469
最大胸径（Max DBH）= 84.20 cm
所在林层（Stratum）= 乔木层 (Canopy)
重要值排序（Importance value rank）= 7/50

落叶小乔木或灌木。小枝暗褐色。叶倒卵状椭圆形或长圆形，先端渐尖，基部楔形至圆形或近心形，叶背有黄色或灰白色腺鳞。雄花序长5～12 cm，雌花单生或每壳斗数个，总苞密生锐刺，内有坚果2～3。花期5～7月，果期9～11月。

树干　Trunk
摄影：卢志军　Photo by: Zhijun Lu

Deciduous small trees or shrubs. Branchlet dark brown. Leaf blade obovate-elliptic or oblong, apex acuminate, base cuneate to rounded or subcordate, abaxially covered with yellow or grayish white glandular scales. Male inflorescences 5-12 cm, female flowers solitary or few per cupule. Cupule densely covered with spinelike bracts, 2-3 nuts inside. Fl. May–Jul., fr. Sept.–Nov..

花枝　Flowering branches
摄影：卢志军　Photo by: Zhijun Lu

果枝　Fruiting branches
摄影：江明喜　Photo by: Mingxi Jiang

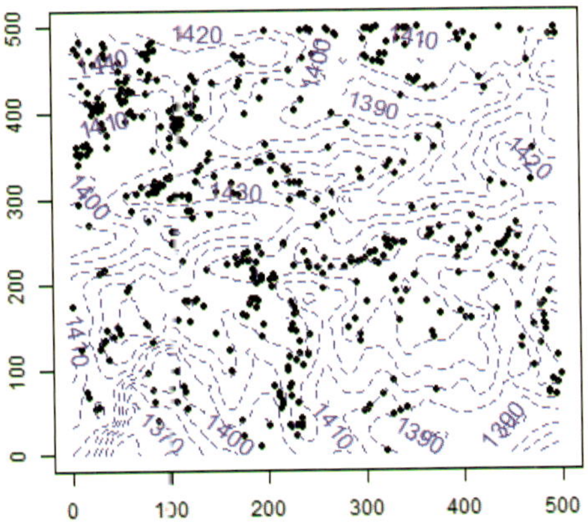

个体分布图　Distribution of individuals

径级分布表　DBH class

胸径区间 (Diameter class) (cm)	个体数 (No. of individuals in the plot)	比例 (Proportion) (%)
1～2.5	6	1.3
2.5～5	4	0.9
5～10	15	3.2
10～25	170	36.2
25～40	196	41.8
40～100	67	14.3
≥100	11	2.3

7 多脉青冈

duō mài qīng gāng | Many-nerve Oak

Cyclobalanopsis multinervis，青冈属

壳斗科 | Fagaceae

代码（SpCode）= CYCMUL

个体数（Individual number/25 hm^2）= 11373

最大胸径（Max DBH）= 46.45 cm

所在林层（Stratum）= 乔木层 (Canopy)

重要值排序（Importance value rank）= 1/50

树干 Trunk

摄影：卢志军 Photo by: Zhijun Lu

常绿乔木。叶革质，狭长圆状椭圆形、倒卵状椭圆形或椭圆状披针形，先端突渐尖或渐尖，基部楔形、宽楔形或近圆形，边缘下部以上有细尖锯齿，仅近基部全缘，下面有毛及易脱落的灰白色粉霜，侧脉10～15对；叶柄长1～3 cm。果序长1～2 cm，上有果2～6个；总苞有鳞片环7层，被毛。果翌年成熟。

Evergreen trees. Leaf blade leathery, long and narrow cone oval, oblong-elliptic to elliptic-lanceolate, apex acute or acuminate, base cuneate, broadly cuneate or subrounded, sharply serrate from lower margin, only entire near base, abaxially covered with hairs and caducous grayish white pruina, secondary veins 10-15, petiole 1-3 cm. Infructescences 1-2 cm, 2-6 fruited, bracts in 7 rings, tomentose. Fr. maturity in following yr..

花枝 Flowering branches

摄影：路俊盟 Photo by: Junmeng Lu

果枝 Fruiting branch

摄影：鲍大川 Photo by: Dachuan Bao

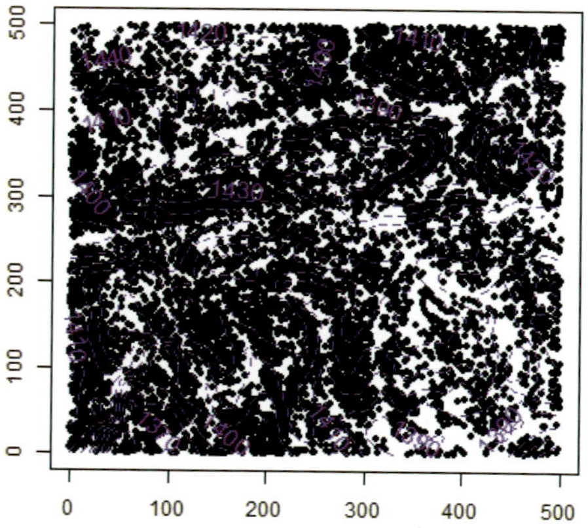

个体分布图 Distribution of individuals

径级分布表 DBH class

胸径区间 (Diameter class) (cm)	个体数 (No. of individuals in the plot)	比例 (Proportion) (%)
1～2.5	3919	34.5
2.5～5	2739	24.1
5～10	1936	17.0
10～25	2329	20.5
25～40	415	3.6
40～100	33	0.3
≥100	2	0.0

8 细叶青冈

xì yè qīng gāng | Thin-leaf Oak

Cyclobalanopsis gracilis，青冈属
壳斗科 | Fagaceae

代码（SpCode）= CYCGRA
个体数（Individual number/25 hm^2） = 5268
最大胸径（Max DBH）= 79.52 cm
所在林层（Stratum）= 乔木层 (Canopy)
重要值排序（Importance value rank）= 3/50

树干 Trunk
摄影：卢志军 Photo by: Zhijun Lu

常绿乔木。幼枝紫棕色，初被微毛，后渐脱落，皮孔显著。叶革质，卵圆状长圆形或长圆形，或披针状长圆形，先端渐尖，基部楔形，或几乎为圆形，边缘有少数明显的齿牙，叶下面有丁字毛，呈灰白色。雄花序5～7 cm，雌花序1～1.5 cm。总苞被较密的平伏毛，上部带灰黄色，鳞片环常有缺齿。花期4月，果熟期9～10月。

Evergreen trees. Young branches purplish brown, tomentose when young, glabrescent, distinct lenticels. Leaf blade leathery, oblong-ovate, elliptic to ovate-lanceolate, apex acuminate, base cuneate to or nearly rounded, margin sometimes sharply serrulate, abaxially whitish and with prostrate simple hairs. Male inflorescences 5-7 cm, female inflorescences 1-1.5 cm. Female inflorescence bracts with prostrate tomentose hairs and upper grayish yellow; bract rings usually dentate. Fl. Apr., fr. maturity Sept.–Oct..

小枝和叶片 Branchlet and leaves
摄影：卢志军 Photo by: Zhijun Lu

小枝和叶背 Branchlet and leaf abaxial surface
摄影：卢志军 Photo by: Zhijun Lu

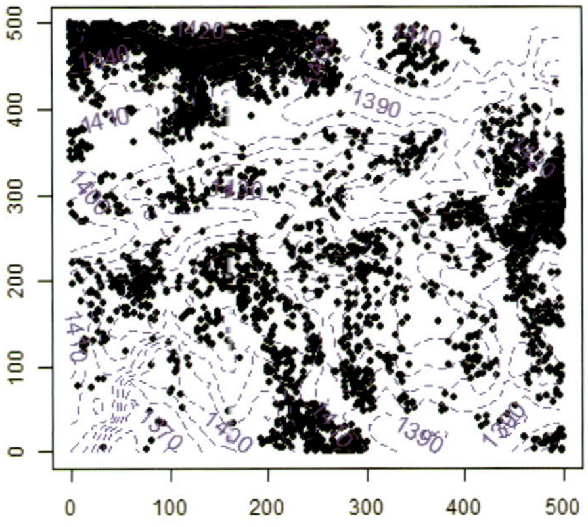

个体分布图 Distribution of individuals

径级分布表 DBH class

胸径区间 (Diameter class) (cm)	个体数 (No. of individuals in the plot)	比例 (Proportion) (%)
1～2.5	2395	45.5
2.5～5	1231	23.4
5～10	802	15.2
10～25	548	10.4
25～40	217	4.1
40～100	68	1.3
≥100	7	0.1

9 曼青冈

màn qīng gāng | Sharpdent Oak

Cyclobalanopsis oxyodon，青冈属
壳斗科 | Fagaceae

代码（SpCode）= CYCOXY
个体数（Individual number/25 hm^2）= 10
最大胸径（Max DBH）= 18.72 cm
所在林层（Stratum）= 乔木层 (Canopy)
重要值排序（Importance value rank）= 30/50

树干 Trunk
摄影：卢志军 Photo by: Zhijun Lu

常绿乔木。小枝灰色，皮孔明显。叶革质，长圆状披针形至长圆形，先端渐尖，基部楔形，上面绿色，下面带苍白色，疏被毛，边缘有疏锯齿。雄花序6～10 cm，雌花序2～5 cm。总苞杯状，直径1.5～2.0 cm，鳞片结合成环，约5～8环，环带边缘有少数圆齿裂。坚果近球形。花期5月，果熟期9～10月。

Evergreen trees. Brachlets gray, distinct lenticels. Leaf blade leathery, oblong-lanceolate to elliptic, apex acuminate, base cuneate, adaxially green, abaxially whitish, with sparse hairs, margin coarsely serrate. Male inflorescences 6-10 cm, female inflorescences 2-5 cm. Cupule cupular, 1.5-2.0 cm in diam, circular bracts in 5-8 rings, margin sometimes dentate. Nut subglobose. Fl. May, fr. maturity Sept.–Oct..

小枝和叶片 Branchlet and leaves
摄影：卢志军 Photo by: Zhijun Lu

小枝和叶背 Branchlet and leaf abaxial surface
摄影：卢志军 Photo by: Zhijun Lu

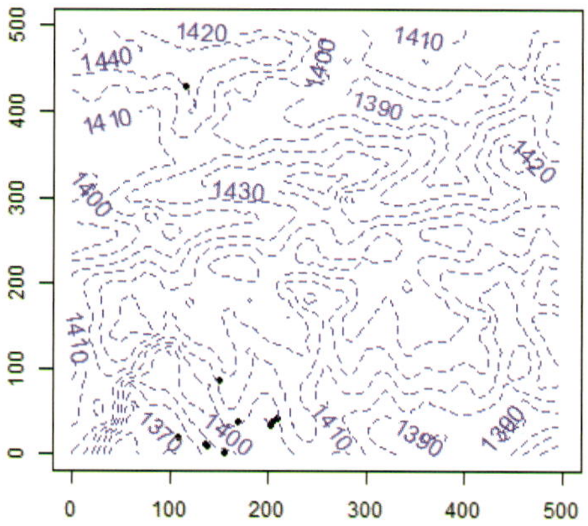

个体分布图 Distribution of individuals

径级分布表 DBH class

胸径区间 (Diameter class) (cm)	个体数 (No. of individuals in the plot)	比例 (Proportion) (%)
1～2.5	1	10.0
2.5～5	3	30.0
5～10	1	10.0
10～25	5	50.0
25～40	0	0.0
40～100	0	0.0
⩾100	0	0.0

10 亮叶水青冈 liàng yè shuǐ qīng gāng | Shining Beech

Fagus lucida，水青冈属
壳斗科 | Fagaceae

代码（SpCode）= FAGLUC
个体数（Individual number/25 hm^2）= 2765
最大胸径（Max DBH）= 89.58 cm
所在林层（Stratum）= 乔木层 (Canopy)
重要值排序（Importance value rank）= 2/50

落叶乔木。幼枝紫褐色，无毛。叶纸质，卵形至椭圆状卵形，先端急尖或短渐尖，基部宽楔形、近圆形或近心形，边缘浅波状，在弯缺外有短三角形的牙齿，下面除脉上被长伏毛外无毛。总苞外被三角状的疣状突起；坚果有三棱。有黄褐色微柔毛。果期8～10月。

Deciduous trees. Young branchlets purplish brown, glabrous. Leaf blade papery, ovate to elliptic-ovate, apex acute to shortly acuminate, base broadly cuneate, subrounded or subcordate, margin slightly sinuate with minute teeth, abaxially glabrous except for silky pubescence on veins. Cupule tuberculate, triangular mucronate. Nuts prismatoid, yellowish brown puberulose. Fr. Aug.–Oct..

树干 Trunk
摄影：江明喜 Photo by: Mingxi Jiang

果枝 Fruiting branch
摄影：卢志军 Photo by: Zhijun Lu

总苞和坚果 Cupule and nuts
摄影：卢志军 Photo by: Zhijun Lu

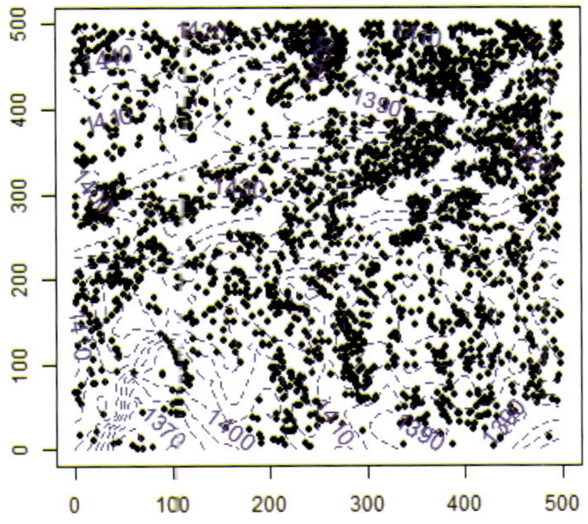

个体分布图 Distribution of individuals

径级分布表 DBH class

胸径区间 (Diameter class) (cm)	个体数 (No. of individuals in the plot)	比例 (Proportion) (%)
1～2.5	552	20.0
2.5～5	487	17.6
5～10	539	19.5
10～25	691	25.0
25～40	355	12.8
40～100	123	4.4
≥100	18	0.7

11 水青冈

shuǐ qīng gāng | Long-petiole Beech

Fagus longipetiolata，水青冈属
壳斗科 | Fagaceae

代码（SpCode）= FAGLON
个体数（Individual number/25 hm^2）= 10
最大胸径（Max DBH）= 34.68 cm
所在林层（Stratum）= 乔木层 (Canopy)
重要值排序（Importance value rank）= 28/50

树干 Trunk
摄影：卢志军 Photo by: Zhijun Lu

落叶乔木。幼枝紫褐色，无毛。叶厚纸质，卵形至卵状长圆形，先端急尖至短渐尖，基部宽楔形至近圆形，边缘有疏浅锯齿，下面密被柔毛并带苍白色。总苞密生锈色茸毛，有极多向外弯的钝刺。坚果三棱，有黄褐色微柔毛，顶端有长柔毛，花柱宿存。花期4月，果熟期8～10月。

Deciduous trees. Branchlets purplish brown, glabrous. Leaf blade papery, ovate to ovate-oblong, apex acute to shortly acuminate, base broadly cuneate to subrounded, margin remotely serrate, abaxially finely densely pubescent and glaucescent. Cupule pubescent, bracts linear and recurved. Nuts prismatoid, yellowish brown puberulose, villose apex, style persistent. Fl. Apr., fr. maturity Aug.–Oct..

叶片 Leaves
摄影：江明喜 Photo by: Mingxi Jiang

果枝 Fruiting branch
摄影：鲍大川 Photo by: Dachuan Bao

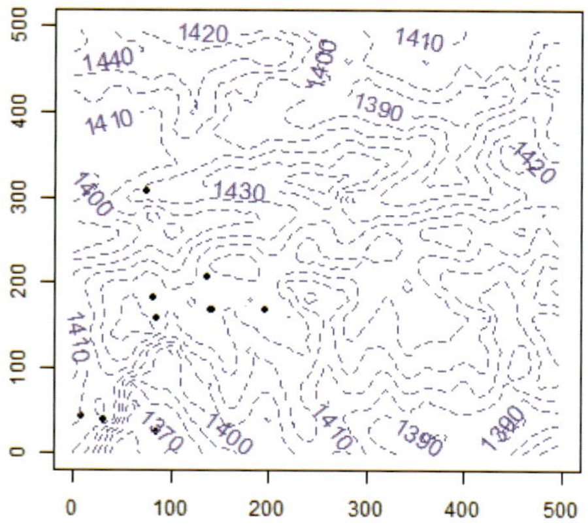

个体分布图 Distribution of individuals

径级分布表 DBH class

胸径区间 (Diameter class) (cm)	个体数 (No. of individuals in the plot)	比例 (Proportion) (%)
1～2.5	1	10.0
2.5～5	4	40.0
5～10	2	20.0
10～25	1	10.0
25～40	2	20.0
40～100	0	0.0
⩾100	0	0.0

12 绵柯

mián kē | Henry Tan-oak

Lithocarpus henryi，柯属
壳斗科 | Fagaceae

代码（SpCode）= LITHEN
个体数（Individual number/25 hm^2）= 127
最大胸径（Max DBH）= 28.97 cm
所在林层（Stratum）= 乔木层 (Canopy)
重要值排序（Importance value rank）= 16/50

树干 Trunk
摄影：卢志军 Photo by: Zhijun Lu

常绿乔木。小枝初被灰色微毛，后无毛，有皮孔。叶革质，长圆状披针形或狭长椭圆形，先端渐尖，略歪斜，基部楔形。雄穗状花序单生叶腋；雌花序20 cm，雌花3朵一簇。总苞盘状，鳞片三角形；坚果扁圆球形，有白粉。花期7月，果熟期翌年8～10月。

Evergreen trees. Branchlets grayish microtriche when young, then glabrous, with lenticels. Leaf blade leathery, oblong-lanceolate or narrowly oblong, apex shortly acuminate, slightly asymmetric, base cuneate. Male inflorescences spicate, solitary and axillary; female inflorescences 20 cm, 3 flowers per cluster. Cupule discoid, bracts triangular. Nuts depressed globose, with white farinose. Fl. Jul., fr. maturity Aug.–Oct. of following yr..

花枝 Flowering branches
摄影：鲍大川 Photo by: Dachuan Bao

果枝 Fruiting branches
摄影：鲍大川 Photo by: Dachuan Bao

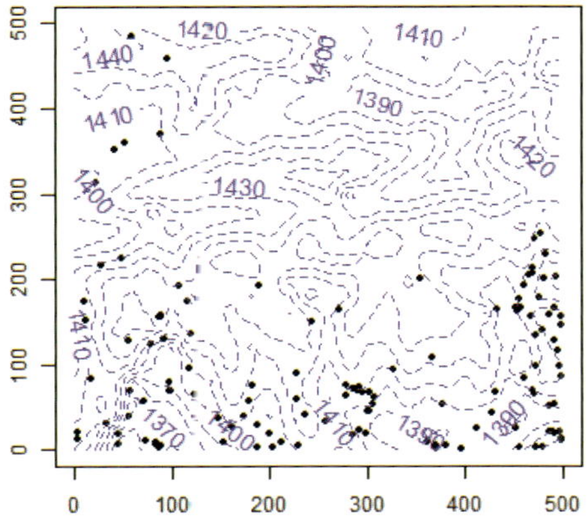

个体分布图 Distribution of individuals

径级分布表 DBH class

胸径区间 (Diameter class) (cm)	个体数 (No. of individuals in the plot)	比例 (Proportion) (%)
1～2.5	49	38.6
2.5～5	31	24.4
5～10	31	24.4
10～25	12	9.4
25～40	2	1.6
40～100	1	0.8
≥100	1	0.8

13 巴东栎

bā dōng lì | Engler Oak

Quercus engleriana，栎属

壳斗科 | Fagaceae

代码（SpCode）= QUEENG

个体数（Individual number/25 hm^2）= 978

最大胸径（Max DBH）= 61.40 cm

所在林层（Stratum）= 乔木层 (Canopy)

重要值排序（Importance value rank）= 13/50

常绿乔木。枝条初被黄色茸毛，后无毛，变褐色。叶革质，卵圆形、卵状椭圆形、卵状披针形或披针形，先端急尖，或有时稍钝，基部圆形，不大对称，雄花序生于新枝茎部，长7 cm；雌花序生于新枝上端叶腋，长1-3 cm。总苞半球形，无梗，鳞片褐色，被茸毛；坚果卵圆状长圆形。花期4～5月，果熟期9～10月。

树干 Trunk

摄影：卢志军 Photo by: Zhijun Lu

Evergreen trees. Young branchlets yellowish tomentose, then glabrescent and brown. Leaf blade leathery, ovate, ovate-elliptic, ovate-lanceolate or lanceolate, apex acute, sometimes slightly blunt, base rounded, slightly asymmetric, male inflorescences on new branch dasis, 7 cm; female inflorescences axillaryon upperend of new branch, 1-3 cm. Cupule hemispherical, sessile, brown scales and pubescent. Nut ovoid-elliptic. Fl. Apr.–May, fr. maturity Sept.–Oct..

小枝和叶片 Branchlets and leaves

摄影：卢志军 Photo by: Zhijun Lu

小枝和叶背 Branchlets and leaf abaxial surface

摄影：卢志军 Photo by: Zhijun Lu

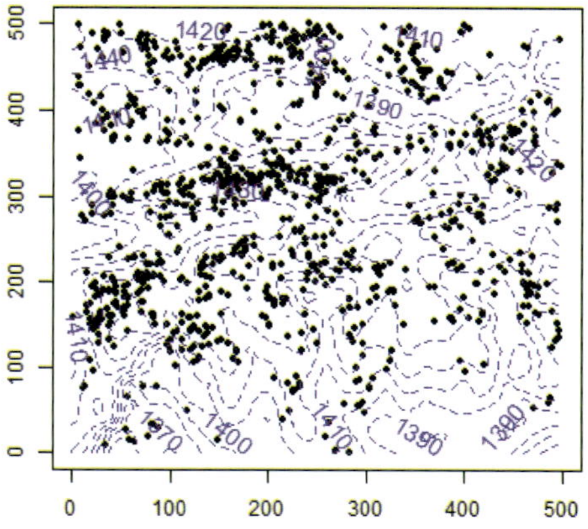

个体分布图 Distribution of individuals

径级分布表 DBH class

胸径区间 (Diameter class) (cm)	个体数 (No. of individuals in the plot)	比例 (Proportion) (%)
1～2.5	299	30.6
2.5～5	238	24.3
5～10	209	21.4
10～25	201	20.6
25～40	24	2.5
40～100	6	0.6
≥100	1	0.1

14 枹栎

bāo lì | Serrate Oak

Quercus serrata-，栎属
壳斗科 | Fagaceae

代码（SpCode）= QUESER
个体数（Individual number/25 hm^2）= 703
最大胸径（Max DBH）= 70.50 cm
所在林层（Stratum）= 乔木层 (Canopy)
重要值排序（Importance value rank）= 9/50

树干 Trunk
摄影：卢志军 Photo by: Zhijun Lu

落叶乔木。叶片薄革质，狭椭圆卵形，卵状披针形，或倒卵形，先端渐尖至急尖，基部楔形至近圆形，叶缘具腺齿。雄花序8～12 cm，雌花序1.5～3.0 cm。壳斗杯状，包被坚果1/4～1/3；坚果卵形或卵球形。花期3～4月，果期9～10月。

Deciduous trees. Leaf blade thinly leathery, narrowly elliptic-ovate, ovate-lanceolate, or obovate, apex acuminate to acute, base cuneate to nearly rounded, margin glandular serrate. Male inflorescences 8-12 cm, female inflorescences 1.5-3.0 cm. Cupule cupular, enclosing 1/4-1/3 of nut. Nut ovoid to ovoid-globose. Fl. Mar.–Apr., fr. Sept.–Oct..

花枝 Flowering branch
摄影：刘海波 Photo by: Haibo Liu

果枝 Fruiting branches
摄影：卢志军 Photo by: Zhijun Lu

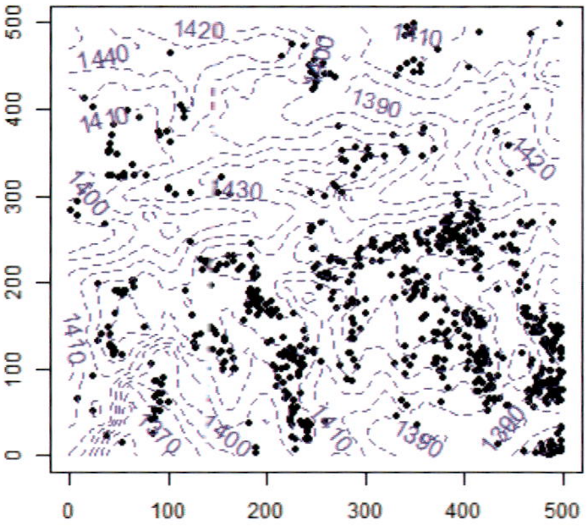

个体分布图 Distribution of individuals

径级分布表 DBH class

胸径区间 (Diameter class) (cm)	个体数 (No. of individuals in the plot)	比例 (Proportion) (%)
1～2.5	20	2.8
2.5～5	24	3.4
5～10	64	9.1
10～25	383	54.5
25～40	185	26.3
40～100	23	3.3
⩾100	4	0.6

15 香桦 xiāng huà | Fragrant Birch

Betula insignis，桦木属
桦木科 | Betulaceae

树干 Trunk
摄影：卢志军 Photo by: Zhijun Lu

代码（SpCode）= BETINS
个体数（Individual number/25 hm^2）= 434
最大胸径（Max DBH）= 94.50 cm
所在林层（Stratum）= 乔木层 (Canopy)
重要值排序（Importance value rank）= 8/50

落叶乔木。小枝黄灰色，白色皮孔显著。叶长圆卵形，先端渐尖，基部圆形或近于心形，边缘有不规则细锯齿，上面深绿色，幼时疏被毛，后无毛，下面淡绿色，无毛，密被褐色或黑色腺点，下面各脉密被平伏白色丝状毛，脉腋通常无毛或有细须状毛。果序单生，长圆形，几乎无果序总梗；果苞基部楔形，背面密被短柔毛，边缘疏被短纤毛，脱落后常以纤维与果序轴相连；有3片披针形裂片，侧裂片直立；翅果狭长圆形，膜质翅极狭。果期6～8月。

Deciduous trees. Branchlets grayish yellow, white lenticels distinct. Leaf blade elliptic-ovate, apex acuminate, base rounded or subcordate, margin irregularly mucronate serrate, adaxially dark green, sparsely tomentulose when young then glabrous, abaxially light green, glabrous, densely brown or black resinous punctate, villous along veins, bearded or glabrous in axils of lateral veins. Female inflorescence solitary, oblong, peduncle obscure, bract base cuneate, back densely pubescent, margin sparsely ciliolate, usually connected by fiber with inflorescence axis after falling off, with three lanceolate lobes, lateral lobes erect. Nutlets narrowly oblong, with extremely narrow membranous wings. Fr. Jun.–Aug..

花枝 Flowering branch
摄影：刘海波 Photo by: Haibo Liu

果枝 Fruiting branches
摄影：江明喜 Photo by: Mingxi Jiang

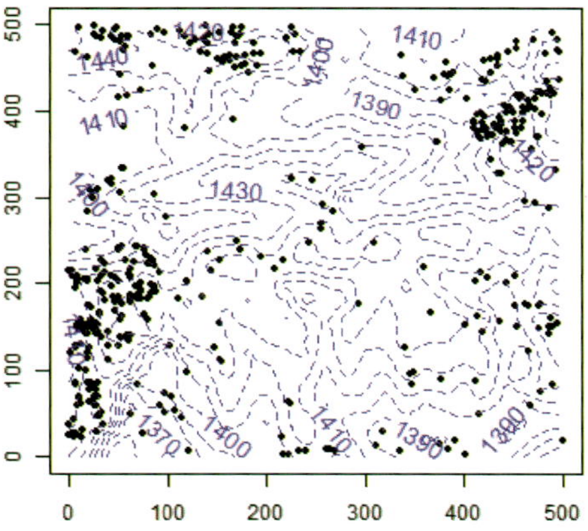

个体分布图 Distribution of individuals

径级分布表 DBH class

胸径区间 (Diameter class) (cm)	个体数 (No. of individuals in the plot)	比例 (Proportion) (%)
1～2.5	4	0.9
2.5～5	5	1.2
5～10	19	4.4
10～25	170	39.2
25～40	176	40.6
40～100	50	11.5
≥100	10	2.3

16 亮叶桦 liàng yè huà | Shiningleaf Birch

Betula luminifera，桦木属
桦木科 | Betulaceae

代码（SpCode）= BETLUM
个体数（Individual number/25 hm^2）= 2
最大胸径（Max DBH）= 20.63 cm
所在林层（Stratum）= 乔木层 (Canopy)
重要值排序（Importance value rank）= 46/50

落叶乔木。小枝褐色，被微柔毛。叶卵形、长卵形或长圆形，先端渐尖，基部圆形，有时近心形或宽楔形，边缘有不规则的刺毛状锯齿。雄花序2～5枚簇生于小枝顶端或单生于小枝上部叶腋。果序单生，长圆柱形；果苞被短柔毛及缘毛，中裂片长圆形、披针形或倒披针形，先端圆或渐尖，侧裂片极小；翅果倒卵形，翅伸出果苞之外。花期4～6月，果期5～8月。

树干 Trunk
摄影：卢志军 Photo by: Zhijun Lu

Deciduous trees. Branchlets brown, pubescent. Leaf blade ovate, oblong-ovate or elliptic, apex acuminate, base rounded, sometimes subcordate, or broadly cuneate, margin irregularly and doubly setiform serrate. Male inflorescences, 2-5 per cluster terminal on young branch or axillary on upper end of young branch. Female inflorescence solitary, narrowly cylindric, bracts pubescent and ciliolate, middle lobe oblong, lanceolate or oblanceolate, apex rounded or acuminate, lateral lobes tiny. Nutlets obovate, wings beyond bracts. Fl. Apr.–Jun., fr. May–Aug..

小枝和叶片 Branchlet and leaves
摄影：江明喜 Photo by: Mingxi Jiang

小枝和叶背 Branchlet and leaf abaxial surface
摄影：卢志军 Photo by: Zhijun Lu

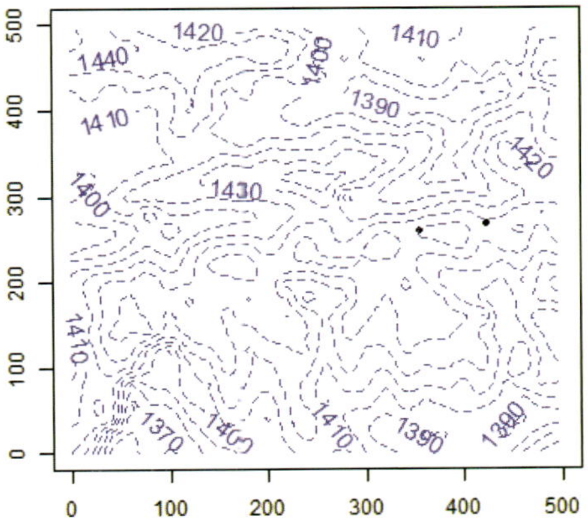

个体分布图 Distribution of individuals

径级分布表 DBH class

胸径区间 (Diameter class) (cm)	个体数 (No. of individuals in the plot)	比例 (Proportion) (%)
1～2.5	0	0.0
2.5～5	0	0.0
5～10	0	0.0
10～25	2	100.0
25～40	0	0.0
40～100	0	0.0
⩾100	0	0.0

17 雷公鹅耳枥 léi gōng é ěr lì | Vimineous Hornbeam

Carpinus viminea，鹅耳枥属
桦木科 | Betulaceae

代码（SpCode）= CARVIM
个体数（Individual number/25 hm^2）= 2430
最大胸径（Max DBH）= 72.73 cm
所在林层（Stratum）= 乔木层 (Canopy)
重要值排序（Importance value rank）= 4/50

树干 Trunk
摄影：卢志军 Photo by: Zhijun Lu

落叶乔木。树皮深灰色。小枝棕褐色，无毛。叶厚纸质，椭圆形、矩圆形、卵状披针形，顶端渐尖、尾状渐尖至长尾状，基部圆楔形、圆形兼有微心形，有时两侧略不等，边缘具规则或不规则的重锯齿。雌花序5～15×2.5～3 cm 果序下垂，序梗疏被短柔毛。小坚果宽卵圆形，无毛。

Deciduous trees. Bark dark gray. Branchlets brown, glabrous. Leaf blade thickly papery, elliptic-oblong, ovate-lanceolate, apex acuminate, caudate acuminate to long caudate, base rounded-cuneate, rounded or subcordate, sometimes slightly asymmetric, margin regularly or irregularly doubly mucronate serrate. Female inflorescence pendulous, 5-15 × 2.5-3 cm peduncle sparsely pubescent. Nutlet broadly ovoid, glabrous.

果枝 Fruiting branch
摄影：刘海波 Photo by: Haibo Liu

小枝和叶背 Branchlet and leaf abaxial surface
摄影：卢志军 Photo by: Zhijun Lu

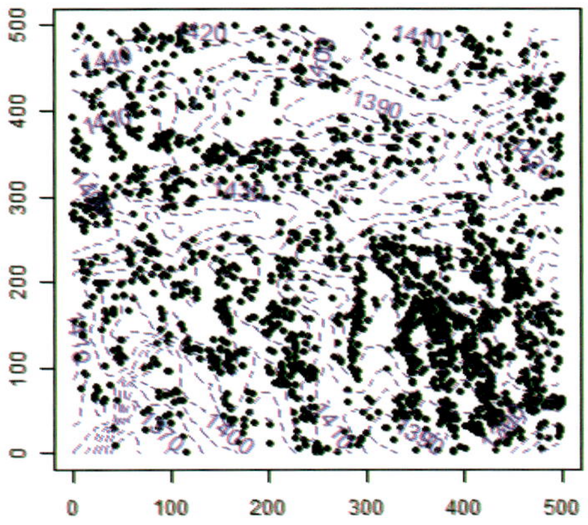

个体分布图 Distribution of individuals

径级分布表 DBH class

胸径区间 (Diameter class) (cm)	个体数 (No. of individuals in the plot)	比例 (Proportion) (%)
1～2.5	733	30.2
2.5～5	275	11.3
5～10	299	12.3
10～25	883	36.3
25～40	211	8.7
40～100	29	1.2
≥100	0	0.0

18 糙叶树

cāo yè shù | Scabrous Aphananthe

Aphananthe aspera，糙叶树属
榆科 | Ulmaceae

代码（SpCode）= APHASP
个体数（Individual number/25 hm^2）= 3
最大胸径（Max DBH）= 10.78 cm
所在林层（Stratum）= 乔木层 (Canopy)
重要值排序（Importance value rank）= 40/50

落叶乔木。小枝褐色，被糙毛。叶纸质，卵形至狭卵形，先端长渐尖，基部圆楔形，边缘有锐锯齿，基出脉3条，上面被糙毛，下面被短伏毛。雄聚伞花序生于新枝下部叶腋；雌花单生于新枝的上部叶腋。核果卵球形，被短伏毛，成熟时黑色。花期5月，果期5～6月。

Deciduous trees. Branchlets brown, coarsely tomentulose. Leaf blade papery, ovate to ovate-elliptic, apex narrowly acuminate, base rounded-cuneate, margin sharply serrate, 3-veined from base, adaxially coarsely tomentulose, abaxially pubescent. Male cyme axillary on new branch bottom; female flower solitary and axillary on new branch top. Drupes ovoid-globose, pubescent, black at maturity. Fl. May, fr. May–Jun..

树干 Trunk
摄影：卢志军 Photo by: Zhijun Lu

果枝 Fruiting branches
摄影：徐文秀、徐文斌 Photo by: Wenxiu Xu, Wenbin Xu

小枝和叶背 Branchlets and leaf abaxial surface
摄影：徐文秀 Photo by: Wenxiu Xu

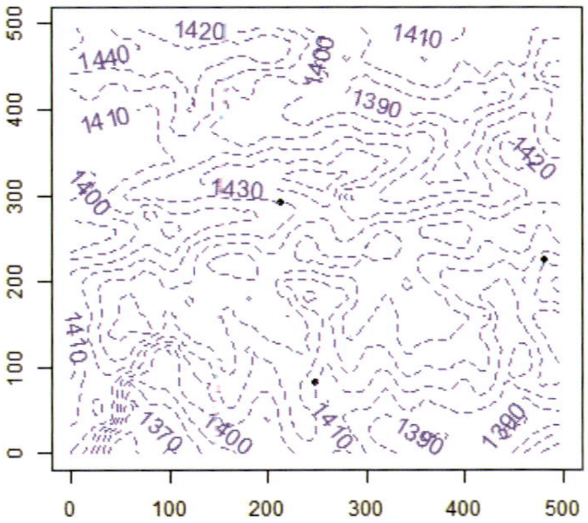

个体分布图 Distribution of individuals

径级分布表 DBH class

胸径区间 (Diameter class) (cm)	个体数 (No. of individuals in the plot)	比例 (Proportion) (%)
1～2.5	1	33.3
2.5～5	1	33.3
5～10	0	0.0
10～25	1	33.3
25～40	0	0.0
40～100	0	0.0
⩾100	0	0.0

19 朴树 pǔ shù | Chinese Hackberry

Celtis sinensis，朴属
榆科 | Ulmaceae

代码（SpCode）= CELSIN
个体数（Individual number/25 hm^2）= 10
最大胸径（Max DBH）= 13.70 cm
所在林层（Stratum）= 乔木层 (Canopy)
重要值排序（Importance value rank）= 31/50

落叶乔木。小枝密被柔毛。叶卵形至长圆状卵形，先端急尖，基部圆形，偏斜，上部边缘有粗锯齿，幼时两面均被柔毛，老后无毛，下面网脉明显。花聚生于叶腋或小枝基部。核果近球形，果核稍有凹点及棱脊；果梗与叶柄等长或稍长，被疏柔毛。花期5月，果熟期7~9月。

Deciduous trees. Branchlet densely pubescent. Leaf blade ovate to ovate-elliptic, apex acute, base rounded, oblique, margin coarsely dentate on apical half, usually puberulent when young, glabrous and abaxially conspicuous veins. Flowers fascicled in leaf axils or stem bases. Drupe subglobose. Stone slightly foveolate, ribbed. Fruit pedicel similar to or longer than petiole, pubescent. Fl. May, fr. maturity Jul.–Sept..

树干 摄影：卢志军
Trunk Photo by: Zhijun Lu

小枝和叶片 摄影：卢志军
Branchlets and leaves Photo by: Zhijun Lu

果枝 摄影：卢志军
Fruiting branches Photo by: Zhijun Lu

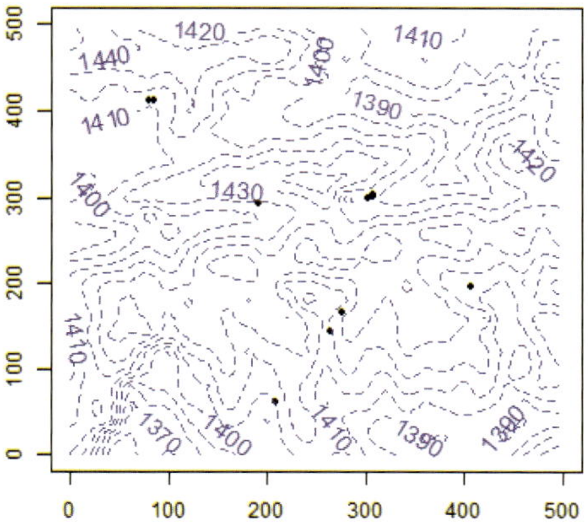

个体分布图 Distribution of individuals

径级分布表 DBH class

胸径区间 (Diameter class) (cm)	个体数 (No. of individuals in the plot)	比例 (Proportion) (%)
1~2.5	4	40.0
2.5~5	4	40.0
5~10	1	10.0
10~25	1	10.0
25~40	0	0.0
40~100	0	0.0
≥100	0	0.0

20 紫弹树　　zǐ dàn shù | Biond Hackberry

Celtis biondii，朴属
榆科 I Ulmaceae

代码（SpCode）= CELBIO
个体数（Individual number/25 hm^2）= 4
最大胸径（Max DBH）= 10.95 cm
所在林层（Stratum）= 亚乔木层 (Understory)
重要值排序（Importance value rank）= 50/54

树干　　Trunk
摄影：卢志军　　Photo by: Zhijun Lu

落叶灌木或小乔木。小枝被锈色茸毛。叶卵形至狭椭圆形，先端渐尖，基部宽楔形，不对称，边缘在中部以上有圆齿状锯齿，有时近全缘，上面被毛，粗糙，老时几乎无毛，下面干后棕色或棕黄色，初被毛，后无毛。核果卵形，果核有微细点。花期3～4月，果期5～7月。

Deciduous shrubs or small trees. Branchlets rusty pubescent. Leaf blade ovate to ovate-elliptic, apex acuminate, base broadly-cuneate, asymmetric, margin ovate dentate on apical half, sometimes subentire, adaxially coarsely tomentulose, nearly glabrous when aged, abaxially brown or brownish yellow when dried, tomentulose then glabrescent. Drupes ovate. Stones slightly foveolate. Fl. Mar.–Apr., fr. May–Jul..

小枝和叶片　　Branchlets and leaves
摄影：陈彬　　Photo by: Bin Chen

果枝和叶背　　Fruiting branches and leaf abaxial surface
摄影：陈彬　　Photo by: Bin Chen

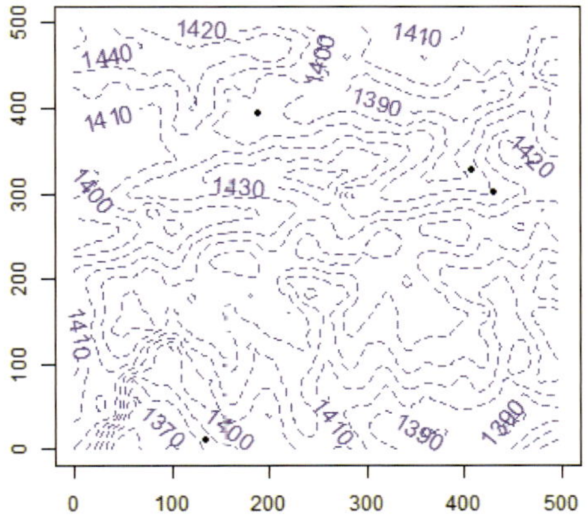

个体分布图 Distribution of individuals

径级分布表 DBH class

胸径区间 (Diameter class) (cm)	个体数 (No. of individuals in the plot)	比例 (Proportion) (%)
1～2.5	0	0.0
2.5～5	1	25.0
5～8	0	0.0
8～11	3	75.0
11～15	0	0.0
15～20	0	0.0
≥20	0	0.0

21 西川朴

xī chuān pǔ | Vandervoet Hackberry

Celtis vandervoetiana，朴属
榆科 | Ulmaceae

代码（SpCode）= CELVAN
个体数（Individual number/25 hm^2）= 3
最大胸径（Max DBH）= 12.28 cm
所在林层（Stratum）= 乔木层 (Canopy)
重要值排序（Importance value rank）= 39/50

落叶乔木。一年生枝无毛。叶革质，卵状椭圆形或卵形，近基部或中部以上边缘有单锯齿，上面无毛，下面脉隆起，网脉稍明显，除脉腋有疏毛外无毛，灰绿色。核果单生，卵状椭圆形，橙黄色，无毛；果核脐部有齿和穴，他处微有穴和肋。花期4月，果熟期8月。

树干 Trunk
摄影：卢志军 Photo by: Zhijun Lu

Deciduous trees. 1-yr branchlets glabrous. Leaf blade leathery, ovate-elliptic or ovate, margin dentate near base or on apical half, adaxially glabrescent, abaxially veins prominent and slightly conspicuous, glabrous except puberulous in vein axils, grayish green. Drupe solitary, ovate-elliptic, orange yellow, glabrous. Stone with teeth and cavity on umbilicus, cavity and ribs for the rest. Fl. Apr., fr. maturity Aug..

小枝和叶片 Branchlets and leaves
摄影：刘海波 Photo by: Haibo Liu

小枝和叶背 Branchlets and leaf abaxial surface
摄影：刘海波 Photo by: Haibo Liu

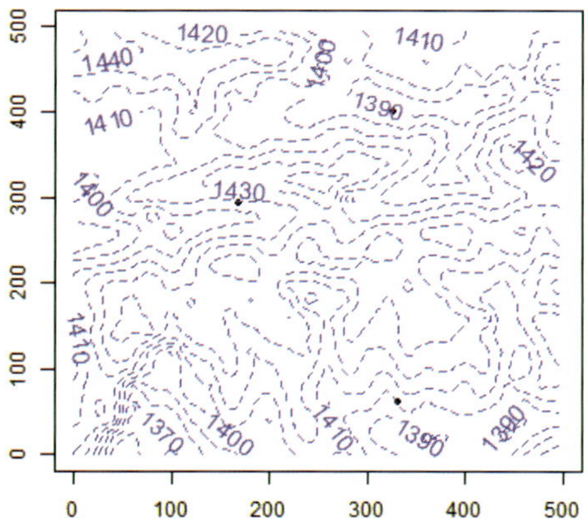

个体分布图 Distribution of individuals

径级分布表 DBH class

胸径区间 (Diameter class) (cm)	个体数 (No. of individuals in the plot)	比例 (Proportion) (%)
1～2.5	0	0.0
2.5～5	1	33.3
5～10	1	33.3
10～25	1	33.3
25～40	0	0.0
40～100	0	0.0
≥100	0	0.0

22 小果朴

xiǎo guǒ pǔ | Cherryfruit Hackberry

Celtis cerasifera，朴属
榆科 | Ulmaceae

代码（SpCode）= CELCER
个体数（Individual number/25 hm²）= 2
最大胸径（Max DBH）= 4.12 cm
所在林层（Stratum）= 乔木层 (Canopy)
重要值排序（Importance value rank）= 47/50

落叶乔木。一年生枝有角棱，无毛，或稍被柔毛，黄棕色，皮孔多。叶大，宽披针形、宽卵形或卵状椭圆形，先端短渐尖，基部圆形或近心形，不大对称。果黑色，有粉，球形，无毛。花期4月，果期9～10月。

Deciduous trees. 1-yr branchlets with edges and corners, glabrous, or slightly puberscent, yellowish brown, with many lenticels. Leaf blade large, broad-lanceolate, broad-ovate or ovate-elliptic, apex shortly acuminate, base rounded or subcordate, asymmetric. Drupes black, glaucous, globose, glabrous. Fl. Apr., fr. Sept.–Oct..

树干　Trunk
摄影：卢志军　Photo by: Zhijun Lu

小枝和叶片　Branchlets and leaves
摄影：卢志军　Photo by: Zhijun Lu

小枝和叶背　Branchlets and leaf abaxial surface
摄影：卢志军　Photo by: Zhijun Lu

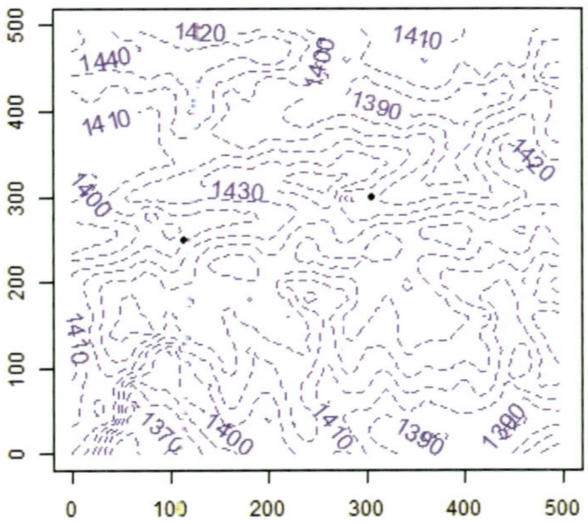

个体分布图 Distribution of individuals

径级分布表 DBH class

胸径区间 (Diameter class) (cm)	个体数 (No. of individuals in the plot)	比例 (Proportion) (%)
1～2.5	1	50.0
2.5～5	1	50.0
5～10	0	0.0
10～25	0	0.0
25～40	0	0.0
40～100	0	0.0
⩾100	0	0.0

23 兴山榆 xīng shān yú | Bergman Elm

Ulmus bergmanniana，榆属
榆科 | Ulmaceae

树干 Trunk
摄影：卢志军 Photo by: Zhijun Lu

代码（SpCode）= ULMBER
个体数（Individual number/25 hm^2）= 1
最大胸径（Max DBH）= 16.21 cm
所在林层（Stratum）= 乔木层 (Canopy)
重要值排序（Importance value rank）= 49/50

落叶乔木。枝条无毛，二年枝橄榄棕色或红棕色。叶倒卵形、倒卵状长圆形或椭圆形，少数为卵形，基部渐狭，基部圆形，不对称，先端尾状渐尖，上面多少被糙毛，下面脉腋被须状毛，其他各处被糙毛，或几乎无毛，边缘有重锯齿，无毛或上面被长柔毛。聚伞花序。翅果宽倒卵形或倒卵状圆形，基部突缩，先端有时有微缺，无毛；种子着生在果翅中央。果期5月。

Deciduous trees. Branchlets glabrous, 2-yr branchlets olive-brown or reddish brown. Leaf blade obovate, obovate-oblong or elliptic, rarely ovate, base attenuate and rounded, asymmetric, apex caudate acuminate, adaxially slightly coarsely tomentulose, abaxially bearded in vein axils and coarsely tomentulose for the rest or nearly glabrescent, margin doubly dentate, glabrous or adaxially pubescent. Cymes. Samaras broad-obovate or obovate-oblong, base suddenly shrunken, apex sometimes incised, glabrous. Seed at center of samara. Fr. May.

小枝和果序 Branchlets and infructescence
摄影：卢志军 Photo by: Zhijun Lu

小枝和叶背 Branchlets and leaf abaxial surface
摄影：卢志军 Photo by: Zhijun Lu

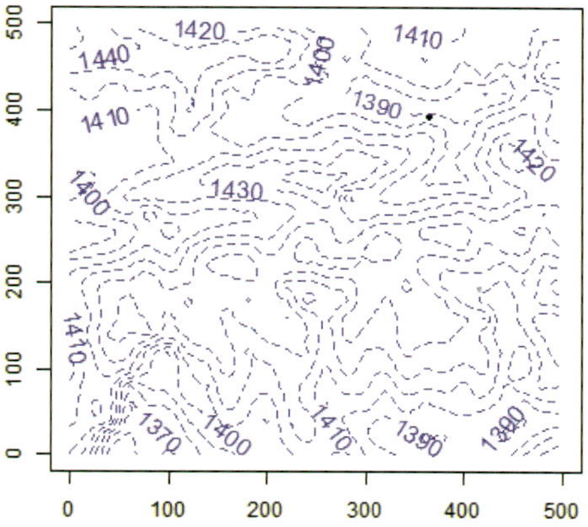

个体分布图 Distribution of individuals

径级分布表 DBH class

胸径区间 (Diameter class) (cm)	个体数 (No. of individuals in the plot)	比例 (Proportion) (%)
1～2.5	0	0.0
2.5～5	0	0.0
5～10	0	0.0
10～25	1	100.0
25～40	0	0.0
40～100	0	0.0
≥100	0	0.0

24 异叶榕 yì yè róng | Diverse-leaf Fig

Ficus heteromorpha，榕属
桑科 | Moraceae

代码（SpCode）= FICHET
个体数（Individual number/25 hm^2）= 2182
最大胸径（Max DBH）= 12.11 cm
所在林层（Stratum）= 灌木层 (Shrub)
重要值排序（Importance value rank）= 16/128

落叶灌木或小乔木，含白色乳汁。树皮红褐色，小枝无毛。叶互生，叶形多样，变化很大，通常为长椭圆形、倒卵状长椭圆形、三角状椭圆披针形，或有3裂，先端短钝尖或长渐尖，基部圆形或心形，全缘，上面深绿色，常有细毛，下面淡绿色，脉上有稀疏短毛。隐花果球形，无梗，上端有一嘴尖稍突出，光滑，成熟时呈紫红色。花期4～5月，果期7～11月。

Deciduous shrubs or small trees, with white milk. Bark reddish brown. Branchlets glabrous. Leaves alternate, shape diversified, oblong, obovate-oblong, triangular elliptic to lanceolate, or 3-lobed, apex shortly obtuse or narrowly acuminate, base rounded or cordate, margin entire, adaxially dark green, usually pubescent, abaxially light green, veins sparsely pubescent. Figs globose, pedicel absent, apical pore navel-like, smooth, purplish red when mature. Fl. Apr.–May, fr. Jul.–Nov..

树干 Trunk
摄影：卢志军 Photo by: Zhijun Lu

小枝和叶片 Branchlet and leaves
摄影：刘海波 Photo by: Haibo Liu

果枝 Fruiting branch
摄影：卢志军 Photo by: Zhijun Lu

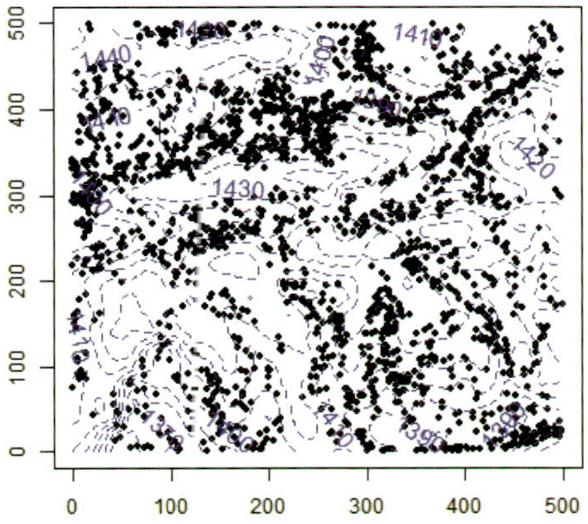

个体分布图 Distribution of individuals

径级分布表 DBH class

胸径区间 (Diameter class) (cm)	个体数 (No. of individuals in the plot)	比例 (Proportion) (%)
1～2	896	40.6
2～3	442	20.0
3～4	277	12.5
4～5	196	8.9
5～7	186	8.4
7～10	146	6.6
⩾10	39	1.8

25 鸡桑

jī sāng | Japanese Mulberry

Morus australis，桑属

桑科 | Moraceae

代码（SpCode）= MORAUS

个体数（Individual number/25 hm^2）= 13

最大胸径（Max DBH）= 10.62 cm

所在林层（Stratum）= 灌木层 (Shrub)

重要值排序（Importance value rank）= 92/128

落叶小乔木或灌木。树皮灰黑色，纵长裂，小枝光滑或幼嫩时有毛。叶卵形，先端渐尖，基部圆形或近心形，边缘有锐利或钝圆的锯齿，不分裂或分裂，上面稍粗糙，下面幼时稍有细毛。雌雄异株；雄花序淡黄色；雌花序绿色，柱头2裂，与花柱等长，宿存。桑葚果黄色，成熟时呈暗红色。花期4月，果期5~6月。

Deciduous small trees or shrubs. Bark grayish black, peeling in strips. Branchlets glabrous or tomentulose when young. Leaf blade ovate, apex acuminate, base rounded or subcordate, margin sharply or obtusely dentate, ± lobes, adaxially slightly coarse, abaxially slightly pubescent when young. Dioecious. Male catkins light yellow; female flowers green, stigma 2-branched, similar height to persistent style. Syncarp yellow, dark red when mature. Fl. Apr., fr. May–Jun..

树干 Trunk

摄影：卢志军 Photo by: Zhijun Lu

小枝和叶片 Branchlets and leaves

摄影：刘海波 Photo by: Haibo Liu

花枝 Flowering branches

摄影：卢志军 Photo by: Zhijun Lu

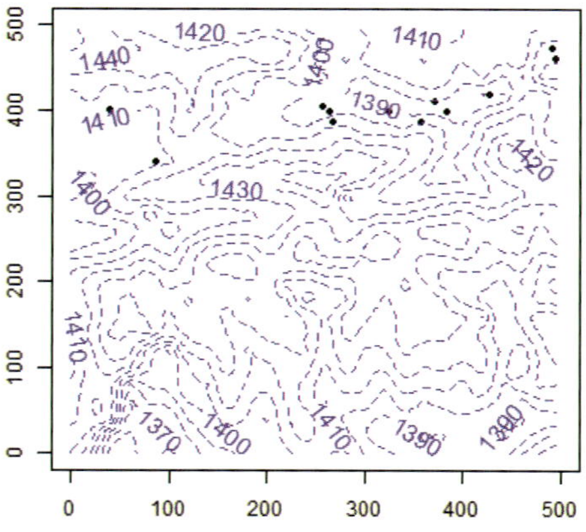

个体分布图 Distribution of individuals

径级分布表 DBH class

胸径区间 (Diameter class) (cm)	个体数 (No. of individuals in the plot)	比例 (Proportion) (%)
1~2	6	46.2
2~3	4	30.8
3~4	1	7.7
4~5	0	0.0
5~7	1	7.7
7~10	0	0.0
≥10	1	7.7

26 青皮木

qīng pí mù | Common Grey-twig

Schoepfia jasminodora，青皮木属
铁青树科 | Olacaceae

代码（SpCode）= SCHJAS
个体数（Individual number/25 hm^2）= 100
最大胸径（Max DBH）= 27.93 cm
所在林层（Stratum）= 灌木层 (Shrub)
重要值排序（Importance value rank）= 39/128

落叶灌木或小乔木。树皮灰色或灰褐色。叶卵形，先端尾状长渐尖，基部近圆形，全缘，两面光滑无毛。花两性，少数单性同株，由几朵小花集成腋生穗状花序或复穗状花序，花冠黄色，裂片长卵形，反卷。核果椭圆形，成熟时紫黑色，开裂。花期春季，6月果熟。

Deciduous shrubs or small trees. Bark gray or grayish brown. Leaf blade ovate, apex caudate narrowly acuminate, base subrounded, margin entire, glabrous on both surfaces. Flowers bisexual, rarely monoecious. Axillary spikes or compound spike inflorescences with several flowerlets; corolla yellow, lobes oblong, recurved. Drupe ellipsoid, purplish black when mature, dehiscence. Fl. Spring, fr. maturity Jun..

树干 Trunk
摄影：卢志军 Photo by: Zhijun Lu

小枝和叶片 Branchlets and leaves
摄影：段来军 Photo by: Laijun Duan

果枝 Fruiting branch
摄影：段来军 Photo by: Laijun Duan

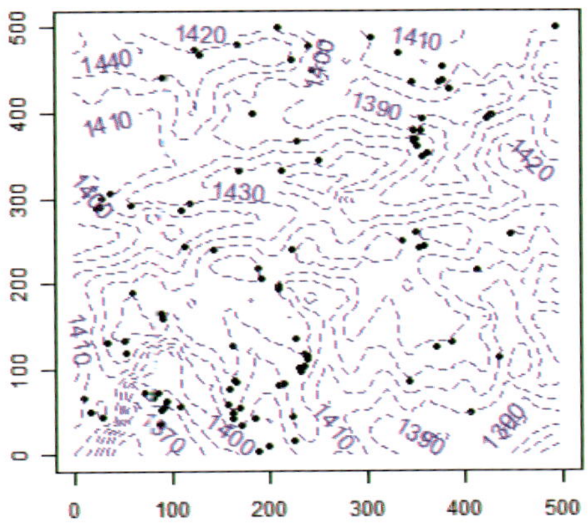

个体分布图 Distribution of individuals

径级分布表 DBH class

胸径区间 (Diameter class) (cm)	个体数 (No. of individuals in the plot)	比例 (Proportion) (%)
1～2	30	30.0
2～3	7	7.0
3～4	4	4.0
4～5	1	1.0
5～7	10	10.0
7～10	7	7.0
≥10	41	41.0

27 多花含笑

duō huā hán xiào | Many-flower Michelia

Michelia floribunda，含笑属
木兰科 I Magnoliaceae

代码（SpCode）= MICFLO
个体数（Individual number/25 hm²）= 1
最大胸径（Max DBH）= 5.38 cm
所在林层（Stratum）= 乔木层 (Canopy)
重要值排序（Importance value rank）= 50/50

树干 Trunk
摄影：卢志军 Photo by: Zhijun Lu

常绿乔木或小乔木。小枝褐色，被绢毛或近无毛。叶革质，披针形至长圆形，先端突渐尖，基部宽楔形，两面无毛，下面苍白色，两面细网脉明显。花单生枝顶，白色，总梗短，苞片密被长柔毛；萼片与花瓣倒披针形或楔形，先端急尖；雄蕊多数，有突尖，花药室平行，侧裂；雌蕊柄被长柔毛，雌蕊群短穗状，心皮卵形。果穗长7～8 cm，蓇葖果，种子1～3。花期2～4月，果期8～9月。

Trees or small trees. Branchlets brown, sericeous or nearly glabrous. Leaf blade leathery, lanceolate to oblong, apex acute, base broadly cuneate, glabrescent with conspicuous veins, abaxially pallescent. Flowers solitary on branch top white, short peduncle, bracts densely pubescent; sepal and petal oblanceolate or cuneate, apex acute; polyandrous, apex acute, anther parallel, lateral cleft; gynophore pubescent, female inflorescence shortly spicate, carpel ovate. Follicle fruit 7-8 cm, seeds 1-3. Fl. Feb.–Apr., fr. Aug.–Sept..

叶片 Leaves
摄影：卢志军 Photo by: Zhijun Lu

叶背 Leaf abaxial surface
摄影：卢志军 Photo by: Zhijun Lu

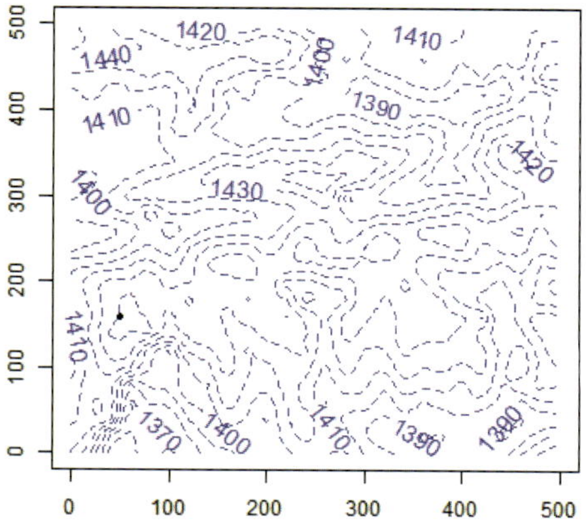

个体分布图 Distribution of individuals

径级分布表 DBH class

胸径区间 (Diameter class) (cm)	个体数 (No. of individuals in the plot)	比例 (Proportion) (%)
1～2.5	0	0.0
2.5～5	0	0.0
5～10	1	100.0
10～25	0	0.0
25～40	0	0.0
40～100	0	0.0
⩾100	0	0.0

28 乐东拟单性木兰 lè dōng nǐ dān xìng mù lán | Lotung Parakmeria

Parakmeria lotungensis，拟单性木兰属
木兰科 | Magnoliaceae

代码（SpCode）= PARLOT
个体数（Individual number/25 hm²）= 3
最大胸径（Max DBH）= 30.98 cm
所在林层（Stratum）= 乔木层 (Canopy)
重要值排序（Importance value rank）= 37/50

树干 Trunk
摄影：卢志军 Photo by: Zhijun Lu

常绿乔木。树皮灰白色；当年生枝绿色。叶革质，狭倒卵状椭圆形、倒卵状椭圆形或狭椭圆形，先端尖而头钝，基部楔形，或狭楔形；上面深绿色，有光泽。雄花两性花异株，花杂性。聚合果卵状长圆形或椭圆状卵圆形，很少倒卵形；种子椭圆形或椭圆状卵圆形，外种皮红色。花期4～5月，果期8～9月。

Evergreen trees. Bark grayish white. Annual twigs green. Leaf blade leathery, narrowly obovate-elliptic, obovate-elliptic, or narrowly elliptic, apex acuminate with an obtuse tip, base cuneate to narrowly cuneate, adaxially green and glossy. Male flower and bisexual flower. Polygamous. Fruit ellipsoid-ovoid or ellipsoidal, rarely obovoid. Seeds ellipsoid to ellipsoid-ovoid; testa red. Fl. Apr.–May, fr. Aug.–Sept..

花枝 Flowering branch
摄影：江明喜 Photo by: Mingxi Jiang

果枝 Fruiting branch
摄影：鲍大川 Photo by: Dachuan Bao

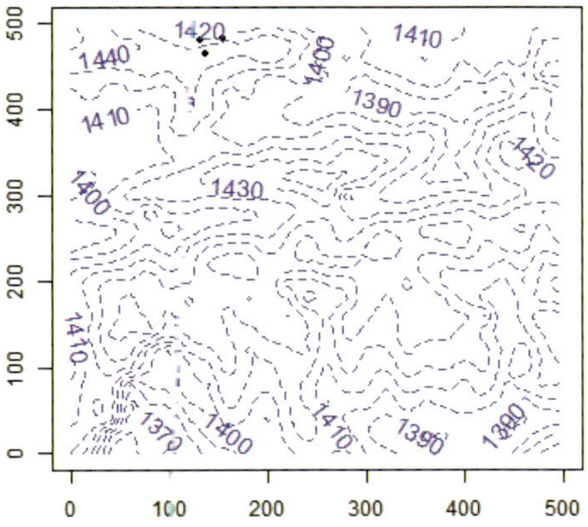

个体分布图 Distribution of individuals

径级分布表 DBH class

胸径区间 (Diameter class) (cm)	个体数 (No. of individuals in the plot)	比例 (Proportion) (%)
1～2.5	0	0.0
2.5～5	0	0.0
5～10	0	0.0
10～25	2	66.7
25～40	1	33.3
40～100	0	0.0
≥100	0	0.0

29 望春玉兰

wàng chūn yù lán | Biond Magnolia

Yulania biondii，玉兰属

木兰科 | Magnoliaceae

代码（SpCode）= YULBIO

个体数（Individual number/25 hm^2）= 7

最大胸径（Max DBH）= 27.68 cm

所在林层（Stratum）= 亚乔木层 (Understory)

重要值排序（Importance value rank）= 40/54

树干 Trunk

摄影：卢志军 Photo by: Zhijun Lu

落叶乔木。小枝无毛，二年生枝紫褐色。叶纸质，长圆状披针形，少数为倒披针状长圆形或卵状披针形，先端渐尖，基部圆形或宽楔形，上面无毛，下面沿中脉及侧脉基部被柔毛，两面网脉明显。花先叶开放；萼片线形；花瓣白色，圆匙形。蓇葖果圆柱形。花期3～5月，果期6～8月。

Deciduous trees. Branchlets glabrous; 2-yr branchlets purplish brown. Leaf blade papery, oblong-lanceolate, rarely lanceolate-oblong or ovate-lanceolate, apex acuminate, base rounded or broadly cuneate, veins conspicuous, adaxially grabrous, abaxially pubescent on midvein and secondary vein base. Flowers appearing before leaves; sepal linear; petal white, spatulate. Follicle cylindric. Fl. Mar.–May, fr. Jun.–Aug..

花 Flowers

摄影：江明喜 Photo by: Mingxi Jiang

蓇葖果与种子 Follicles and seeds

摄影：陈彬 Photo by: Bin Chen

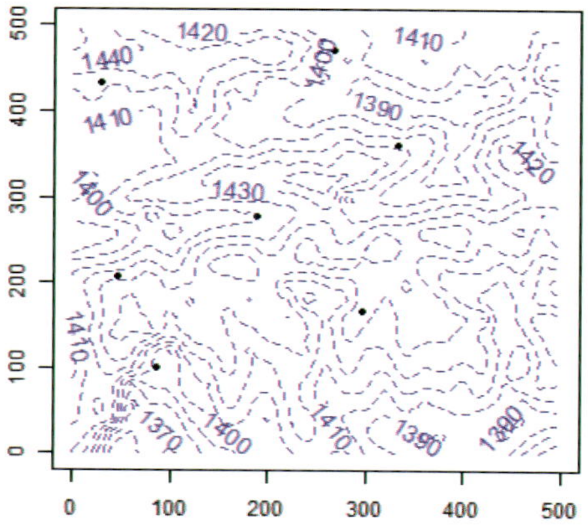

个体分布图 Distribution of individuals

径级分布表 DBH class

胸径区间 (Diameter class) (cm)	个体数 (No. of individuals in the plot)	比例 (Proportion) (%)
1～2.5	2	28.6
2.5～5	0	0.0
5～8	0	0.0
8～11	0	0.0
11～15	1	14.3
15～20	1	14.3
≥20	3	42.9

30 武当玉兰　wǔ dāng yù lán | Sprenger Magnolia

Yulania sprengeri，玉兰属
木兰科 | Magnoliaceae

代码（SpCode）= YULSPR
个体数（Individual number/25 hm^2）= 6
最大胸径（Max DBH）= 38.66 cm
所在林层（Stratum）= 亚乔木层 (Understory)
重要值排序（Importance value rank）= 37/54

落叶乔木。树皮苍灰棕色或黑棕色。幼枝无毛，有皮孔散生。花先叶开放，花芽有丝状长柔毛，花梗短，无毛；萼片3，不等大，外面的三角形，较小，先端渐尖，内面的较大，先端圆；花瓣长圆状匙状，先端圆，外面有深紫色纵纹，花柱先端弯曲。果实圆柱形，成熟后心皮棕色。花期3～4月，果期8～9月。

Deciduous trees. Bark pale grayish brown or blackish brown. Young branchlets glabrous, with scattered lenticels. Blossom bud filiform pubescent. Flowers appearing before leaves, peduncle short glabrous; 3-sepaled, asymmetric, outer triangular, smaller, apex acuminate, inner larger, apex rounded; petal oblong-elliptic spatulate, apex rounded, outside dark purple stria ted; style apex curved. Fruit cylindric; mature carpels brown. Fl. Mar.–Apr., fr. Aug.–Sept..

树干　Trunk
摄影：卢志军　Photo by: Zhijun Lu

花　Flowers
摄影：江明喜　Photo by: Mingxi Jiang

叶背　Leaf abaxial surface
摄影：卢志军　Photo by: Zhijun Lu

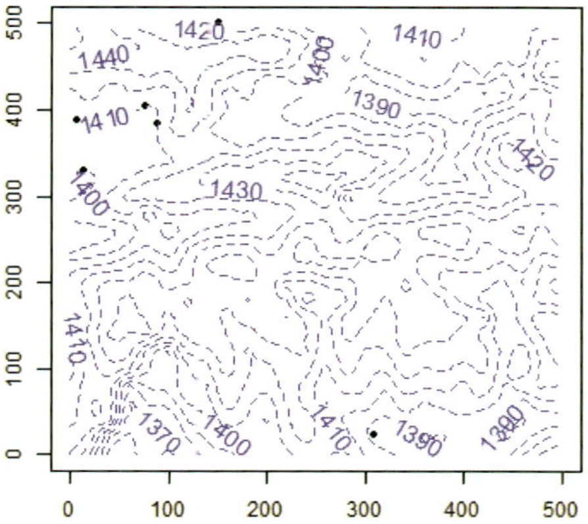

个体分布图 Distribution of individuals

径级分布表 DBH class

胸径区间 (Diameter class) (cm)	个体数 (No. of individuals in the plot)	比例 (Proportion) (%)
1～2.5	1	16.7
2.5～5	0	0.0
5～8	1	16.7
8～11	0	0.0
11～15	1	16.7
15～20	0	0.0
≥20	3	50.0

31 玉兰 yù lán | Yulan Magnolia

Yulania denudata，玉兰属
木兰科 | Magnoliaceae

代码（SpCode）= MAGDEN
个体数（Individual number/25 hm^2）= 3
最大胸径（Max DBH）= 34.33 cm
所在林层（Stratum）= 亚乔木层 (Understory)
重要值排序（Importance value rank）= 45/54

落叶乔木。小枝淡灰褐色或灰色，时有柔毛。叶宽倒卵形、倒卵形至长圆状倒卵形，先端短急尖，基部楔形或宽楔形，全缘。花单一顶生，先叶开放，白色，有芳香。蓇葖果圆柱状，淡褐色，果梗有毛。花期3～4月，果期5～11月。

Deciduous trees. Branchlets grayish brown or gray, pubescent when young. Leaf blade broadly obovate, obovate to oblong-obovate, apex shortly acute, base cuneate or broadly cuneate, margin entire. Flower solitary on branch apex, appearing prior to foliage, white, fragrant. Fruit cylindric, light brown, pedicel tomentulose. Fl. Mar.–Apr., fr. May–Nov..

树干 Trunk
摄影：卢志军 Photo by: Zhijun Lu

花 Flowers
摄影：江明喜 Photo by: Mingxi Jiang

小枝和叶背 Branchlet and leaf abaxial surface
摄影：卢志军 Photo by: Zhijun Lu

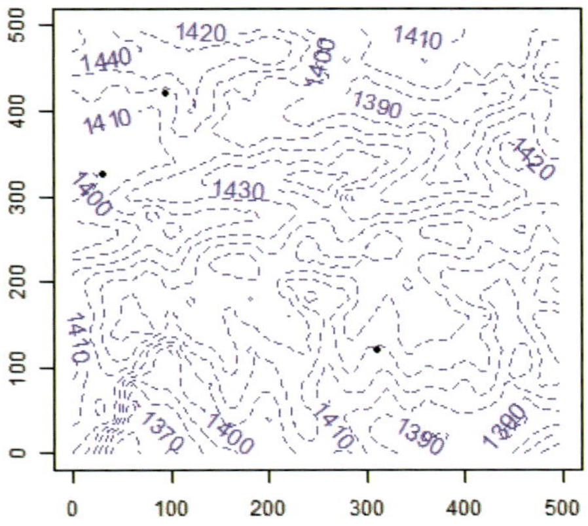

个体分布图 Distribution of individuals

径级分布表 DBH class

胸径区间 (Diameter class) (cm)	个体数 (No. of individuals in the plot)	比例 (Proportion) (%)
1～2.5	0	0.0
2.5～5	1	33.3
5～8	0	0.0
8～11	0	0.0
11～15	0	0.0
15～20	0	0.0
⩾20	2	66.7

32 野八角

yě bā jiǎo | Simons Anise-tree

Illicium simonsii，八角属
八角科 | Illiciaceae

树干 Trunk
摄影：卢志军 Photo by: Zhijun Lu

代码（SpCode）= ILLSIM
个体数（Individual number/25 hm^2）= 96
最大胸径（Max DBH）= 22.66 cm
所在林层（Stratum）= 灌木层 (Shrub)
重要值排序（Importance value rank）= 54/128

常绿灌木或乔木。幼枝褐色。叶近对生或互生，少数聚生节上，革质，披针形至椭圆形，或长圆状椭圆形，先端短渐尖，基部渐狭，边缘稍反卷，上面暗绿色，下面棕褐色，中脉在上面下凹，在下面隆起。花腋生，聚生枝顶端，淡黄色，芳香。花期2～5月，果期6～10月。

Evergreen shrubs or trees. Young branchlets brown. Leaf blade subopposite or alternate, rarely on nodes, leathery, lanceolate to elliptic, or oblong-elliptic, apex shortly acuminate, base decurrent, margin slightly revolute, adaxially dark green and midvein concave, abaxially brown and midvein prominent. Flowers axillary, compound on branch top, light yellow, fragrant. Fl. Feb. –May, fr. Jun.–Oct..

果枝 Fruiting branches
摄影：刘海波 Photo by: Haibo Liu

花 Flowers
摄影：卢志军 Photo by: Zhijun Lu

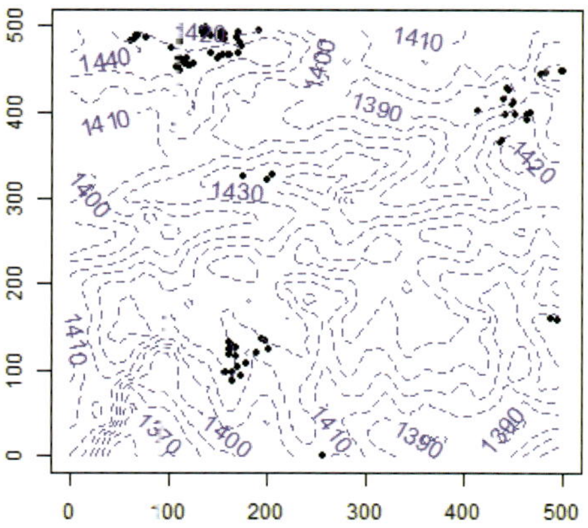

个体分布图 Distribution of individuals

径级分布表 DBH class

胸径区间 (Diameter class) (cm)	个体数 (No. of individuals in the plot)	比例 (Proportion) (%)
1～2	51	53.1
2～3	23	24.0
3～4	11	11.5
4～5	2	2.1
5～7	4	4.2
7～10	0	0.0
≥10	5	5.2

33 红果黄肉楠 hóng guǒ huáng ròu nán | Redfruit Actinodaphne

Actinodaphne cupularis，黄肉楠属
樟科 | Lauraceae

代码（SpCode）= ACTCUP
个体数（Individual number/25 hm^2）= 21
最大胸径（Max DBH）= 15.52 cm
所在林层（Stratum）= 灌木层 (Shrub)
重要值排序（Importance value rank）= 74/128

常绿灌木或小乔木。一年生枝被灰褐色毛，老枝褐灰色。叶近轮生，长椭圆形至倒披针形，先端急尖，基部楔形，全缘，稍呈波状。花序密束生，花梗短。浆果卵圆形，成熟时红色。花期11月至翌年3月，果期7～10月。

Evergreen shrubs or small trees. 1-yr branchlets puberulent, old brownish gray. Leaf blade subverticillate, oblong to oblanceolate, apex acute, base cuneate, margin entire, slightly crenate. Umbel inflorescence dense, peduncle short. Fruit ovoid, red when mature. Fl. Nov.–Mar. of following yr, fr. Jul.–Oct..

树干 Trunk
摄影：卢志军 Photo by: Zhijun Lu

叶片 Leaves
摄影：卢志军 Photo by: Zhijun Lu

叶背 Leaf abaxial surface
摄影：卢志军 Photo by: Zhijun Lu

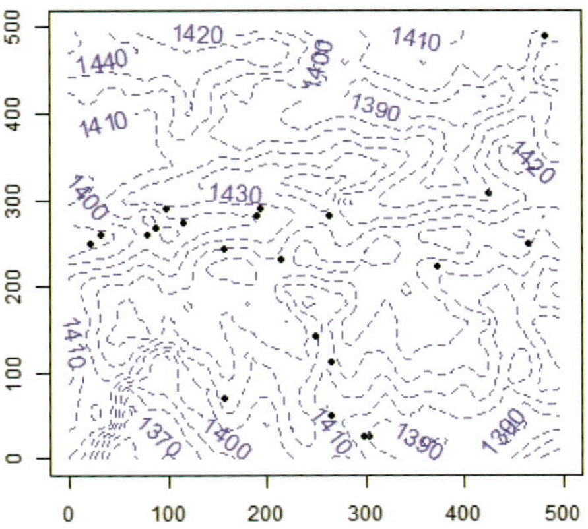

个体分布图 Distribution of individuals

径级分布表 DBH class

胸径区间 (Diameter class) (cm)	个体数 (No. of individuals in the plot)	比例 (Proportion) (%)
1～2	6	28.6
2～3	2	9.5
3～4	4	19.0
4～5	3	14.3
5～7	1	4.8
7～10	2	9.5
⩾10	3	14.3

34 川桂

chuān guì | Wilson Cinnamon

Cinnamomum wilsonii，樟属
樟科 | Lauraceae

代码（SpCode）= CINWIL
个体数（Individual number/25 hm^2）= 66
最大胸径（Max DBH）= 11.88 cm
所在林层（Stratum）= 乔木层 (Canopy)
重要值排序（Importance value rank）= 19/50

常绿乔木或小乔木。小枝圆，深棕色或带暗棕色。叶互生或近于对生，革质，椭圆状长圆形，先端渐尖，钝头，基部渐狭，沿叶柄下延，边缘反卷，带骨质，离基三出脉。花黄白色，横列成伞形或总状花序。果托杯状，截头。花期5～6月。

Evergreen trees or small trees. Branchlets rounded, nut brown or dark brown. Leaf blade alternate or subopposite, leathery, elliptic-oblong, apex acuminate, obtuse, base attenuate, decurrent along petiole, margin revolute, cartilaginous, 3-veined from base. Flowers yellowish white, umbels or racemes. Perianth cup scyphoid, obtuse. Fl. May–Jun..

树干 Trunk
摄影：卢志军 Photo by: Zhijun Lu

小枝和叶片 Branchlets and leaves
摄影：卢志军 Photo by: Zhijun Lu

小枝和叶背 Branchlets and leaf abaxial surface
摄影：卢志军 Photo by: Zhijun Lu

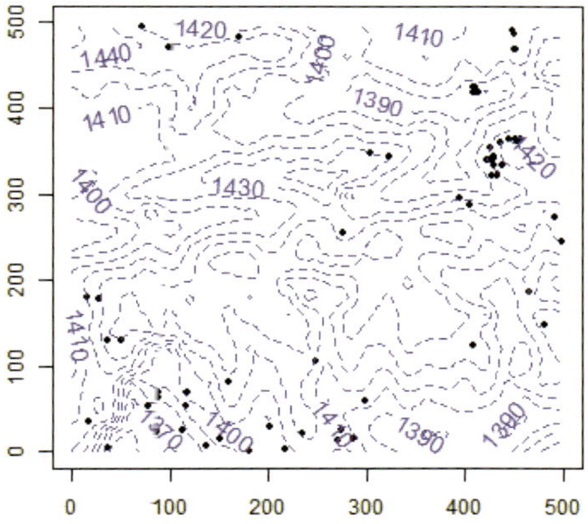

个体分布图 Distribution of individuals

径级分布表 DBH class

胸径区间 (Diameter class) (cm)	个体数 (No. of individuals in the plot)	比例 (Proportion) (%)
1～2.5	38	57.6
2.5～5	19	28.8
5～10	7	10.6
10～25	2	3.0
25～40	0	0.0
40～100	0	0.0
≥100	0	0.0

35 香粉叶　xiāng fěn yè | Pretty Spice-bush

Lindera pulcherrima var. *attenuata*，山胡椒属
樟科 | Lauraceae

代码（SpCode）= LINPUL
个体数（Individual number/25 hm^2）= 1512
最大胸径（Max DBH）= 11.18 cm
所在林层（Stratum）= 亚乔木层 (Understory)
重要值排序（Importance value rank）= 14/54

常绿乔木。小枝绿色，光滑。叶互生；叶片狭卵形、倒卵形、长圆形、长圆披针形或椭圆形；上面绿色，下面蓝灰色；幼时两面被白色茸毛，后无毛或近无毛；三出脉，上面略凸起，下面显著凸起；基部圆形或宽楔形；顶端渐尖或尾状渐尖。伞形花序。雄花被片椭圆形，等长。果实椭圆形。果期6～8月。

Evergreen trees. Branchlets green, smooth. Leaves alternate; leaf blade narrowly ovate, obovate, oblong, oblong-lanceolate, or elliptic, green adaxially, blue-gray abaxially, white pubescent on both surfaces when young, later glabrous or nearly glabrous, trinerved, distinctly convex abaxially, slightly convex adaxially, base rounded or broadly cuneate, apex acuminate or caudate-acuminate. Umbels. Male flower tepals elliptic, equal in length. Fruits ellipsoid. Fr. Jun.–Aug..

树干　Trunk
摄影：卢志军　Photo by: Zhijun Lu

果枝　Fruiting branch
摄影：刘海波　Photo by: Haibo Liu

小枝和叶背　Branchlets and leaf abaxial surface
摄影：卢志军　Photo by: Zhijun Lu

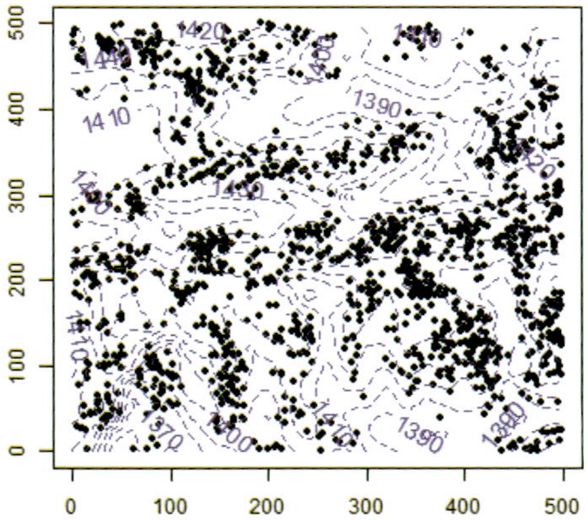

个体分布图　Distribution of individuals

径级分布表　DBH class

胸径区间 (Diameter class) (cm)	个体数 (No. of individuals in the plot)	比例 (Proportion) (%)
1～2.5	993	65.1
2.5～5	409	26.8
5～8	79	5.2
8～11	21	1.4
11～15	4	0.3
15～20	4	0.3
≥20	2	0.1

36 绿叶甘橿 lǜ yè gān jiāng | Nees Spice-bush

Lindera neesiana，山胡椒属
樟科 | Lauraceae

代码（SpCode）= LINNEE
个体数（Individual number/25 hm^2）= 831
最大胸径（Max DBH）= 18.69 cm
所在林层（Stratum）= 灌木层 (Shrub)
重要值排序（Importance value rank）= 33/128

落叶乔木。树皮绿色或绿棕色。幼枝光滑。叶互生，纸质，宽卵形，先端急尖或渐尖，下面带绿苍白色，无毛或稍被柔毛，有离基三出脉。伞形花序，总梗短。果短卵圆形至近球形，成熟时红色。花期4月，果期7～8月。

Deciduous trees. Bark green or green-brown. Bark green or green-brown. Young branchlets smooth. Leaves alternate, papery, broadly ovate, apex acute or acuminate, greenish pale abaxially, glabrous or slightly pubescent, 3-veined from base. Umbel with short peduncles. Fruits short ovoid to subglobose, red when mature. Fl. Apr., fr. Jul.–Aug..

树干 Trunk
摄影：卢志军 Photo by: Zhijun Lu

果枝 Fruiting branches
摄影：刘海波 Photo by: Haibo Liu

种子和叶背 Seeds and leaf abaxial surface
摄影：卢志军 Photo by: Zhijun Lu

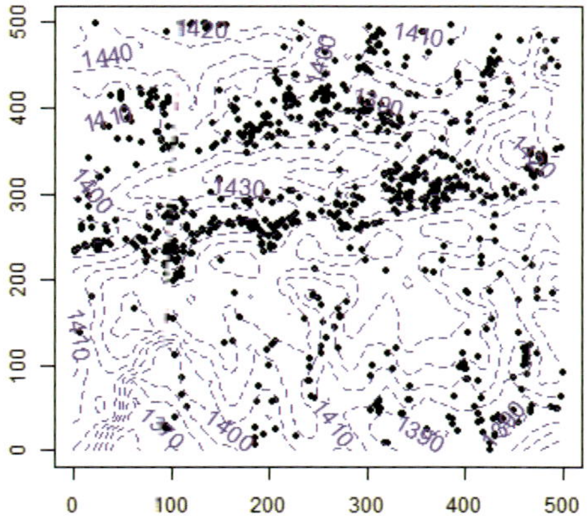

个体分布图 Distribution of individuals

径级分布表 DBH class

胸径区间 (Diameter class) (cm)	个体数 (No. of individuals in the plot)	比例 (Proportion) (%)
1～2	538	63.9
2～3	178	21.1
3～4	68	8.1
4～5	27	3.2
5～7	14	1.7
7～10	2	0.2
≥10	4	0.5

37 三桠乌药

sān yā wū yào | Obtuse-lobe Spice-bush

Lindera obtusiloba，山胡椒属
樟科 I Lauraceae

代码（SpCode）= LINOBT
个体数（Individual number/25 hm^2）= 801
最大胸径（Max DBH）= 55.04 cm
所在林层（Stratum）= 灌木层 (Shrub)
重要值排序（Importance value rank）= 10/128

落叶灌木或小乔木。树皮黑棕色。叶卵圆形至近圆形，基部截形至圆形，或为心形，先端通常为三裂，裂片钝，卵圆状三角形，基部三出脉。花先叶开放，伞形花序。果球形，暗红色或灰黑色。花期4月，果期8～9月。

Deciduous shrubs or small trees. Bark black-brown. Leaf blade ovate to subrounded, base truncate to rounded, or cordate, apex 3-lobed, lobes obtuse and ovoid-triangular, 3-veined from base. Umbels, flowers appearing before leaves. Fruit globose; dark red or grayish black. Fl. Apr., fr. Aug.–Sept..

树干 Trunk
摄影：卢志军 Photo by: Zhijun Lu

小枝和叶片 Branchlets and leaves
摄影：刘海波 Photo by: Haibo Liu

小枝和叶背 Branchlets and leaf abaxial surface
摄影：刘海波 Photo by: Haibo Liu

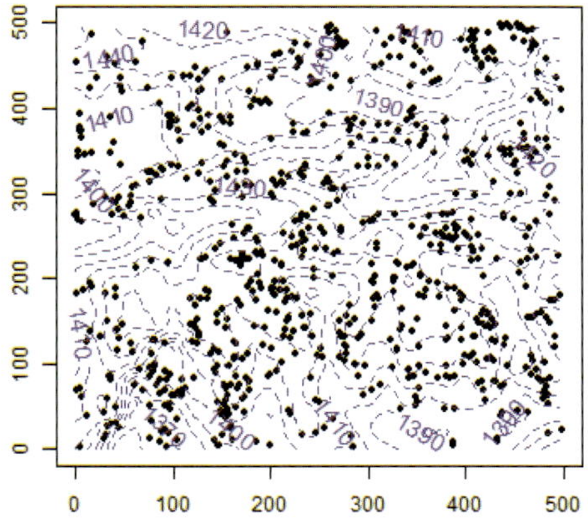

个体分布图 Distribution of individuals

径级分布表 DBH class

胸径区间 (Diameter class) (cm)	个体数 (No. of individuals in the plot)	比例 (Proportion) (%)
1～2.5	143	17.4
2.5～5	102	12.4
5～8	56	6.8
8～11	31	3.8
11～15	61	7.4
15～20	85	10.3
⩾20	323	39.2

38 红果山胡椒 hóng guǒ shān hú jiāo | Red-fruit Spice-bush

Lindera erythrocarpa，山胡椒属
樟科 | Lauraceae

代码（SpCode）= LINERY
个体数（Individual number/25 hm^2）= 720
最大胸径（Max DBH）= 59.95 cm
所在林层（Stratum）= 灌木层 (Shrub)
重要值排序（Importance value rank）= 5/128

落叶乔木。树皮灰棕色。小枝无毛，或幼时稍被柔毛。叶纸质，倒披针形或倒披针状长圆形，先端长渐尖，基部楔形，下延，上面无毛而稍发亮，下面稍被粉，无毛，或在叶脉以外各处稍被长柔毛；叶脉羽状，在下面稍隆起。花序伞形，总苞片早落，伞形花半圆形。果球形，紫红色。花期5月，果期7～10月。

树干 Trunk
摄影：卢志军 Photo by: Zhijun Lu

Deciduous trees. Bark gray-brown. Branchlets glabrous, or slightly pubescent when young. Leaf blade papery, oblanceolate or oblanceolate-oblong, apex long acuminate, base cuneate, decurrent, adaxially glabrouse and slightly glossy, abaxially slightly glaucous and glabrous, or slightly pubescent except veins; pinninerved, abaxially slightly prominent. Umbels, phyllary caducous, corrola semicircular. Fruit globose, purplish red. Fl. May, fr. Jul.–Oct..

小枝和叶片 Branchlets and leaves
摄影：覃必伍 Photo by: Biwu Qin

小枝和叶背 Branchlets and leaf abaxial surface
摄影：覃必伍 Photo by: Biwu Qin

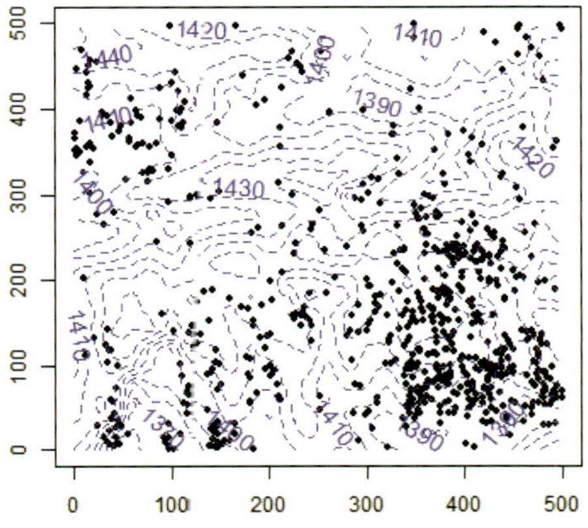

个体分布图 Distribution of individuals

径级分布表 DBH class

胸径区间 (Diameter class) (cm)	个体数 (No. of individuals in the plot)	比例 (Proportion) (%)
1～2	32	4.4
2～3	17	2.3
3～4	7	1.0
4～5	19	2.6
5～7	38	5.2
7～10	80	11.0
≥10	527	72.4

39 香叶树　xiāng yè shù | Obtuse-lobe Spice-bush

Lindera communis，山胡椒属
樟科 | Lauraceae

代码（SpCode）= LINCOM
个体数（Individual number/25 hm^2）= 149
最大胸径（Max DBH）= 6.63 cm
所在林层（Stratum）= 灌木层 (Shrub)
重要值排序（Importance value rank）= 55/128

常绿灌木或乔木。树皮棕色。一年生小枝密被茸毛，老枝灰褐色，被微柔毛或近无毛。叶互生，厚革质，椭圆状披针形，有时为披针形，少数为卵形，先端急尖至渐尖，有时为长尾状渐尖，基部渐狭或楔形，上面绿色带棕褐色，无毛而发亮，下面浅绿色而带浅棕色，有毛或有时无毛。伞形花序，总梗短，花黄色。果红色，卵形，位于小杯状的花被筒上。花期3～4月，果期9～10月。

Evergreen shrubs or trees. Bark brown. 1-yr branchlets densely pubescent; old grayish brown, pubescent or subglabrous. Leaf blade alternate, thickly leathery, elliptic-lanceolate, sometimes lanceolate, rarely ovate, apex acute to acuminate, sometimes long caudate acuminate, base tapering or cuneate, adaxially green with nut-brown, glabrous and glossy, abaxially light green with light brown, tomentulose or sometimes glabrous. Umbels, peduncles short, flowers yellow. Fruit red, ovoid, on cotyloid perianth tube. Fl. Mar.–Apr., fr. Sept.–Oct..

树干　Trunk
摄影：卢志军　Photo by: Zhijun Lu

花枝　Flowering branch
摄影：徐耀粘　Photo by: Yaozhan Xu

果枝　Fruiting branch
摄影：刘海波　Photo by: Haibo Liu

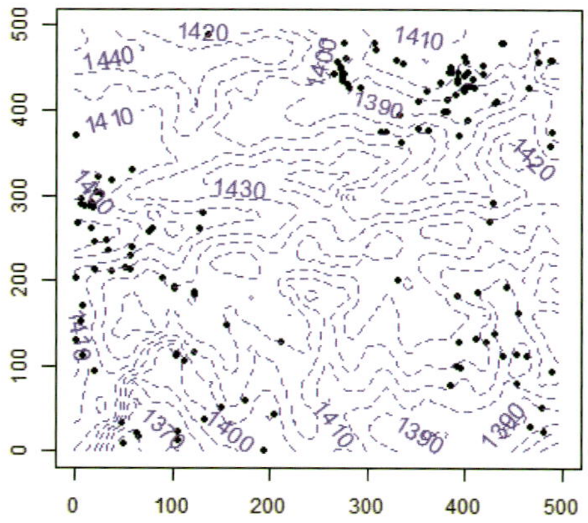

个体分布图 Distribution of individuals

径级分布表 DBH class

胸径区间 (Diameter class) (cm)	个体数 (No. of individuals in the plot)	比例 (Proportion) (%)
1～2	117	78.5
2～3	21	14.1
3～4	6	4.0
4～5	2	1.3
5～7	3	2.0
7～10	0	0.0
≥10	0	0.0

40 山胡椒 shān hú jiāo | Grey-blue Spice-bush

Lindera glauca，山胡椒属
樟科 | Lauraceae

代码（SpCode）= LINGLA
个体数（Individual number/25 hm^2）= 73
最大胸径（Max DBH）= 19.29 cm
所在林层（Stratum）= 灌木层 (Shrub)
重要值排序（Importance value rank）= 59/128

树干 Trunk
摄影：卢志军 Photo by: Zhijun Lu

落叶灌木或乔木。树支灰或灰白色。小枝深灰色或灰棕色，被微柔毛，后转为无毛，皮孔稍显。叶互生，薄革质，长圆状椭圆形，先端宽急尖，基部圆形或渐尖。雌雄异株，伞形花序，近无总梗，先叶或与叶同时开放；雌花着生二年生枝条上，腋生，花梗被柔毛，花绿黄色，无毛，子房无毛。果球形，黑色。花期4月，果期7～8月。

Deciduous shrubs or trees. Bark gray or gray-white. Young branchlets dark gray or grayish brown, pubescent, later glabrate, lenticels slightly prominent. Leaf blade alternate, thinly leathery, oblong-elliptic, apex broadly acute, base rounded or acuminate. Dioecious. Umbels, peduncles nearly absent, appearing before leaves or simultaneously; Female flowers on 2-yr branchlets, axillary, peduncles pubescent; flowers greenish yellow, glabrous; ovary glabrescent. Fruit globose, black. Fl. Apr., fr. Jul.–Aug..

果枝 Fruiting branch
摄影：卢志军、江明喜 Photo by: Zhijun Lu, Mingxi Jiang

果枝 Fruiting branch
摄影：卢志军 Photo by: Zhijun Lu

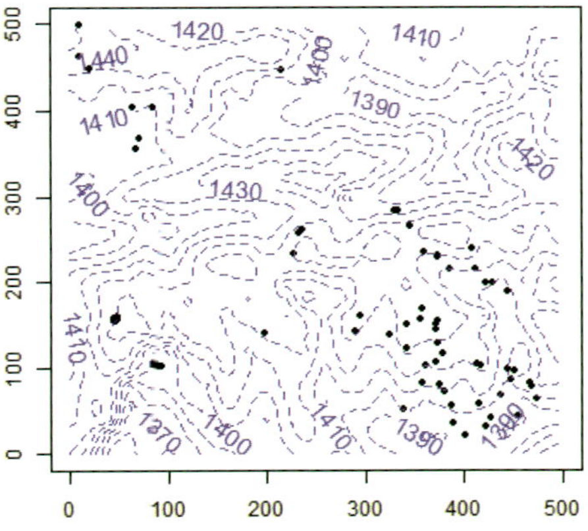

个体分布图 Distribution of individuals

径级分布表 DBH class

胸径区间 (Diameter class) (cm)	个体数 (No. of individuals in the plot)	比例 (Proportion) (%)
1～2	23	31.5
2～3	14	19.2
3～4	14	19.2
4～5	6	8.2
5～7	9	12.3
7～10	3	4.1
≥10	4	5.5

41 香叶子

xiāng yè zǐ | Fragrant Spice-bush

Lindera fragrans，山胡椒属
樟科 | Lauraceae

代码（SpCode）= LINFRA
个体数（Individual number/25 hm^2）= 48
最大胸径（Max DBH）= 7.24 cm
所在林层（Stratum）= 灌木层 (Shrub)
重要值排序（Importance value rank）= 73/128

常绿小乔木。树皮黄褐色。幼枝青绿或棕黄色。叶互生；披针形至长狭卵形，先端渐尖，基部楔形或宽楔形；上面绿色，无毛；下面绿带苍白色，无毛或被白色微柔毛；三出脉。伞形花序腋生。果长卵形，成熟时紫黑色，有疏柔毛，果托膨大。

Evergreen small trees. Bark yellow-brown. Young branchlets blue-green or brown-yellow. Leaves alternate; leaf blade lanceolate or narrowly ovate, apex acuminate, base cuneate or broadly cuneate, green adaxially, glabrous, pale green abaxially, glabrous or white pubescent, trinerved. Umbels axillary. Fruits ovate, purple-black at maturity, laxly pubescent, hypocarpium inflated.

树干 Trunk
摄影：卢志军 Photo by: Zhijun Lu

花枝 Flowering branch
摄影：徐文斌 Photo by: Wenbin Xu

花枝 Flowering branch
摄影：徐文斌 Photo by: Wenbin Xu

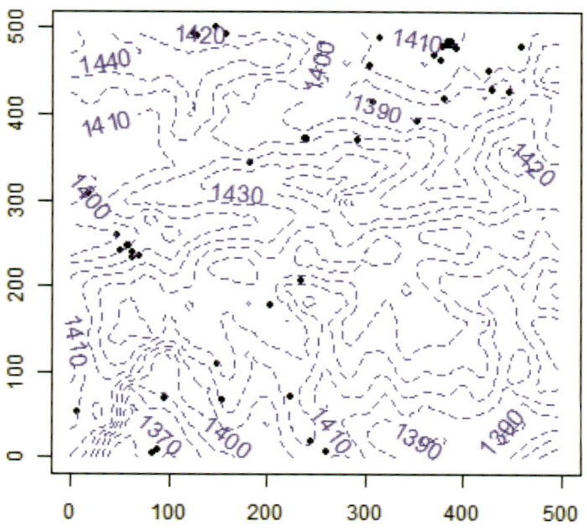

个体分布图 Distribution of individuals

径级分布表 DBH class

胸径区间 (Diameter class) (cm)	个体数 (No. of individuals in the plot)	比例 (Proportion) (%)
1～2	35	72.9
2～3	9	18.8
3～4	3	6.3
4～5	0	0.0
5～7	0	0.0
7～10	1	2.1
≥10	0	0.0

42 黑壳楠 hēi ké nán | Large-leaf Spice-bush

Lindera megaphylla，山胡椒属
樟科 | Lauraceae

代码（SpCode）= LINMEG
个体数（Individual number/25 hm^2）= 2
最大胸径（Max DBH）= 28.35 cm
所在林层（Stratum）= 乔木层 (Canopy)
重要值排序（Importance value rank）= 45/50

常绿乔木。树皮灰黑色。小枝紫黑色，光滑或被毛。叶互生，厚革质，披针形，或披针状长圆形，先端急尖，基部渐尖，无毛或近无毛，上面暗绿色，有光亮，下面灰绿色或稍带绿苍白色。伞形花序，花多数，黄绿色，总苞片广宽，盔状，被细绢毛。果大，黑色，卵圆形，生在杯状果托上。花期1～3月，果熟期8～10月。

树干 Trunk
摄影：卢志军 Photo by: Zhijun Lu

Evergreen trees. Bark gray-black. Branchlets purple-black, glabrous or pubescent. Leaves alternate; leaf blade thickly leathery, lanceolate or lanceolate-oblong, apex acute, base attenuate, glabrous or subgrabrous, adaxially dark green, glossy, abaxially grayish green or with greenish pale. Umbel, polymerous flowers, yellowish green, phyllary broad, helmet shaped, pubescent. Fruit large, black, ovoid, on cup-shaped hypocarpium. Fl. Jan.–Mar., fr. maturity Aug.–Oct..

花 Flower
摄影：刘海波 Photo by: Haibo Liu

叶背 Leaf abaxial surface
摄影：刘海波 Photo by: Haibo Liu

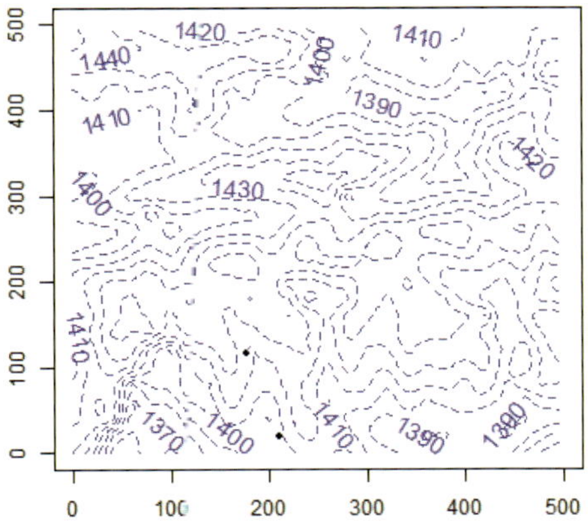

个体分布图 Distribution of individuals

径级分布表 DBH class

胸径区间 (Diameter class) (cm)	个体数 (No. of individuals in the plot)	比例 (Proportion) (%)
1～2.5	0	0.0
2.5～5	1	50.0
5～10	0	0.0
10～25	0	0.0
25～40	1	50.0
40～100	0	0.0
≥100	0	0.0

43 黄丹木姜子 huáng dān mù jiāng zǐ | Elongate Litse

Litsea elongata，木姜子属
樟科 | Lauraceae

代码（SpCode）= LITELO
个体数（Individual number/25 hm^2）= 20905
最大胸径（Max DBH）= 45.1 cm
所在林层（Stratum）= 亚乔木层 (Understory)
重要值排序（Importance value rank）= 2/54

树干 Trunk
摄影：卢志军 Photo by: Zhijun Lu

常绿灌木或小乔木。枝条密被锈色柔毛。叶互生，常聚生枝顶，革质，长圆形、倒披针形或椭圆状披针形，先端钝急尖或渐尖，基部楔形或圆形。花序伞形；总苞片革质。果长椭圆形，位于杯状花被筒上，熟时黑紫色。花期8月。

Evergreen shrubs or small trees. Branchlets densely brown tomentose. Leaves alternate, usually on branchlet top, leathery; leaf blade oblong, oblanceolate or elliptic-lanceolate, apex obtusely acute or acuminate, base cuneate or rounded. Umbels, phyllary leathery. Fruit oblong, on cup-shaped perianth tube, black-purple at maturity. Fl. Aug..

花枝 Flowering branch
摄影：刘海波 Photo by: Haibo Liu

果枝 Fruiting branch
摄影：刘海波 Photo by: Haibo Liu

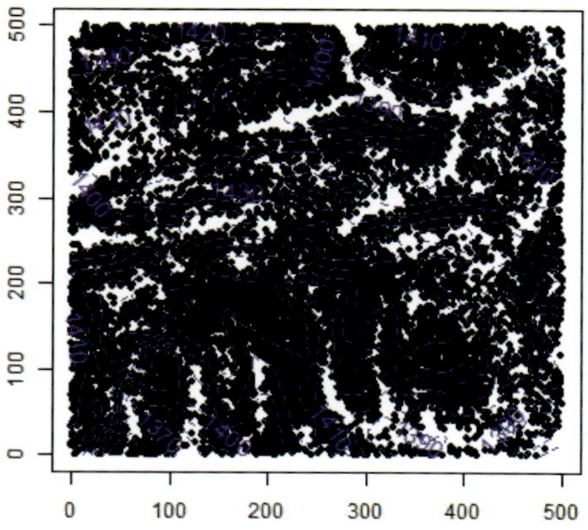

个体分布图 Distribution of individuals

径级分布表 DBH class

胸径区间 (Diameter class) (cm)	个体数 (No. of individuals in the plot)	比例 (Proportion) (%)
1～2.5	10728	50.8
2.5～5	8276	39.2
5～8	1236	5.9
8～11	411	1.9
11～15	199	0.9
15～20	33	0.2
≥20	22	0.1

44 钝叶木姜子 dùn yè mù jiāng zǐ | Veitch Litse

Litsea veitchiana，木姜子属
樟科 | Lauraceae

代码（SpCode）= LITVEI
个体数（Individual number/25 hm^2）= 322
最大胸径（Max DBH）= 35.57 cm
所在林层（Stratum）= 灌木层 (Shrub)
重要值排序（Importance value rank）= 28/128

落叶灌木或小乔木。树皮灰褐色或黑褐色。幼枝被黄白色长绢毛，后脱落无毛。叶互生，倒卵形或倒卵状长圆形，先端急尖或钝，基部楔形或宽楔形，纸质，下面有稀疏长绢毛。伞形花序生于去年枝顶，单生，淡黄色。果球形，成熟时黑色。花期4～5月，果期8～9月。

Deciduous shrubs or small trees. Bark grayish brown or blackish brown. Young branchlets yellow-white long sericeous then glabrous. Leaves alternate, leaf blade obovate or obovate-oblong, apex acute or obtuse, base cuneate or broadly cuneate, papery, abaxially sparsely pubescent. Umbels solitary on apex of previous yr branchlets, flowers light yellow. Fruit globose, black at maturity. Fl. Apr.–May, fr. Aug.–Sept..

树干 Trunk
摄影：卢志军 Photo by: Zhijun Lu

花和果枝 Flowers and fruiting branch
摄影：刘海波 Photo by: Haibo Liu

果枝 Fruiting branch
摄影：刘海波 Photo by: Haibo Liu

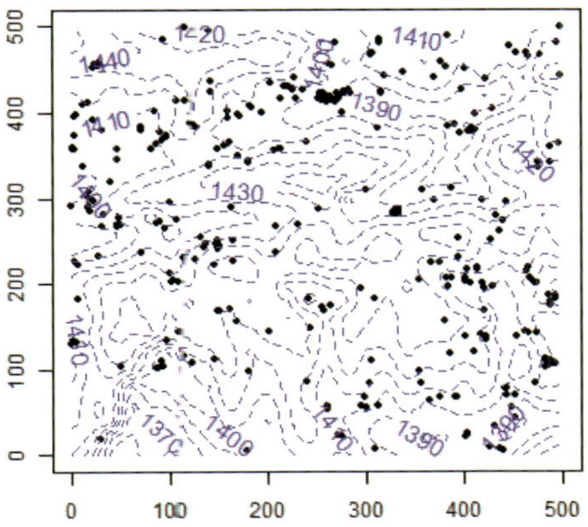

个体分布图 Distribution of individuals

径级分布表 DBH class

胸径区间 (Diameter class) (cm)	个体数 (No. of individuals in the plot)	比例 (Proportion) (%)
1～2	145	44.6
2～3	52	16.0
3～4	12	3.7
4～5	6	1.8
5～7	11	3.4
7～10	24	7.4
≥10	72	22.2

45 山鸡椒

shān jī jiāo | Mountain Spicy-tree

Litsea cubeba，木姜子属
樟科 | Lauraceae

代码（SpCode）= LITCUB
个体数（Individual number/25 hm^2）= 13
最大胸径（Max DBH）= 6.00 cm
所在林层（Stratum）= 灌木层 (Shrub)
重要值排序（Importance value rank）= 96/128

落叶灌木或小乔木。小枝光滑或被丝状毛。叶及果实有芳香。叶互生，膜质，披针形或长圆状披针形，先端渐尖，基部楔形，上面绿而发亮，下面绿苍白色，干后变黑。花与叶同时或先叶开放，伞形花；花黄白色。果近球形，黑色。花期4月，果期6~7月。

Deciduous shrubs or small trees. Branchlets glabrous or sericeous-pubescent. Leaves and fruits fragrant. Leaves alternate; leaf blade membranous, lanceolate or oblanceolate, apex acuminate, base cuneate, adaxially green and glossy, abaxially greenish pale, black when dried. Umbels, flower yellowish white, appearing with or before leaves. Fruit subglobose, black. Fl. Apr., fr. Jun.–Jul..

树干 Trunk
摄影：卢志军 Photo by: Zhijun Lu

果枝 Fruiting branch
摄影：鲍大川 Photo by: Dachuan Bao

小枝和叶背 Branchlets and leaf abaxial surface
摄影：卢志军 Photo by: Zhijun Lu

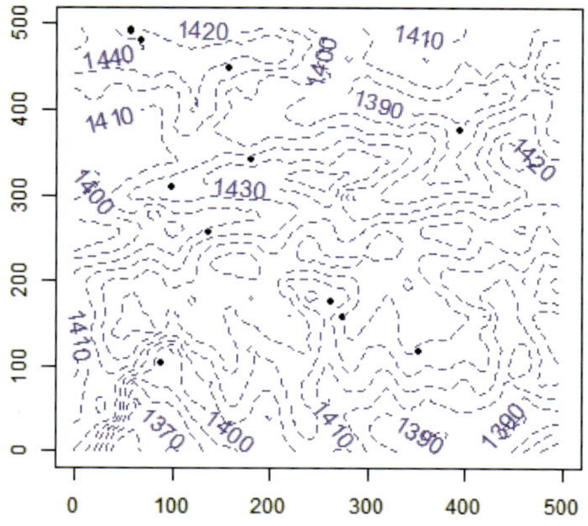

个体分布图 Distribution of individuals

径级分布表 DBH class

胸径区间 (Diameter class) (cm)	个体数 (No. of individuals in the plot)	比例 (Proportion) (%)
1~2	9	69.2
2~3	2	15.4
3~4	0	0.0
4~5	1	7.7
5~7	1	7.7
7~10	0	0.0
≥10	0	0.0

46 毛叶木姜子

máo yè mù jiāng zǐ | Hairyleaf Litse

Litsea mollis，木姜子属
樟科 | Lauraceae

代码（SpCode）= LITMOL
个体数（Individual number/25 hm^2）= 12
最大胸径（Max DBH）= 3.08 cm
所在林层（Stratum）=灌木层 (Shrub)
重要值排序（Importance value rank）= 97/128

落叶灌木或小乔木，树皮绿色，光滑，有黑斑。叶纸质，互生或聚生枝顶，长圆形或椭圆形，先端突尖，基部楔形，下面羽状脉，突起。伞形花序腋生。果球形，成熟时蓝黑色。花期3～4月，果期9～10月。

Deciduous shrubs or small trees. Bark green, smooth, black-spotted. Leaf blade papery, alternate or clustered at apex of branchlet, oblong or elliptic, apex abruptly acute, base cuneate, abaxially with pinnate and prominent veins. Umbels axillary. Fruit globose, blue-black at maturity. Fl. Mar.–Apr., fr. Sept.–Oct..

树干 Trunk
摄影：卢志军 Photo by: Zhijun Lu

小枝和叶片 Branchlets and leaves
摄影：卢志军 Photo by: Zhijun Lu

小枝和叶背 Branchlets and leaf abaxial surface
摄影：卢志军 Photo by: Zhijun Lu

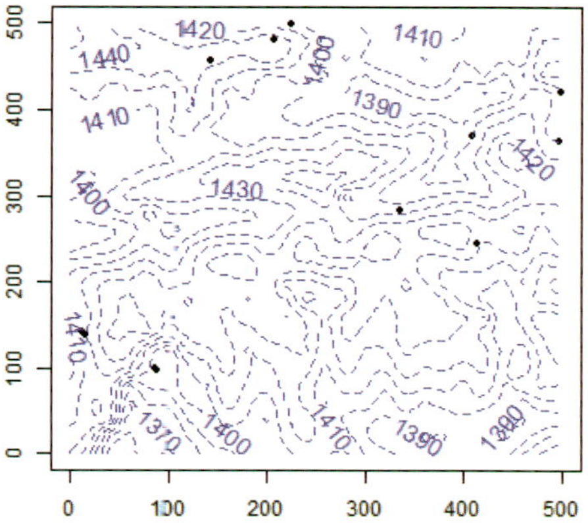

个体分布图 Distribution of individuals

径级分布表 DBH class

胸径区间 (Diameter class) (cm)	个体数 (No. of individuals in the plot)	比例 (Proportion) (%)
1～2	8	66.7
2～3	3	25.0
3～4	1	8.3
4～5	0	0.0
5～7	0	0.0
7～10	0	0.0
≥10	0	0.0

47 木姜子 mù jiāng zǐ | Pungent Litse

Litsea pungens，木姜子属
樟科 | Lauraceae

代码（SpCode）= LITPUN
个体数（Individual number/25 hm^2）= 4
最大胸径（Max DBH）= 2.06 cm
所在林层（Stratum）= 灌木层 (Shrub)
重要值排序（Importance value rank）= 110/128

落叶小乔木。幼枝有毛。叶互生，叶常聚生枝顶，膜质，初被柔毛，后变为无毛；叶披针形，或倒披针形至卵圆状倒披针形，顶端短急尖，基部渐尖。花黄色，伞形花序。果黑色，球形。花期3～4月，果期7～8月。

Deciduous small trees. Young branchlets pubescent. Leaves alternate, often clustered at apex of branchlet, membranous, initially pubescent, then glabrous; leaf blade lanceolate or obovate-lanceolate to ovoid-lanceolate, apex shortly acute, base cuneate. Umbels, yellow. Fruit black, globose. Fl. Mar.–Apr., fr. Jul.–Aug..

树干 Trunk
摄影：卢志军 Photo by: Zhijun Lu

果枝 Fruiting branch
摄影：鲍大川 Photo by: Dachuan Bao

果枝 Fruiting branch
摄影：卢志军 Photo by: Zhijun Lu

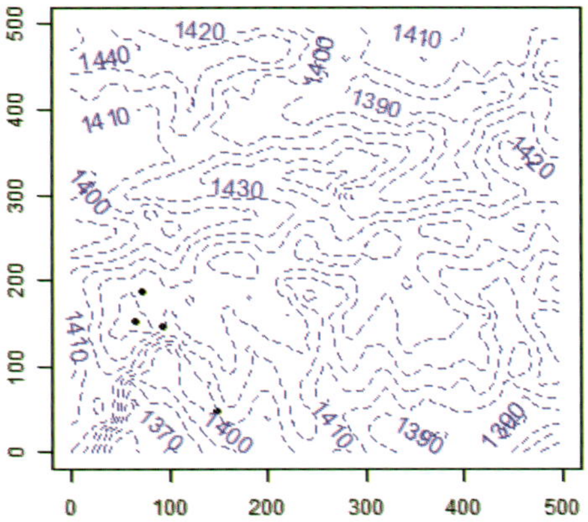

个体分布图 Distribution of individuals

径级分布表 DBH class

胸径区间 (Diameter class) (cm)	个体数 (No. of individuals in the plot)	比例 (Proportion) (%)
1～2	3	75.0
2～3	1	25.0
3～4	0	0.0
4～5	0	0.0
5～7	0	0.0
7～10	0	0.0
≥10	0	0.0

48 宜昌润楠 yí chāng rùn nán | Yichang Machilus

Machilus ichangensis，润楠属
樟科 | Lauraceae

代码（SpCode）= MACICH
个体数（Individual number/25 hm²）= 2799
最大胸径（Max DBH）= 27.80 cm
所在林层（Stratum）=乔木层 (Canopy)
重要值排序（Importance value rank）= 11/50

常绿乔木。枝条细弱，暗赤色，疏生皮孔，无毛。叶互生，革质，长圆状披针形或披针形，长渐尖，基部渐狭，少数为宽楔形，无毛，上面亮黄绿色，下面稍被粉，无毛，少数稍被绢毛，后变为无毛。圆锥花序，花黄白色。核果近圆形，花被片宿存。花期4月，果期8～9月。

Evergreen trees. Branchlets slender, dark red, lenticel sparse, glabrous. Leaves alternate, leathery; leaf blade oblanceolate or lanceolate, apex long acuminate, base tapering, rarely broadly cuneate, glabrous, adaxially bright yellowish green, abaxially glaucous, glabrous, rarely pubescent, then glabrous. Panicle, flowers yellowish white. Drupes subglobose, tepals persistent. Fl. Apr., fr. Aug.–Sept..

树干 Trunk
摄影：卢志军 Photo by: Zhijun Lu

花枝 Flowering branches
摄影：江明喜 Photo by: Mingxi Jiang

果枝 Fruiting branches
摄影：江明喜 Photo by: Mingxi Jiang

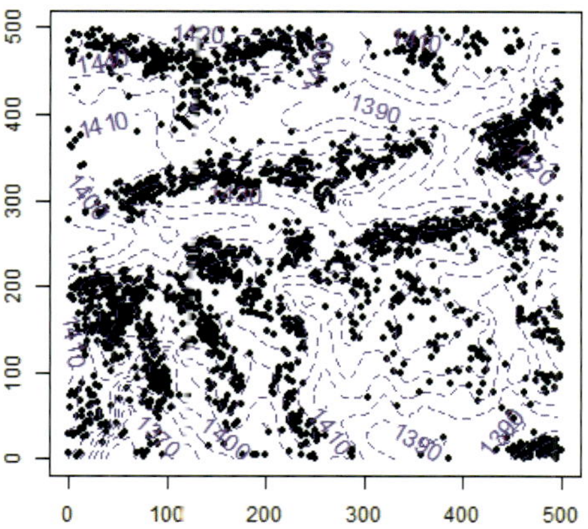

个体分布图 Distribution of individuals

径级分布表 DBH class

胸径区间 (Diameter class) (cm)	个体数 (No. of individuals in the plot)	比例 (Proportion) (%)
1～2.5	1579	55.8
2.5～5	354	12.5
5～10	303	10.7
10～25	552	19.5
25～40	11	0.4
40～100	0	0.0
≥100	0	0.0

49 巫山新木姜子 wū shān xīn mù jiāng zǐ | Wushan Newlitse

Neolitsea wushanica，新木姜子属
樟科 | Lauraceae

代码（SpCode）= NEOWUS
个体数（Individual number/25 hm^2）= 739
最大胸径（Max DBH）= 26.65 cm
所在林层（Stratum）= 灌木层 (Shrub)
重要值排序（Importance value rank）= 32/128

常绿小乔木。树皮黄绿色，平滑。叶薄革质，互生或聚生于枝顶，椭圆形或长圆状披针形，先端急尖或近于渐尖，基部多少有点渐尖，下面苍绿色，具白粉。伞形花序腋生或侧生。果球形，成熟时紫黑色。花期10月，果期翌年6～7月。

Evergreen small trees. Bark yellow-green, smooth. Leaf blade thin-leathery, alternate or clustered at apex of branchlet, elliptic or oblong-lanceolate, apex acute or subacuminate, base somewhat acuminate, abaxially glaucous with white pubescent. Umbels axillary or lateral. Fruit globose, purple-black at maturity. Fl. Oct., fr. Jun. to Jul. of following yr..

树干 Trunk
摄影：卢志军 Photo by: Zhijun Lu

小枝和叶片 Branchlets and leaves
摄影：卢志军 Photo by: Zhijun Lu

叶背 Leaf abaxial surface
摄影：卢志军 Photo by: Zhijun Lu

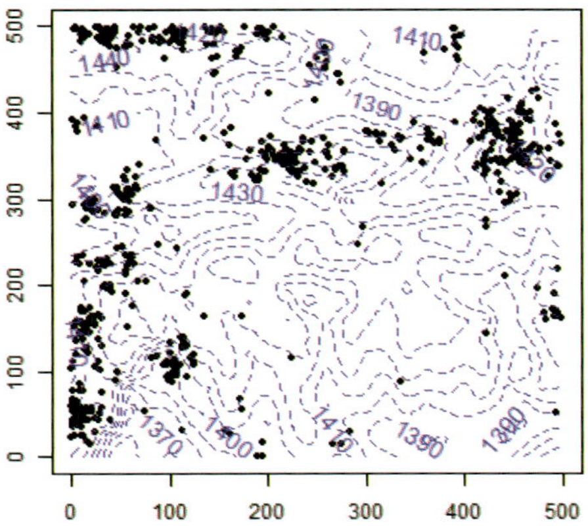

个体分布图 Distribution of individuals

径级分布表 DBH class

胸径区间 (Diameter class) (cm)	个体数 (No. of individuals in the plot)	比例 (Proportion) (%)
1～2	416	55.8
2～3	146	19.6
3～4	82	11.0
4～5	34	4.6
5～7	37	5.0
7～10	16	2.1
≥10	8	1.1

50 新木姜子

xīn mù jiāng zǐ | Common Newlitse

Neolitsea aurata，新木姜子属
樟科 I Lauraceae

代码（SpCode）= NEOAUR
个体数（Individual number/25 hm^2）= 163
最大胸径（Max DBH）= 12.45 cm
所在林层（Stratum）= 灌木层 (Shrub)
重要值排序（Importance value rank）= 48/128

常绿灌木或小乔木。枝条褐色，初密被短毛。叶近轮生，革质，倒卵状长圆形至长圆形，先端渐尖，有时有尾，基部楔形，离基三出脉，上面无毛，下面密被金黄色茸毛。花序伞形，几乎无总梗。果球形，位于浅碟形的果托上。花期4月，果期9～10月。

Evergreen shrubs or small trees. Branchlets brown, initially pubescent. Leaves subverticillate, leathery, leaf blade obovate-oblong to oblong, apex acuminate, sometimes caudate, base cuneate, 3-verned from base, adaxially glabrous, abaxially densely golden yellow sericeous. Umbels, pedicel nearly absent. Fruit ellipsoid, on shallowly discoid perianth tube. Fl. Apr., fr. Sept.–Oct..

树干 Trunk
摄影：卢志军 Photo by: Zhijun Lu

花枝和叶片 Flowering branch and leaves
摄影：徐耀粘 Photo by: Yaozhan Xu

叶背 Leaf abaxial surface
摄影：徐耀粘 Photo by: Yaozhan Xu

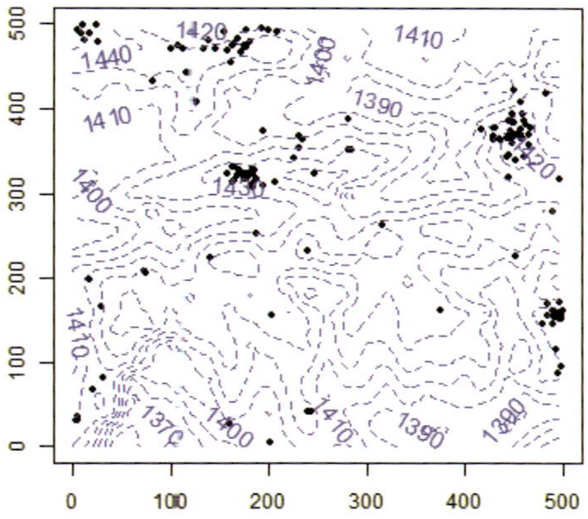

个体分布图 Distribution of individuals

径级分布表 DBH class

胸径区间 (Diameter class) (cm)	个体数 (No. of individuals in the plot)	比例 (Proportion) (%)
1～2	63	38.4
2～3	39	23.8
3～4	19	11.6
4～5	14	8.5
5～7	16	9.8
7～10	8	4.9
≥10	4	2.4

51 簇叶新木姜子

cù yè xīn mù jiāng zǐ | Dense-leaf Newlitse

Neolitsea confertifolia，新木姜子属

樟科 | Lauraceae

代码（SpCode）= NEOCON

个体数（Individual number/25 hm^2）= 159

最大胸径（Max DBH）= 11.39 cm

所在林层（Stratum）= 灌木层 (Shrub)

重要值排序（Importance value rank）= 53/128

常绿灌木或小乔木。全株无毛。幼枝被灰棕毛，后无毛。叶密集近轮生枝顶及分枝处，革质，狭披针形，先端急尖，基部宽楔形，边缘波状，上面灰绿色，下面绿苍白色。伞形花序，着生叶腋，近无柄。果黑色，卵圆形，生于扁平盘状花筒上。花期4月，果期8月。

Evergreen shrubs or small trees. Glabrous. Young branchlets gray-brown pubescence then glabrous. Leaves conferted, subverticillate on branchlet apex and node, leathery; leaf blade narrowly lanceolate, apex acute, base broadly cuneate, margin crenate, adaxially grayish green, abaxially greenish pale.Umbels in leaf axils, subsessile. Fruit black and ovoid, seated on flat discoid perianth tube. Fl. Apr., fr. Aug..

树干 Trunk

摄影：卢志军 Photo by: Zhijun Lu

果枝 Fruiting branches

摄影：刘海波 Photo by: Haibo Liu

叶背 Leaf abaxial surface

摄影：卢志军 Photo by: Zhijun Lu

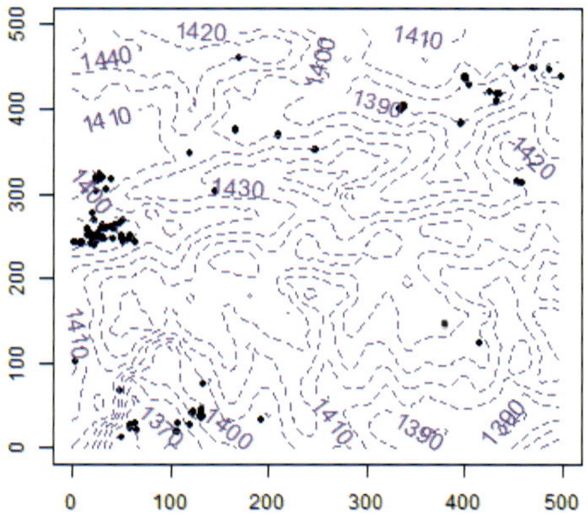

个体分布图 Distribution of individuals

径级分布表 DBH class

胸径区间 (Diameter class) (cm)	个体数 (No. of individuals in the plot)	比例 (Proportion) (%)
1～2	83	52.2
2～3	32	20.1
3～4	21	13.2
4～5	11	6.9
5～7	7	4.4
7～10	3	1.9
≥10	2	1.3

52 利川润楠 lì chuān rùn nán | Lichuan Machilus

Machilus lichuanensis，润楠属
樟科 | Lauraceae

代码（SpCode）= MACLIC
个体数（Individual number/25 hm^2）= 7
最大胸径（Max DBH）= 5.36 cm
所在林层（Stratum）= 乔木层 (Canopy)
重要值排序（Importance value rank）= 33/50

常绿高大乔木。枝紫褐色或紫黑色。叶革质，椭圆形或狭倒卵形，先端短渐尖至急尖，基部楔形，上面绿色，稍光亮，下面中脉和侧脉两侧密被柔毛。聚伞状圆锥花序生当年生枝下端。果扁球形。花期5月，果期9月。

Evergreen large trees. Branch purple-brown or purple-black. Leaf blade leathery, elliptic or narrowly obovate, apex shortly acuminate to acute, base cuneate, adaxially green, slightly shiny, abaxially densely pubescent along midvein and lateral veins. Cymose panicles on lower part of 1-yr branchlet. Fruit compressed globose. Fl. May, fr. Sept..

树干 Trunk
摄影：卢志军 Photo by: Zhijun Lu

叶片 Leaves
摄影：卢志军 Photo by: Zhijun Lu

叶背 Leaf abaxial surface
摄影：卢志军 Photo by: Zhijun Lu

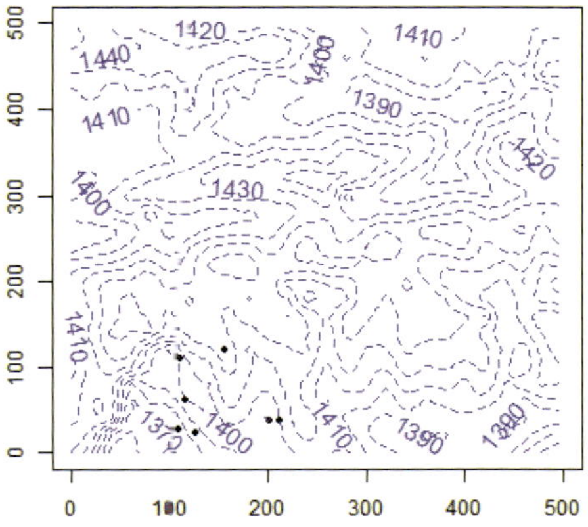

个体分布图 Distribution of individuals

径级分布表 DBH class

胸径区间 (Diameter class) (cm)	个体数 (No. of individuals in the plot)	比例 (Proportion) (%)
1～2.5	2	28.6
2.5～5	4	57.1
5～10	1	14.3
10～25	0	0.0
25～40	0	0.0
40～100	0	0.0
≥100	0	0.0

53 檫木　chá mù | Common Sassafras

Sassafras tzumu，檫木属
樟科 I Lauraceae

代码（SpCode）= SASTZU
个体数（Individual number/25 hm^2）= 472
最大胸径（Max DBH）= 79.00 cm
所在林层（Stratum）= 乔木层 (Canopy)
重要值排序（Importance value rank）= 6/50

落叶乔木。树皮灰棕色。叶卵圆形或倒卵形，全缘或三裂，叶脉羽状或三出。总状花序，花黄色，早春开花。果蓝黑色，被白粉。花期3～4 月，果期8月。

Deciduous trees. Bark grayish brown. Leaf blade ovate or obovate, margin entire or 3-lobed, veins netted or trinerved. Raceme, flower yellow, blossom early Spring. Fruit bluish black, white waxy. Fl. Mar.–Apr., fr. Aug..

树干　Trunk
摄影：卢志军　Photo by: Zhijun Lu

果枝　Fruiting branch
摄影：路俊盟　Photo by: Junmeng Lu

叶背　Leaf abaxial surface
摄影：卢志军　Photo by: Zhijun Lu

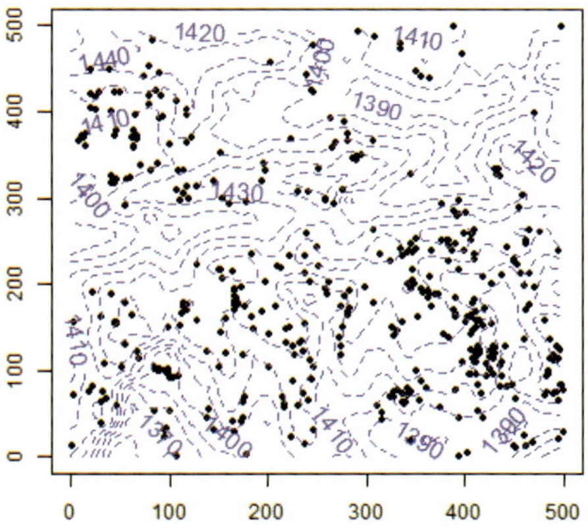

个体分布图 Distribution of individuals

径级分布表 DBH class

胸径区间 (Diameter class) (cm)	个体数 (No. of individuals in the plot)	比例 (Proportion) (%)
1～2.5	29	6.1
2.5～5	24	5.1
5～10	8	1.7
10～25	87	18.4
25～40	209	44.3
40～100	115	24.4
⩾100	0	0.0

54 水青树 shuǐ qīng shù | Tetracentron

Tetracentron sinense 水青树属
水青树科 | Tetracentraceae

代码（SpCode）= TETSIN
个体数（Individual number/25 hm^2）= 3
最大胸径（Max DBH）= 9.21 cm
所在林层（Stratum）= 乔木层 (Canopy)
重要值排序（Importance value rank）= 41/50

树干 Trunk
摄影：卢志军 Photo by: Zhijun Lu

落叶乔木。树皮灰褐色，皮孔凸起。叶纸质，卵圆形，先端渐尖，基部心形，边缘密生细小的钝锯齿，两面光滑无毛。穗状花序，花小，黄绿色，腋生。蒴果褐色，室背开裂。花期7月，果期8～10月。

Deciduous trees. Bark grayish brown, lenticels embossing. Leaf blade papery, ovoid, apex acuminate, base cordate, margin serrulate, glabrous on both surfaces. Spikes, small flowers yellowish green, axillary. Capsule brown, dehiscence loculicidal. Fl. Jul., fr. Aug.–Oct..

花枝 Flowering branch
摄影：刘海波 Photo by: Haibo Liu

花枝和叶背 Flowering branch and leaf abaxial surface
摄影：刘海波 Photo by: Haibo Liu

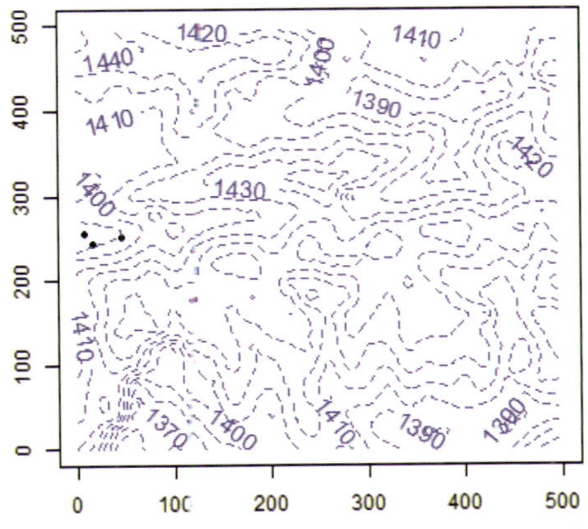

个体分布图 Distribution of individuals

径级分布表 DBH class

胸径区间 (Diameter class) (cm)	个体数 (No. of individuals in the plot)	比例 (Proportion) (%)
1～2.5	0	0.0
2.5～5	2	66.7
5～10	1	33.3
10～25	0	0.0
25～40	0	0.0
40～100	0	0.0
≥100	0	0.0

55 豪猪刺

háo zhū cì | Wintergreen Barberry

Berberis julianae，小檗属
小檗科 | Berberidaceae

代码（SpCode）= BERJUL
个体数（Individual number/25 hm^2）= 17
最大胸径（Max DBH）= 5.00 cm
所在林层（Stratum）= 灌木层 (Shrub)
重要值排序（Importance value rank）= 91/128

常绿灌木。幼枝有条棱，老枝黄褐色或灰褐色，稍有黑色疣点；针刺坚硬，三叉。叶革质，椭圆形、披针形或倒披针形，先端渐尖，基部楔形，边缘有刺齿，上面深绿色，下面淡绿色。花簇生；花瓣黄色，长椭圆形。浆果长圆形，蓝黑色，被白粉，花柱宿存。花期3～6月，果期5～11月。

Evergreen shrubs. Young branchlets sulcate, old ones yellow-brown or gray-brown, slightly black verruculose; spines stout and 3-fid. Leaf blade leathery, elliptic, lanceolate or oblanceolate, apex acuminate, base cuneate, margin spinose-serrate, adaxially deep green, abaxially pale green. Flowers fascicled. Petals yellow, oblong-elliptic. Berry blue-black, oblong, white pruinose, style persistent. Fl. Mar.–Jun., fr. May–Nov..

树干 Trunk
摄影：卢志军 Photo by: Zhijun Lu

花枝 Flowering branch
摄影：江明喜 Photo by: Mingxi Jiang

果枝 Fruiting branch
摄影：江明喜 Photo by: Mingxi Jiang

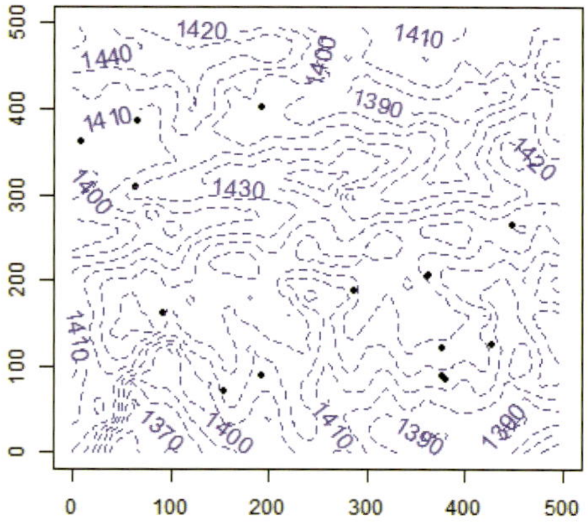

个体分布图 Distribution of individuals

径级分布表 DBH class

胸径区间 (Diameter class) (cm)	个体数 (No. of individuals in the plot)	比例 (Proportion) (%)
1～2	11	64.7
2～3	2	11.8
3～4	2	11.8
4～5	2	11.8
5～7	0	0.0
7～10	0	0.0
≥10	0	0.0

56 阔叶十大功劳

kuò yè shí dà gōng láo | Leather-leaf Mahonia

Mahonia bealei，十大功劳属
小檗科 | Berberidaceae

代码（SpCode）= MAHBEA
个体数（Individual number/25 hm^2）= 1
最大胸径（Max DBH）= 1.90 cm
所在林层（Stratum）= 灌木层 (Shrub)
重要值排序（Importance value rank）= 126/128

常绿灌木。树皮黄褐色。小叶卵形，基部的侧生小叶较小，由下向上渐次增大，顶生小叶较宽大，边缘有刺齿，基部宽楔形至圆形，有时浅心形。总状花序，花黄色。浆果卵圆形，蓝黑色，被白粉，有短宿存花柱或无。花期11月至翌年3月，果期4～8月。

Evergreen shrubs. Bark yellowish brown. Leaflet blade ovoid, smaller on rachis base, increasing in length but decreasing in width from base to apex, margin serrate, base broadly cuneate to rounded, sometimes shallow cordate. Raceme, flowers yellow. Berry ovoid, dark purple, pruinose, style persistent or absent. Fl. Nov.–Mar. of following year, fr. Apr.–Aug..

树干 Trunk
摄影：卢志军 Photo by: Zhijun Lu

花枝 Flowering branch
摄影：徐耀粘 Photo by: Yaozhan Xu

果枝 Fruiting branches
摄影：鲍大川 Photo by: Dachuan Bao

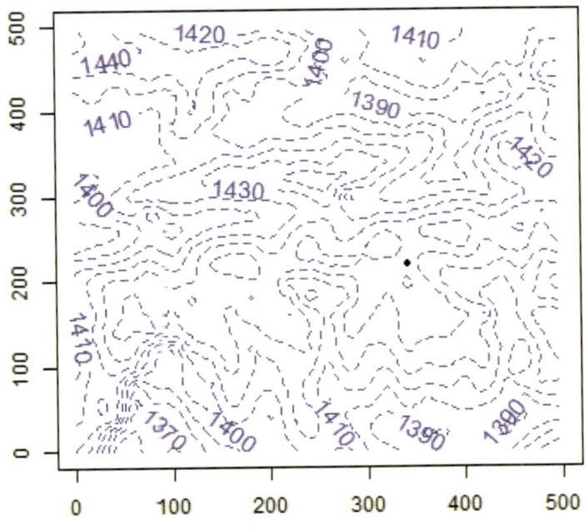

个体分布图 Distribution of individuals

径级分布表 DBH class

胸径区间 (Diameter class) (cm)	个体数 (No. of individuals in the plot)	比例 (Proportion) (%)
1～2	1	100.0
2～3	0	0.0
3～4	0	0.0
4～5	0	0.0
5～7	0	0.0
7～10	0	0.0
≥10	0	0.0

57 猫儿屎

māo ér shǐ | Farges Decaisnea

Decaisnea insignis，猫儿屎属
木通科 | Lardizabalaceae

代码（SpCode）= DECINS
个体数（Individual number/25 hm^2）= 40
最大胸径（Max DBH）= 20.77 cm
所在林层（Stratum）= 灌木层 (Shrub)
重要值排序（Importance value rank）= 58/128

落叶灌木或小乔木。树皮灰褐色。枝稍被白粉。叶互生，奇数羽状复叶，小叶13～25，叶片卵状椭圆形或长椭圆形，先端渐尖，基部近圆形或宽楔形，稍偏斜，全缘。总状花序或圆锥花序，花黄绿色。果实圆柱状，成熟时蓝紫色，含浆汁。种子卵形，扁平，棕黑色。花期4～7月，果期7～10月。

Deciduous shrubs or small trees. Bark grayish brown. Branch apex slightly glaucous. Leaves alternate, imparipinnately compound, leaflets 13-25, leaf blade ovate-elliptic or long-elliptic, apex acuminate, base subrounded or broadly cuneate, slightly oblique, margin entire. Racemes or panicles, flowers yellowish green. Fruits cylindric, blue-purple when mature, with milk. Seeds ovoid, flat, brown-black. Fl. Apr.–Jul., fr. Jul.–Oct..

树干 Trunk
摄影：卢志军 Photo by: Zhijun Lu

花枝 Flowering branches
摄影：卢志军 Photo by: Zhijun Lu

果实和种子 Fruits and seeds
摄影：江明喜 Photo by: Mingxi Jiang

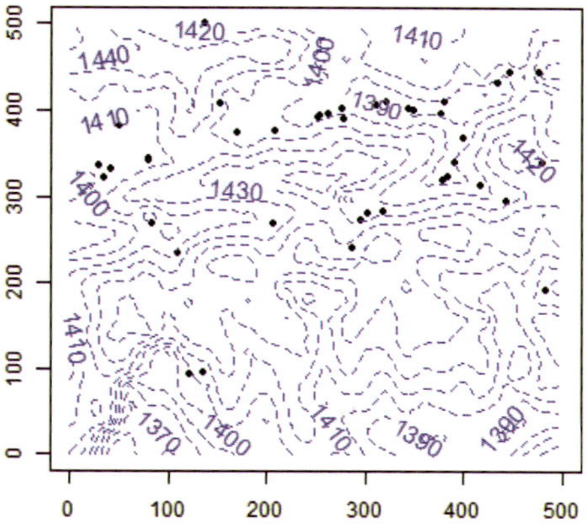

个体分布图 Distribution of individuals

径级分布表 DBH class

胸径区间 (Diameter class) (cm)	个体数 (No. of individuals in the plot)	比例 (Proportion) (%)
1～2	7	17.5
2～3	5	12.5
3～4	3	7.5
4～5	4	10.0
5～7	4	10.0
7～10	7	17.5
≥10	10	25.0

58 西南山茶 xī nán shān chá | Pitard Camellia

Camellia pitardii，山茶属
山茶科 | Theaceae

代码（SpCode）= CAMPIT
个体数（Individual number/25 hm^2）= 3149
最大胸径（Max DBH）= 20.67 cm
所在林层（Stratum）= 灌木层 (Shrub)
重要值排序（Importance value rank）= 7/128

常绿灌木至小乔木。嫩枝灰棕色，无毛。叶革质，披针形或长圆形，先端渐尖或长尾状，基部楔形，上面亮绿色，下面黄绿色，无毛，边缘有尖锐粗锯齿。花顶生，玫瑰色、粉色或白色，无柄。蒴果扁球形，种子棕色，球形或半球形。花期2～5月，果期8～10月。

树干 Trunk
摄影：卢志军 Photo by: Zhijun Lu

Evergreen shrubs or small trees. Young branches grayish brown, glabrous. Leaf blade leathery, oblanceolate or oblong, apex acuminate to caudate, base cuneate, adaxially bright green, abaxially yellowish green, both surfaces glabrous, margin apiculately serrulate. Flowers terminal, rose, pink or white, sessile. Capsule oblate. Seeds brown, globose to semiglobose. Fl. Feb.–May, fr. Aug.–Oct..

花枝 Flowering branch
摄影：徐耀粘 Photo by: Yaozhan Xu

果枝 Fruiting branches
摄影：刘海波 Photo by: Haibo Liu

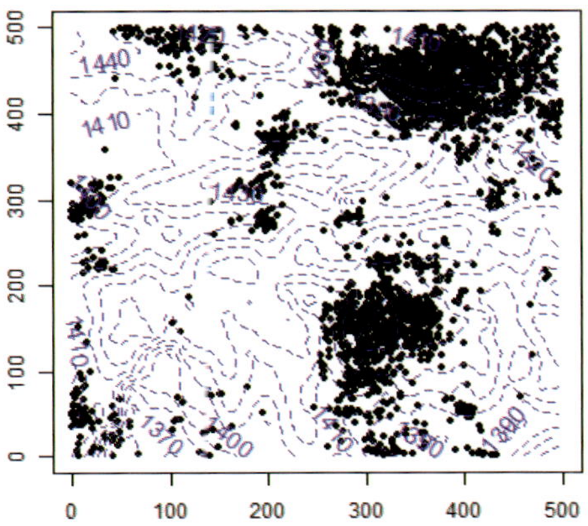

个体分布图 Distribution of individuals

径级分布表 DBH class

胸径区间 (Diameter class) (cm)	个体数 (No. of individuals in the plot)	比例 (Proportion) (%)
1～2	743	23.5
2～3	635	20.1
3～4	467	14.8
4～5	349	11.0
5～7	427	13.5
7～10	287	9.1
≥10	241	7.6

59 连蕊茶　lián ruǐ chá | Taperleaf Camellia

Camellia cuspidata，山茶属
山茶科 | Theaceae

代码（SpCode）= CAMCUS
个体数（Individual number/25 hm^2）= 2721
最大胸径（Max DBH）= 13.50 cm
所在林层（Stratum）= 灌木层 (Shrub)
重要值排序（Importance value rank）= 11/128

常绿灌木。幼枝灰色，稍具细柔毛。叶薄革质，长圆状披针形，先端尾状渐尖，基部楔形，边缘具细锯齿，上面无毛，有光泽，下面淡绿色，散生小点。花顶生或腋生，白色，有郁香。蒴果圆球形；种子淡褐色。花期11月至翌年3月，果期8～11月。

Evergreen shrubs. Young branches brown, slightly puberulent. Leaf blade thinly leathery, oblong-lanceolate, apex caudately acuminate, base cuneate, margin serrulate, adaxially glabrous and glossy, abaxially light green, sparsely verruculose. Flowers terminal or axillary, white, fragrant. Capsule globose. Seeds bright brown. Fl. Nov.–Mar. of following yr, fr. Aug.–Nov..

树干　Trunk
摄影：卢志军　Photo by: Zhijun Lu

花枝　Flowering branch
摄影：刘海波　Photo by: Haibo Liu

果枝　Fruiting branch
摄影：徐耀粘　Photo by: Yaozhan Xu

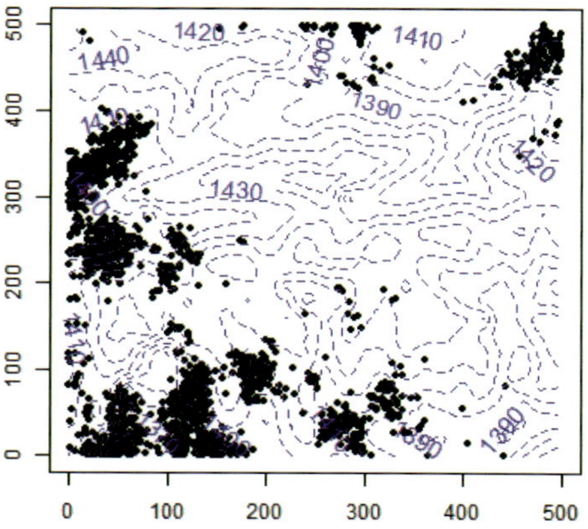

个体分布图 Distribution of individuals

径级分布表 DBH class

胸径区间 (Diameter class) (cm)	个体数 (No. of individuals in the plot)	比例 (Proportion) (%)
1～2	891	32.6
2～3	621	22.7
3～4	415	15.2
4～5	254	9.3
5～7	289	10.6
7～10	177	6.5
⩾10	74	2.7

60 贵州连蕊茶 guì zhōu lián ruǐ chá | Kweichow Camellia

Camellia costei，山茶属
山茶科 | Theaceae

代码（SpCode）= CAMCOS
个体数（Individual number/25 hm^2）= 779
最大胸径（Max DBH）= 17.57 cm
所在林层（Stratum）= 灌木层 (Shrub)
重要值排序（Importance value rank）= 30/128

常绿灌木。幼枝灰棕色，被毛。叶薄革质，椭圆形至椭圆状披针形，先端尾状渐尖，基部楔形，边缘有细锯齿，上面绿色，下面有长柔毛。花单独腋生或顶生，白色或带紫红色。蒴果近球形；种子栗色，近球形。花期4～7月，果期7～10月。

Evergreen shrubs. Young branches grayish brown, glabrescent. Leaf blade thinly leathery, elliptic to elliptic-lanceolate, apex caudately acuminate, base cuneate, margin crenulate-serrulate, adaxially green, abaxially pubescent. Flowers solitary axillary or terminal, white or with purplish red. Capsule subglobose. Seeds chestnut brown, subglobose. Fl. Apr.–Jul., fr. Jul.–Oct..

树干 Trunk
摄影：卢志军 Photo by: Zhijun Lu

花枝 Flowering branch
摄影：刘海波 Photo by: Haibo Liu

小枝和叶背 Branchlet and leaf abaxial surface
摄影：卢志军 Photo by: Zhijun Lu

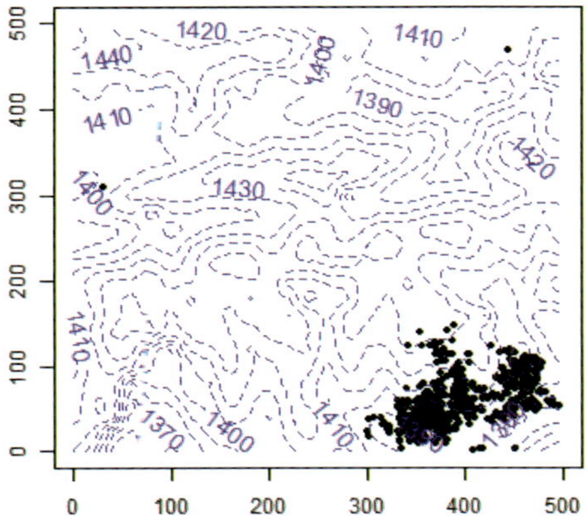

个体分布图 Distribution of individuals

径级分布表 DBH class

胸径区间 (Diameter class) (cm)	个体数 (No. of individuals in the plot)	比例 (Proportion) (%)
1～2	263	33.8
2～3	173	22.2
3～4	100	12.8
4～5	68	8.7
5～7	100	12.8
7～10	57	7.3
≥10	18	2.3

61 红淡比

hóng dàn bǐ | Japanese Cleyera

Cleyera japonica，红淡比属
山茶科 | Theaceae

代码（SpCode）= CLEJAP
个体数（Individual number/25 hm^2）= 142
最大胸径（Max DBH）= 19.90 cm
所在林层（Stratum）= 灌木层 (Shrub)
重要值排序（Importance value rank）= 42/128

常绿灌木或小乔木。树皮灰色，光滑。小枝灰棕色，有角棱。叶革质，椭圆形至倒卵形，先端短尾尖，基部楔形，全缘，上面光亮，下面苍绿色，两面无毛。花单生或簇生于叶腋，白色。浆果球形，成熟时黑色。花期5～7月，果期8～11月。

Evergreen shrubs or small trees. Bark gray, glabrous. Young branches grayish brown, ribbed. Leaf blade leathery, elliptic to obovate, apex caudately acuminate, base cuneate, margin entire, adaxially shiny, abaxially pale green, both surfaces glabrous. Flowers solitary or axillarily fascicled, white. Fruit purplish black when mature, globose. Fl. May–Jul., fr. Aug.–Nov..

树干 Trunk
摄影：卢志军 Photo by: Zhijun Lu

花枝 Flowering branch
摄影：江明喜 Photo by: Mingxi Jiang

果枝 Fruiting branch
摄影：卢志军 Photo by: Zhijun Lu

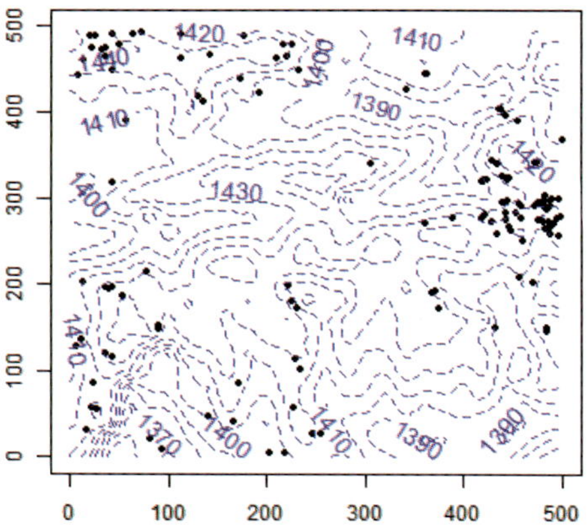

个体分布图 Distribution of individuals

径级分布表 DBH class

胸径区间 (Diameter class) (cm)	个体数 (No. of individuals in the plot)	比例 (Proportion) (%)
1～2	46	32.4
2～3	22	15.5
3～4	19	13.4
4～5	9	6.3
5～7	18	12.7
7～10	10	7.0
≥10	18	12.7

62 短柱柃

duǎn zhù líng | Short-style Eurya

Eurya brevistyla，柃木属
山茶科 | Theaceae

代码（SpCode）= EURBRE
个体数（Individual number/25 hm^2）= 15997
最大胸径（Max DBH）= 12.74 cm
所在林层（Stratum）= 灌木层 (Shrub)
重要值排序（Importance value rank）= 1/128

常绿灌木或小乔木。幼枝灰棕色，有2棱，无毛。叶革质，椭圆形或倒卵状椭圆形，先端急狭缩而锐尖，基部楔形，边缘有锐锯齿，上面暗绿色，有光泽，下面常呈黄绿色。花腋生，白色。果实圆球形，熟时蓝黑色。花期4～11月，果期10～翌年5月。

Evergreen shrubs or small trees. Young branches grayish brown, 2-ribbed, glabrous. Leaf blade leathery, elliptic or obovate-elliptic, apex rounded and retuse with a mucronate tip, base cuneate, margin serrulate, adaxially dark green and shiny, abaxially yellowish green. Flowers axillary, white. Fruit bluish black when mature, globose. Fl. Apr.–Nov., fr. Oct.–May of following yr..

树干 Trunk
摄影：卢志军 Photo by: Zhijun Lu

小枝和叶片 Branchlet and leaves
摄影：卢志军 Photo by: Zhijun Lu

果枝 Fruiting branches
摄影：刘海波 Photo by: Haibo Liu

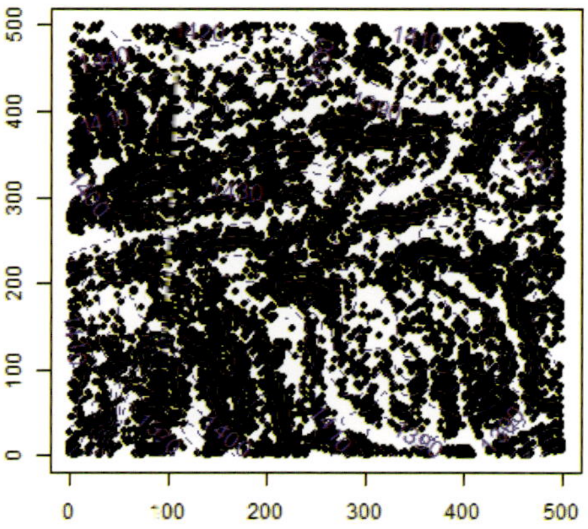

个体分布图 Distribution of individuals

径级分布表 DBH class

胸径区间 (Diameter class) (cm)	个体数 (No. of individuals in the plot)	比例 (Proportion) (%)
1～2	5567	34.6
2～3	4281	26.6
3～4	2689	16.7
4～5	1625	10.1
5～7	1292	8.0
7～10	435	2.7
≥10	108	0.7

63 翅柃

chì líng | Winged Eurya

Eurya alata，柃木属
山茶科 | Theaceae

代码（SpCode）= EURALA
个体数（Individual number/25 hm^2）= 56
最大胸径（Max DBH）= 10.15 cm
所在林层（Stratum）= 灌木层 (Shrub)
重要值排序（Importance value rank）= 65/128

常绿灌木或小乔木。全体无毛。幼枝褐色，显具4棱。叶革质，椭圆形、狭椭圆形或椭圆状倒卵形，先端狭缩而锐尖，基部楔形至宽楔形，边缘密生细锯齿，上面绿色，有光泽，下面带黄绿色。雄花腋生，白色；雌花腋生，萼片卵形。果实圆球形，成熟时黑色。花期9~11月，果期11月至翌年3月。

Evergreen shrubs or small trees. Glabrous. Young branches brown, 4-ribbed. Leaf blade leathery, elliptic, narrow-elliptic or elliptic-obovate, apex acuminate and with an obtuse tip, base cuneate to broadly cuneate, margin densely serrulate, adaxially green, glossy, abaxially yellowish green. Male flowers axillary, white; female flowers axillary, sepals ovate. Fruits globose, black when mature. Fl. Sept.–Nov., fr. Nov.–Mar. of following yr..

树干 Trunk
摄影：卢志军 Photo by: Zhijun Lu

果枝和叶片 Fruiting branch and leaves
摄影：卢志军 Photo by: Zhijun Lu

果枝和叶背 Fruiting branch and leaf abaxial surface
摄影：卢志军 Photo by: Zhijun Lu

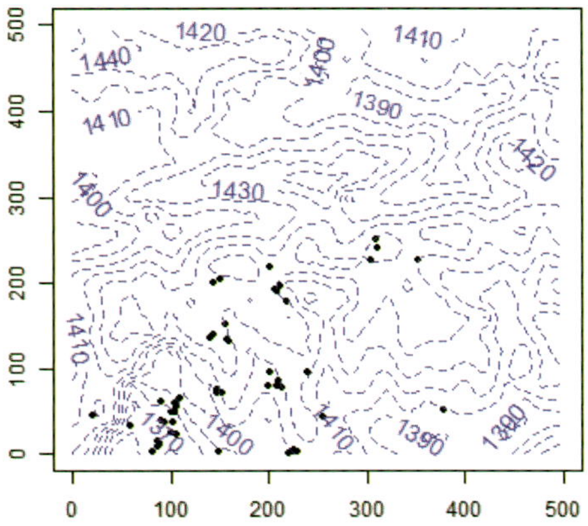

个体分布图 Distribution of individuals

径级分布表 DBH class

胸径区间 (Diameter class) (cm)	个体数 (No. of individuals in the plot)	比例 (Proportion) (%)
1~2	13	23.2
2~3	16	28.6
3~4	11	19.6
4~5	2	3.6
5~7	11	19.6
7~10	2	3.6
≥10	1	1.8

64 细齿叶柃　xì chǐ yè líng | Shining Eurya

Eurya nitida，柃木属
山茶科 | Theaceae

代码（SpCode）= EURNIT
个体数（Individual number/25 hm^2）= 2
最大胸径（Max DBH）= 4.17 cm
所在林层（Stratum）= 灌木层 (Shrub)
重要值排序（Importance value rank）= 115/128

常绿灌木或小乔木。全体无毛。幼枝具2棱。叶薄革质，长圆状椭圆形或倒卵状披针形，先端渐尖，基部楔形，边缘有钝锯齿，上面深绿色，下面淡绿色。花腋生，白色。果实圆球形。花期5～10月，果期8～10月。

Evergreen shrubs or small trees. Glabrous. Young branches 2-ribbed. Leaf blade thinly leathery, oblong-elliptic, obovate-lanceolate, apex acuminate, base cuneate, margin crenate, adaxially dark green, abaxially light green. Flowers axillary, white Fruits globose. Fl. May–Oct., fr. Aug.–Oct..

树干　Trunk
摄影：卢志军　Photo by: Zhijun Lu

小枝和叶片　Branchlets and leaves
摄影：卢志军　Photo by: Zhijun Lu

小枝和叶背　Branchlets and leaf abaxial surface
摄影：卢志军　Photo by: Zhijun Lu

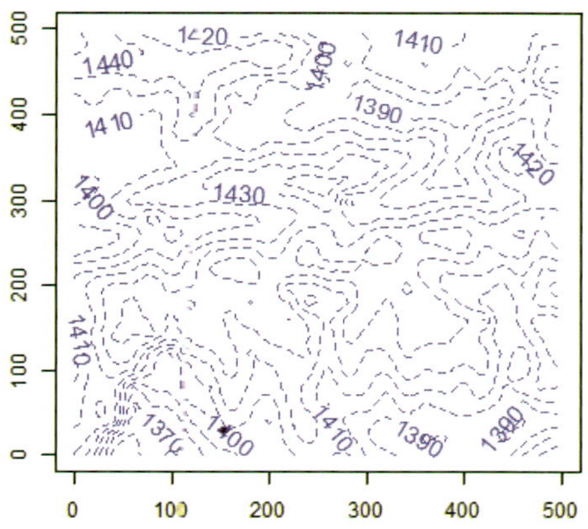

个体分布图 Distribution of individuals

径级分布表 DBH class

胸径区间 (Diameter class) (cm)	个体数 (No. of individuals in the plot)	比例 (Proportion) (%)
1～2	0	0.0
2～3	0	0.0
3～4	1	50.0
4～5	1	50.0
5～7	0	0.0
7～10	0	0.0
≥10	0	0.0

65 小花木荷 xiǎo huā mù hé | Little-flower Guger-tree

Schima parviflora，木荷属
山茶科 | Theaceae

代码（SpCode）= SCHPAR
个体数（Individual number/25 hm²）= 1790
最大胸径（Max DBH）= 117.5 cm
所在林层（Stratum）= 乔木层 (Canopy)
重要值排序（Importance value rank）= 5/50

常绿乔木。树皮灰色，松软而厚，条状深裂。叶革质，长圆形至披针状长圆形，先端渐尖，基部楔形，边缘有不明显的小锯齿，中脉在上面凹陷，在下面稍凸起。花白色，排成短总状花序。蒴果扁球形，褐色。花期8月，果期8～10月。

树干 Trunk
摄影：卢志军 Photo by: Zhijun Lu

Evergreen trees. Bark gray, soft, thick, peeling in strips. Leaf blade leathery, oblong to lanceolate-oblong, apex acuminate, base cuneate, margin crenulate, midvein visible on both surfaces. Flowers white, raceme. Capsule subglobose, brown. Fl. Aug., fr. Aug.–Oct..

花枝 Flowering branches
摄影：江明喜 Photo by: Mingxi Jiang

果枝 Fruiting branches
摄影：徐耀粘 Photo by: Yaozhan Xu

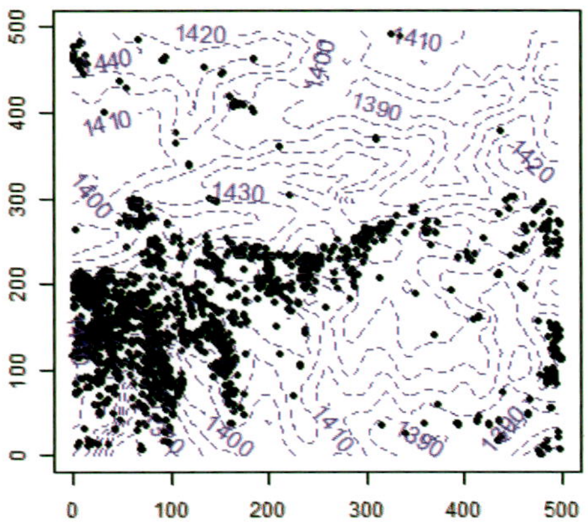

个体分布图 Distribution of individuals

径级分布表 DBH class

胸径区间 (Diameter class) (cm)	个体数 (No. of individuals in the plot)	比例 (Proportion) (%)
1～2.5	801	43.9
2.5～5	310	17.0
5～10	173	9.5
10～25	258	14.1
25～40	171	9.4
40～100	71	3.9
≥100	6	0.3

66 紫茎

zǐ jīng | Chinese Stewartia

Stewartia sinensis，紫茎属
山茶科 | Theaceae

代码（SpCode）= STESIN
个体数（Individual number/25 hm^2）= 1
最大胸径（Max DBH）= 4.27 cm
所在林层（Stratum）= 灌木层 (Shrub)
重要值排序（Importance value rank）= 124/128

树干 Trunk
摄影：卢志军 Photo by: Zhijun Lu

落叶灌木或乔木。树皮褐色，光滑，密生灰白色小粉点，片状脱落。幼枝紫褐色，无毛。叶长圆状椭圆形或倒卵状椭圆形，先端渐尖，基部近圆形或圆楔形，边缘疏生细锯齿，下面疏生平贴细毛，或仅沿中脉有细毛。花白色，单独腋生，花药橘红色。蒴果近卵圆形，顶端略尖。花期7月，果期9～10月。

Deciduous shrubs or trees. Bark brown, glabrous, with dense grayish white pruinose, flake. Young branchlets purplish brown, glabrous. Leaf blade oblong-elliptic to obovate-elliptic, apex acuminate, base subrounded or rounded-cuneate, margin sparsely crenulate, abaxially sparsely pubescent, or only on midvein. Flowers white, solitary and axillary, anther orange-red. Capsule subovoid, apex rostrate. Fl. Jul., fr. Sept.–Oct..

花枝 Flowering branch
摄影：徐耀粘 Photo by: Yaozhan Xu

果枝 Fruiting branch
摄影：江明喜 Photo by: Mingxi Jiang

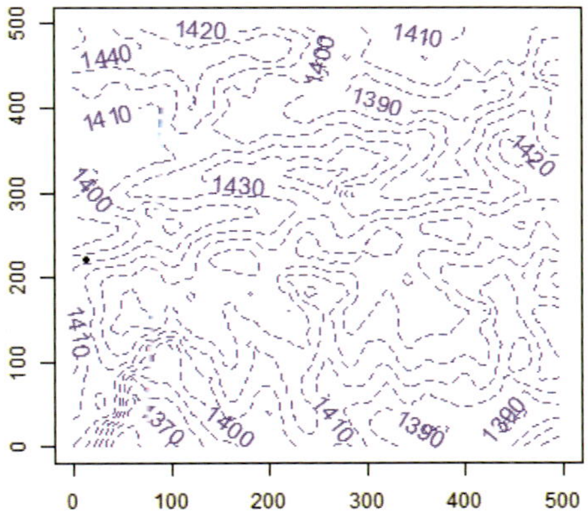

个体分布图 Distribution of individuals

径级分布表 DBH class

胸径区间 (Diameter class) (cm)	个体数 (No. of individuals in the plot)	比例 (Proportion) (%)
1～2	0	0.0
2～3	0	0.0
3～4	0	0.0
4～5	1	100.0
5～7	0	0.0
7～10	0	0.0
⩾10	0	0.0

67 厚皮香

hòu pí xiāng | Naked-anther Ternstroemia

Ternstroemia gymnanthera，厚皮香属

山茶科 | Theaceae

代码（SpCode）= TERGYM

个体数（Individual number/25 hm^2）= 77

最大胸径（Max DBH）= 20.70 cm

所在林层（Stratum）= 亚乔木层 Under-story

重要值排序（Importance value rank）= 35/54

常绿小乔木。树皮灰棕色，光滑。幼枝灰褐色，无毛。叶互生，生于枝顶，革质，长圆状倒卵形，先端急尖，基部楔形，全缘，边缘稍反卷，常带红色。花单独腋生或簇生于小枝顶端，白色或淡黄色。果实球形；种子红色。花期6～8月，果期9～11月。

Evergreen small trees. Bark grayish brown, smooth. Young branches grayish brown, glabrous. Leaves alternate, on branchlet apex, leathery; leaf blade oblong-obovate, apex acute, base cuneate, margin entire, slightly revolute, usually with red. Flowers solitary and axillary, or fascicled on branchlet apex, white or light yellow. Fruit globose; seed red. Fl. Jun.–Aug., fr. Sept.–Nov..

树干 Trunk

摄影：卢志军 Photo by: Zhijun Lu

花枝 Flowering branch

摄影：江明喜 Photo by: Mingxi Jiang

果枝 Fruiting branch

摄影：卢志军 Photo by: Zhijun Lu

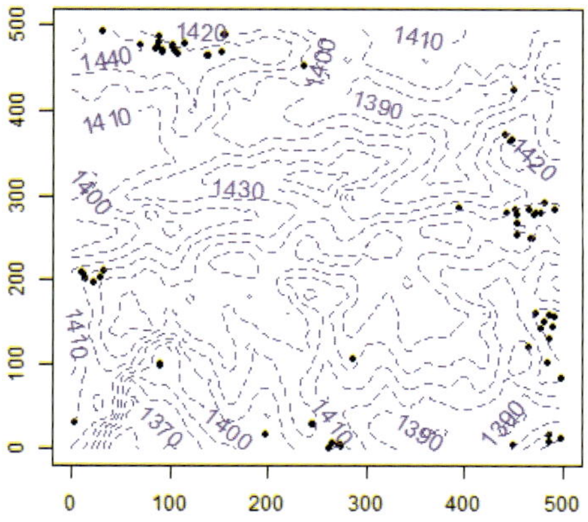

个体分布图 Distribution of individuals

径级分布表 DBH class

胸径区间 (Diameter class) (cm)	个体数 (No. of individuals in the plot)	比例 (Proportion) (%)
1～2.5	40	51.3
2.5～5	28	35.9
5～8	4	5.1
8～11	2	2.6
11～15	1	1.3
15～20	1	1.3
≥20	1	1.3

68 瑞木　　rùi mù | Manyflower Winterhazel

Corylopsis multiflora，蜡瓣花属
金缕梅科 | Hamamelidaceae

代码（SpCode）= CORMUL
个体数（Individual number/25 hm^2）= 19
最大胸径（Max DBH）= 16.98 cm
所在林层（Stratum）= 灌木层 (Shrub)
重要值排序（Importance value rank）= 70/128

落叶或半常绿灌木或小乔木。小枝被星状毛。叶卵形、长圆形至椭圆形，先端渐尖，基部近心形，边缘有刺毛状刺，叶下面有淡蓝色粉霜，疏被毛。总状花序。蒴果。花期4～5月，果期7～8月。

Oeciduous or semievergreen shrubs or small trees. Young branches stellately tomentose. Leaf blade ovate, oblong to elliptic, apex acuminate, base subcordate, margin serrate, teeth mucronate, abaxially light blue pruinose, sparsely pubescent. Raceme. Capsule. Fl. Apr.–May, fr. Jul.–Aug..

树干　　Trunk
摄影：卢志军　　Photo by: Zhijun Lu

花枝　　Flowering branch
摄影：卢志军　　Photo by: Zhijun Lu

小枝和叶背　　Branchlet and leaf abaxial surface
摄影：卢志军　　Photo by: Zhijun Lu

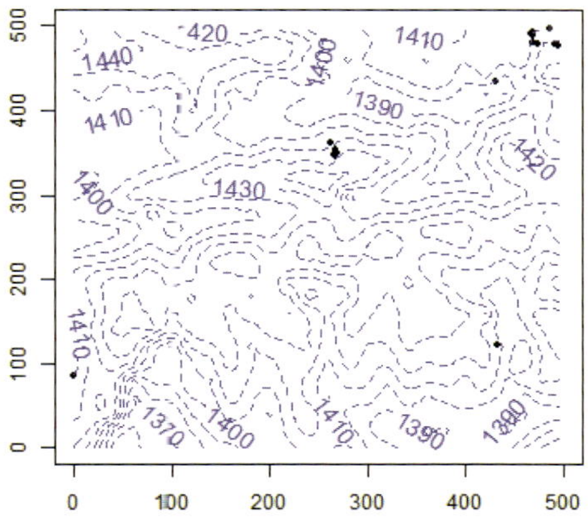

个体分布图　Distribution of individuals

径级分布表　DBH class

胸径区间 (Diameter class) (cm)	个体数 (No. of individuals in the plot)	比例 (Proportion) (%)
1～2	3	15.8
2～3	2	10.5
3～4	1	5.3
4～5	2	10.5
5～7	3	15.8
7～10	4	21.1
⩾10	4	21.1

69 缺萼枫香树 quē è fēng xiāng shù | Calyx-less Sweet-gum

Liquidambar acalycina，枫香树属
金缕梅科 | Hamamelidaceae

代码（SpCode）= LIQACA
个体数（Individual number/25 hm^2）= 324
最大胸径（Max DBH）= 78.49 cm
所在林层（Stratum）= 乔木层 (Canopy)
重要值排序（Importance value rank）= 14/50

树干 Trunk
摄影：卢志军 Photo by: Zhijun Lu

落叶大乔木。树皮黑褐色。小枝无毛，有皮孔。叶宽卵形，掌状3裂，中央裂片较长，先端尾状渐尖，基部心形，侧裂片稍平展，边缘有具腺的锯齿。雌花头状花序。头状果序。花期3～6月，果期7月。

Deciduous trees. Bark black-brown. Branchlets glabrous, lenticellate. Leaf blade broadly ovate, palmately 3-lobed, middle lobe longer, apex caudately acuminate, base cordate, lateral lobes slightly applanate, margin glandular serrate. Head inflorescence and infructescence. Fl. Mar.–Jun., fr. Jul..

果枝 Fruiting branches
摄影：刘海波 Photo by: Haibo Liu

小枝和叶背 Branchlet and leaf abaxial surface
摄影：卢志军 Photo by: Zhijun Lu

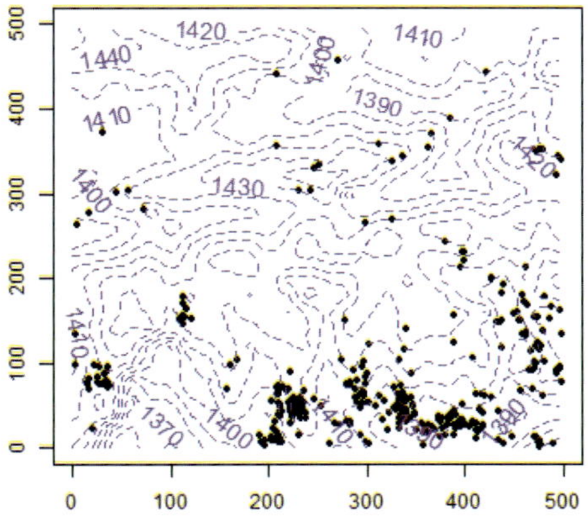

个体分布图 Distribution of individuals

径级分布表 DBH class

胸径区间 (Diameter class) (cm)	个体数 (No. of individuals in the plot)	比例 (Proportion) (%)
1～2.5	26	8.0
2.5～5	16	4.9
5～10	36	11.1
10～25	125	38.6
25～40	85	26.2
40～100	36	11.1
≥100	0	0.0

70 水丝梨 shuǐ sī lí | Chinese Fig-hazel

Sycopsis sinensis，水丝梨属
金缕梅科 I Hamamelidaceae

代码（SpCode）= SYCSIN
个体数（Individual number/25 hm^2）= 24
最大胸径（Max DBH）= 17.42 cm
所在林层（Stratum）= 亚乔木层 (Understory)
重要值排序（Importance value rank）= 43/54

树干 Trunk
摄影：卢志军 Photo by: Zhijun Lu

常绿乔木。小枝灰褐色，近无毛。叶椭圆状卵形至倒卵形，有时披针形，先端渐尖，基部钝圆，全缘。头状穗状花序，总梗短，被苞片。蒴果球形，被褐色茸毛。花期4月，果期7～9月。

Evergreen trees. Branchlets dark brown, subglabrous. Leaf blade elliptic-ovate to obovate, sometimes lanceolate, apex acuminate, base obtusely rounded, margin entire. Inflorescence head spike, peduncle short, with bracts. Capsules globose, brown pubescent. Fl. Apr., fr. Jul.–Sept..

花枝 Flowering branches
摄影：江明喜 Photo by: Mingxi Jiang

果枝 Fruiting branch
摄影：鲍大川 Photo by: Dachuan Bao

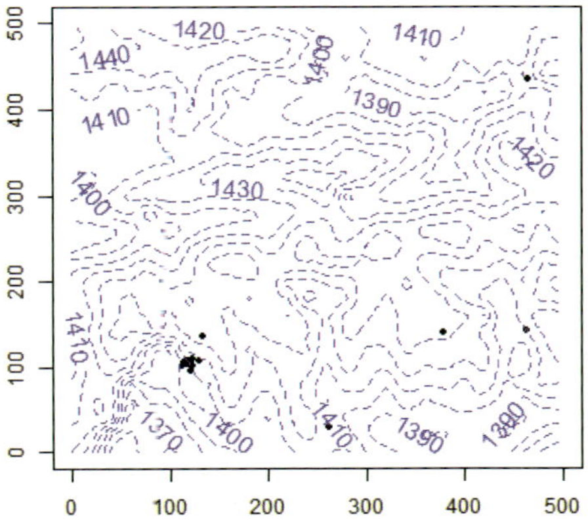

个体分布图 Distribution of individuals

径级分布表 DBH class

胸径区间 (Diameter class) (cm)	个体数 (No. of individuals in the plot)	比例 (Proportion) (%)
1～2.5	12	50.0
2.5～5	8	33.3
5～8	2	8.3
8～11	1	4.2
11～15	0	0.0
15～20	1	4.2
≥20	0	0.0

71 异色溲疏 yì sè sōu shū | Discolor Deutzia

Deutzia discolor，溲疏属
虎耳草科 | Saxifragaceae

代码（SpCode）= DEUDIS
个体数（Individual number/25 hm^2）= 9
最大胸径（Max DBH）= 5.11 cm
所在林层（Stratum）= 灌木层 (Shrub)
重要值排序（Importance value rank）= 98/128

落叶灌木。幼枝被星状毛，后脱落。叶纸质，卵状长圆形至披针形，先端渐尖或急尖，基部宽楔形或楔形，边缘有细锯齿。伞房花序；花白色或粉红色。蒴果近球形。花期5月，果期10月。

Deciduous shrubs. Branchlets stellately tomentulose, then glabrous. Leaf blade papery, ovate-oblong to lanceolate, apex acuminate or acute, base broadly cuneate or cuneate, margin serrulate. Umbels, flowers white or pinkish red. Capsules subglobose. Fl. May, fr. Oct..

树干 Trunk
摄影：卢志军 Photo by: Zhijun Lu

花枝 Flowering branches
摄影：江明喜 Photo by: Mingxi Jiang

小枝和叶背 Branchlet and leaf abaxial surface
摄影：刘海波 Photo by: Haibo Liu

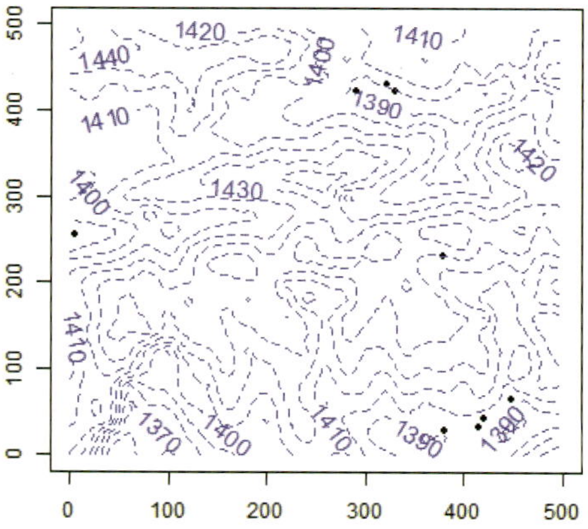

个体分布图 Distribution of individuals

径级分布表 DBH class

胸径区间 (Diameter class) (cm)	个体数 (No. of individuals in the plot)	比例 (Proportion) (%)
1～2	2	22.2
2～3	3	33.3
3～4	2	22.2
4～5	1	11.1
5～7	1	11.1
7～10	0	0.0
≥10	0	0.0

72 蜡莲绣球 là lián xiù qiú | Strigose Hydrangea

Hydrangea strigosa，绣球属
虎耳草科 | Saxifragaceae

代码（SpCode）= HYDSTR
个体数（Individual number/25 hm^2）= 2033
最大胸径（Max DBH）= 35.90 cm
所在林层（Stratum）= 灌木层 (Shrub)
重要值排序（Importance value rank）= 12/128

树干 Trunk
摄影：卢志军 Photo by: Zhijun Lu

落叶灌木。树皮灰褐色，呈薄片状脱落。小枝密生粗伏毛。叶长圆状卵形，披针形或椭圆状披针形，先端渐尖，基部楔形或圆形，边缘有细锯齿。伞房状聚伞花序顶生；花白色或带紫色。蒴果半球形，顶端平截；种子宽椭圆形，两端有短翅。花期8～9月，果期10月。

Deciduous shrubs. Bark grayish brown, usually peeled off into fragments. Branchlets densely strigose. Leaf blade obovate, lanceolate or elliptic-lanceolate, apex acuminate, base cuneate or rounded. Margin serrulate. Inflorescences corymbose cymes, on apex, flowers white or purplish. Capsule semiglobose, apex truncate. Seeds broadly elliptic, winged at both ends. Fl. Aug.–Sept., fr. Oct..

花枝 Flowering branches
摄影：徐耀粘 Photo by: Yaozhan Xu

小枝和叶背 Branchlet and leaf abaxial surface
摄影：卢志军 Photo by: Zhijun Lu

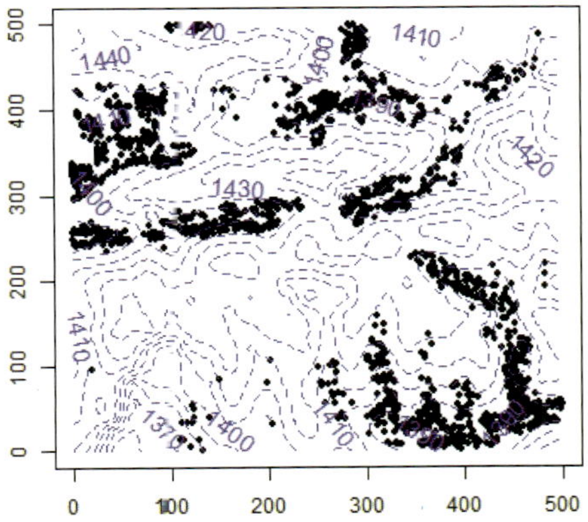

个体分布图 Distribution of individuals

径级分布表 DBH class

胸径区间 (Diameter class) (cm)	个体数 (No. of individuals in the plot)	比例 (Proportion) (%)
1～2	419	20.1
2～3	323	15.5
3～4	282	13.5
4～5	247	11.9
5～7	410	19.7
7～10	233	11.2
≥10	119	5.7

73 绢毛山梅花 juàn máo shān méi huā | Silk Mock-orange

Philadelphus sericanthus，山梅花属
虎耳草科 | Saxifragaceae

代码（SpCode）= PHISER
个体数（Individual number/25 hm²）= 18
最大胸径（Max DBH）= 6.4 cm
所在林层（Stratum）= 灌木层 (Shrub)
重要值排序（Importance value rank）= 86/128

落叶灌木。一年枝近无毛，二年枝灰褐色至赤褐色。叶椭圆状卵形至披针状卵形，先端渐尖，基部宽楔形，边缘有浅锯齿。总状花序；花瓣倒卵形，白色。蒴果倒卵形。花期6~7月，果期7~8月。

Deciduous shrubs. 1-yr branchlets subglabrous; 2-yr branchlets grayish brown to reddish brown. Leaf blade elliptic-ovate to lanceolate-ovate, apex acuminate, base broadly cuneate, margin serrate. Raceme, petals obovate, white. Capsules obovate. Fl. Jun.–Jul., fr. Jul.–Aug..

树干 Trunk
摄影：卢志军 Photo by: Zhijun Lu

花枝 Flowering branch
摄影：鲍大川 Photo by: Dachuan Bao

小枝和叶背 Branchlet and leaf abaxial surface
摄影：刘海波 Photo by: Haibo Liu

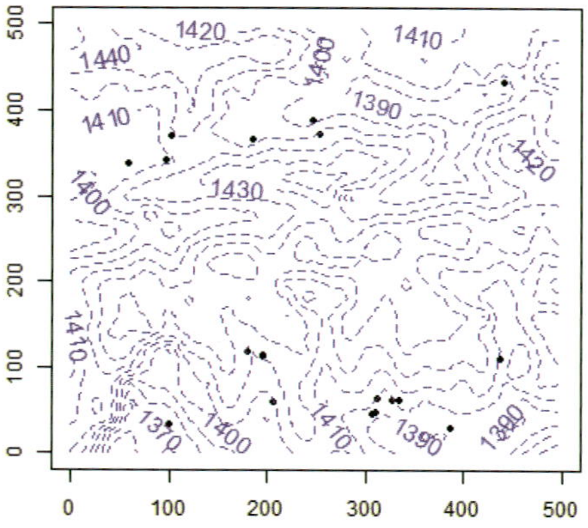

个体分布图 Distribution of individuals

径级分布表 DBH class

胸径区间 (Diameter class) (cm)	个体数 (No. of individuals in the plot)	比例 (Proportion) (%)
1~2	5	27.8
2~3	6	33.3
3~4	4	22.2
4~5	1	5.6
5~7	2	11.1
7~10	0	0.0
≥10	0	0.0

74 棱果海桐 léng guǒ hǎi tóng | Three-rib-fruit Pittosporum

Pittosporum trigonocarpum，海桐花属
海桐花科 | Pittosporaceae

代码（SpCode）= PITTRI
个体数（Individual number/25 hm²）= 667
最大胸径（Max DBH）= 17.71 cm
所在林层（Stratum）= 灌木层 (Shrub)
重要值排序（Importance value rank）= 35/128

常绿灌木或小乔木。幼枝无毛；老枝灰色，皮孔明显。叶互生，薄革质，在枝顶近丛生，叶倒披针形或宽倒披针形。先端渐尖或急尖，基部渐狭，全缘，稍反卷，无毛。伞形花序，顶生，无毛，花有芳香；花瓣黄色。蒴果椭圆形，有3棱；种子小，多数，暗红色，不整齐。花期3～5月，果期6～10月。

Evergreen shrubs or small trees. Young branchlets glabrous; old gray, lenticellate. Leaves alternate, thinly leathery, clustered at branchlet apex; leaf blade oblanceolate or broadly oblanceolate, apex acuminate or acute, base decurrent, margin entire, slightly revolute, glabrous. Umbels, terminal, glabrous, flowers fragrant, petals yellow. Capsules ellipsoid, 3-ribbed. Seeds small and numerous, dark red, irregular. Fl. Mar.–May, fr. Jun.–Oct..

树干 Trunk
摄影：卢志军 Photo by: Zhijun Lu

花枝 Flowering branch
摄影：刘海波 Photo by: Haibo Liu

果枝和种子 Fruiting branch and seeds
摄影：刘海波、卢志军 Photo by: Haibo Liu, Zhijun Lu

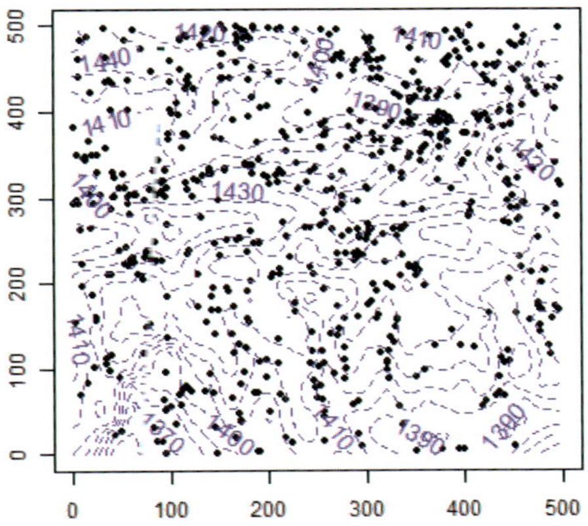

个体分布图 Distribution of individuals

径级分布表 DBH class

胸径区间 (Diameter class) (cm)	个体数 (No. of individuals in the plot)	比例 (Proportion) (%)
1～2	554	83.1
2～3	91	13.6
3～4	8	1.2
4～5	7	1.0
5～7	4	0.6
7～10	1	0.1
≥10	2	0.3

75 线叶柄果海桐 xiàn yè bǐng guǒ hǎi tóng | Narrowleaf Pittosporum

Pittosporum podocarpum var. *angustatum*，海桐花属
海桐花科 | Pittosporaceae

代码（SpCode）= PITANG
个体数（Individual number/25 hm^2）= 202
最大胸径（Max DBH）= 7.50 cm
所在林层（Stratum）= 灌木层 (Shrub)
重要值排序（Importance value rank）= 52/128

常绿灌木，嫩枝无毛。叶簇生枝顶，薄革质，带状或狭窄披针形，先端急尖或渐尖，基部楔形，通常下延，全缘。伞形花序，顶生。蒴果梨形或椭圆形。花期3～5月，果期6～10月。

Evergreen shrubs. Young branchlets glabrous. Leaves clustered at branchlet apex, thinly leathery, leaf blade beltlike or narrowly lanceolate, apex acute or acuminate, base cuneate, usually decurrent, margin entire. Inflorescences umbel, terminal. Capsule pear-shaped or ellipsoid. Fl. Mar.–May, fr. Jun.–Oct..

树干 Trunk
摄影：卢志军 Photo by: Zhijun Lu

花枝 Flowering branches
摄影：鲍大川 Photo by: Dachuan Bao

果枝 Fruiting branches
摄影：刘海波 Photo by: Haibo Liu

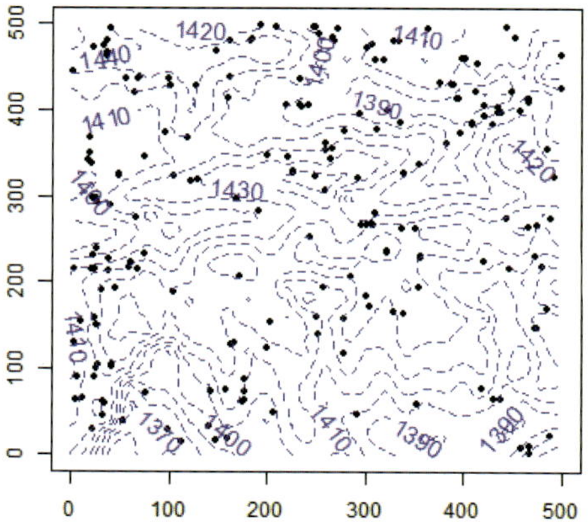

个体分布图 Distribution of individuals

径级分布表 DBH class

胸径区间 (Diameter class) (cm)	个体数 (No. of individuals in the plot)	比例 (Proportion) (%)
1～2	169	83.7
2～3	27	13.4
3～4	4	2.0
4～5	1	0.5
5～7	0	0.0
7～10	1	0.5
⩾10	0	0.0

76 海金子

hǎi jīn zǐ | Anise-tree-like Pittosporum

Pittosporum illicioides，海桐花属
海桐花科 | Pittosporaceae

代码（SpCode）= PITILL
个体数（Individual number/25 hm^2）= 77
最大胸径（Max DBH）= 18.65 cm
所在林层（Stratum）= 灌木层 (Shrub)
重要值排序（Importance value rank）= 62/128

常绿小乔木或灌木。幼枝光滑，老枝皮孔明显。叶互生，薄革质，倒披针形至倒卵形，先端急渐尖，基部渐狭，全缘，无毛。伞形花序，顶生。蒴果绿色或黑色，椭圆形，三棱；种子暗红色。花期4～5月，果期7～9月。

Evergreen small trees or shrubs. Young branchlets glabrous; old lenticellate. Leaves alternate, thinly leathery, oblanceolate to obovate, apex acutely acuminate, base decurrent, margin entire, glabrous. Umbels terminal. Capsules green or black, ellipsoid, 3-lobed. Seeds dark red. Fl. Apr.–May, fr. Jul.–Sept..

树干 Trunk
摄影：卢志军 Photo by: Zhijun Lu

果枝 Fruiting branch
摄影：鲍大川 Photo by: Dachuan Bao

小枝和叶背 Branchlet and leaf abaxial surface
摄影：卢志军 Photo by: Zhijun Lu

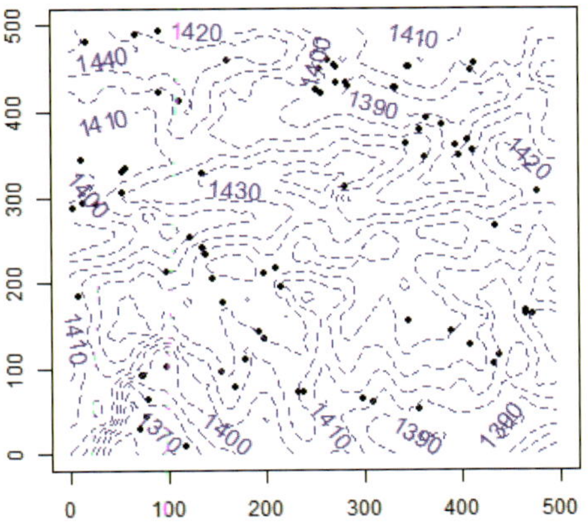

个体分布图 Distribution of individuals

径级分布表 DBH class

胸径区间 (Diameter class) (cm)	个体数 (No. of individuals in the plot)	比例 (Proportion) (%)
1～2	59	76.6
2～3	12	15.6
3～4	2	2.6
4～5	1	1.3
5～7	2	2.6
7～10	0	0.0
≥10	1	1.3

77 崖花子

yá huā zǐ | Truncate Pittosporum

Pittosporum truncatum，海桐花属
海桐花科 | Pittosporaceae

代码（SpCode）= PITTRU
个体数（Individual number/25 hm^2）= 34
最大胸径（Max DBH）= 5.10 cm
所在林层（Stratum）= 灌木层 (Shrub)
重要值排序（Importance value rank）= 77/128

常绿灌木。幼枝被灰茸毛，后无毛。叶聚生枝顶，革质，菱状倒卵形，短渐尖，少数倒卵形而急尖，全缘，稍反卷，无毛。伞形花序。蒴果球形，簇生；黑红色或浅红色。花期3～5月，果期6～10月。

Evergreen shrubs. Young branchlets gray pubescent, then glabrate. Leaves clustered at branchlet apex; leathery, rhombic-obovate, shortly acuminate, rarely obovate and acute, margin entire, slightly revolute, glabrous. Capsule globose, fascicled, blackish red or light red. Fl. Mar.–May, fr. Jun.–Oct..

树干 Trunk
摄影：卢志军 Photo by: Zhijun Lu

花枝 Flowering branch
摄影：刘海波 Photo by: Haibo Liu

小枝和叶背 Branchlet and leaf abaxial surface
摄影：卢志军 Photo by: Zhijun Lu

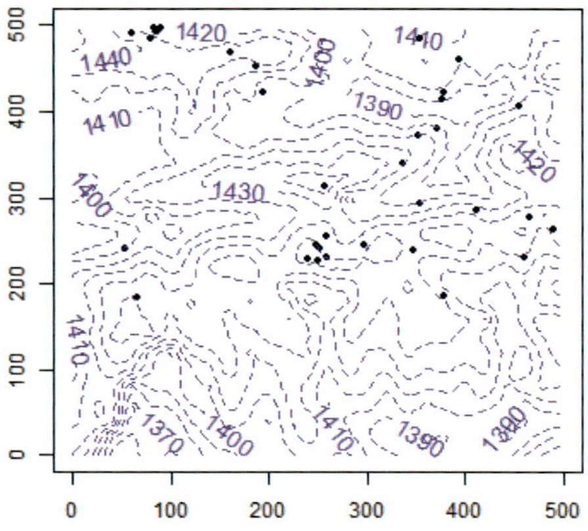

个体分布图 Distribution of individuals

径级分布表 DBH class

胸径区间 (Diameter class) (cm)	个体数 (No. of individuals in the plot)	比例 (Proportion) (%)
1～2	25	73.5
2～3	7	20.6
3～4	1	2.9
4～5	0	0.0
5～7	1	2.9
7～10	0	0.0
≥10	0	0.0

78 柄果海桐

bǐng guǒ hǎi tóng | Stalked-fruit Pittosporum

Pittosporum podocarpum，海桐花属
海桐花科 | Pittosporaceae

代码（SpCode）= PITPOD
个体数（Individual number/25 hm²）= 2
最大胸径（Max DBH）= 2.00 cm
所在林层（Stratum）= 灌木层 (Shrub)
重要值排序（Importance value rank）= 121/128

常绿灌木。幼枝无毛，老枝皮孔明显。叶聚生枝条分叉处及枝顶端，薄革质，倒披针形至椭圆形，先端急渐尖，基部楔形，全缘，稍反卷，无毛。花序伞形或近伞形。蒴果梨形，3瓣裂；种子红色。花期4月，果期6月。

Evergreen shrubs. Young branchlets glabrous; old lenticellate. Leaves clustered at node and branchlet apex, thinly leathery, blade oblanceolate to elliptic, apex acutely acuminate, base cuneate, margin entire, slightly revolute, grabrous. Umbels or nearly umbels. Capsules pyriform, 3-lobed. Seeds red. Fl. Apr., fr. Jun..

树干 Trunk
摄影：卢志军 Photo by: Zhijun Lu

小枝和叶片 Branchlet and leaves
摄影：卢志军 Photo by: Zhijun Lu

小枝和叶背 Branchlet and leaf abaxial surface
摄影：卢志军 Photo by: Zhijun Lu

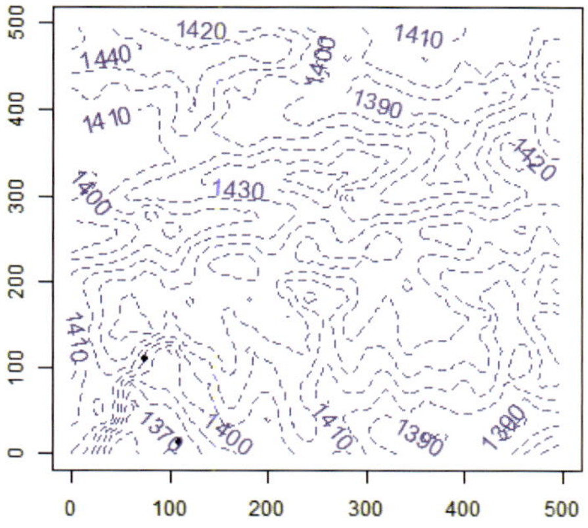

个体分布图 Distribution of individuals

径级分布表 DBH class

胸径区间 (Diameter class) (cm)	个体数 (No. of individuals in the plot)	比例 (Proportion) (%)
1～2	1	50.0
2～3	1	50.0
3～4	0	0.0
4～5	0	0.0
5～7	0	0.0
7～10	0	0.0
≥10	0	0.0

79 厚圆果海桐 hòu yuán guǒ hǎi tóng | Rehder Pittosporum

Pittosporum rehderianum，海桐花属
海桐花科 | Pittosporaceae

代码（SpCode）= PITREH
个体数（Individual number/25 hm²）= 2
最大胸径（Max DBH）= 2.00 cm
所在林层（Stratum）= 灌木层 (Shrub)
重要值排序（Importance value rank）= 120/128

常绿灌木。幼枝干后棕色，无毛；老枝灰棕色，皮孔明显。叶聚生枝条分叉处及顶端，革质，倒披针形至倒卵形，顶端急尖至渐尖，基部渐狭，全缘，两面无毛。花序近伞形，顶生，花瓣黄色。蒴果近球形，3裂；种子多数，暗红色。花期3～5月，果期5～11月。

Evergreen shrubs. Young branchlets brown after dried, glabrous; old branchlets gray-brown, lenticellate. Leaves clustered at branchlet apex, leathery. Leaf blade oblanceolate to obovate, apex acute to acuminate, base decurrent, margin entire, glabrous on both surfaces. Nearly umbels, terminal, petals yellow. Capsules subglobose, 3-lobed. Seeds numerous, dark red. Fl. Mar.–May, fr. May–Nov..

树干 Trunk
摄影：卢志军 Photo by: Zhijun Lu

小枝和叶片 Branchlet and leaves
摄影：卢志军 Photo by: Zhijun Lu

小枝和叶背 Branchlet and leaf abaxial surface
摄影：卢志军 Photo by: Zhijun Lu

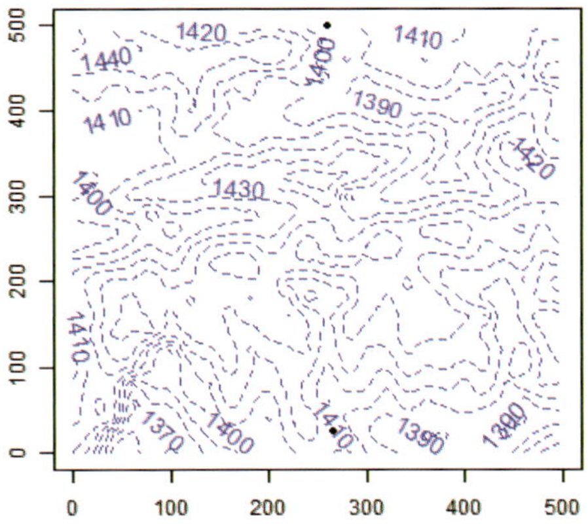

个体分布图 Distribution of individuals

径级分布表 DBH class

胸径区间 (Diameter class) (cm)	个体数 (No. of individuals in the plot)	比例 (Proportion) (%)
1～2	1	50.0
2～3	1	50.0
3～4	0	0.0
4～5	0	0.0
5～7	0	0.0
7～10	0	0.0
≥10	0	0.0

80 尾叶樱桃

wěi yè yīng táo | Diels Cherry

Cerasus dielsiana，樱属
蔷薇科 | Rosaceae

代码（SpCode）= CERDIE
个体数（Individual number/25 hm^2）= 421
最大胸径（Max DBH）= 37.72 cm
所在林层（Stratum）= 亚乔木层 (Understory)
重要值排序（Importance value rank）= 19/54

落叶灌木或乔木。老枝灰色或深棕灰色，无毛。叶纸质，长圆形，先端突狭渐尖，基部急尖或圆，或有微凹，边缘有单及重锯齿，齿端有腺体，上面无毛，下面被柔毛。花先叶开放，伞形花序；花瓣白色或淡红色，卵形。核果球形，红色。花期4～5月，果期6～8月。

Deciduous shrubs or trees. Branchlets grayish or dark brown, glabrous. Leaf blade papery, oblong, apex caudate-acuminate, base acute or rounded, or slightly concave, margin acutely serrate or biserrate and teeth with a apical gland, adaxially glabrous, abaxially pubescent. Flowers appearing before leaves, umbels, patels white or light red, ovate. Drupes globose, red. Fl. Apr.–May, fr.Jun.–Aug..

树干 Trunk
摄影：卢志军 Photo by: Zhijun Lu

果枝 Fruiting branch
摄影：卢志军 Photo by: Zhijun Lu

果枝和叶背 Fruiting branch and leaf abaxial surface
摄影：卢志军 Photo by: Zhijun Lu

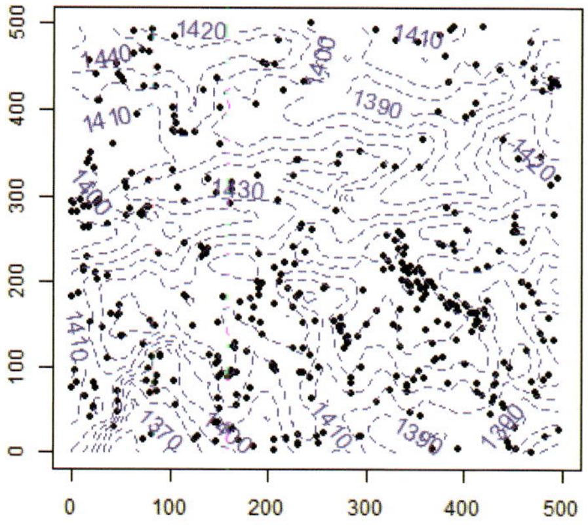

个体分布图 Distribution of individuals

径级分布表 DBH class

胸径区间 (Diameter class) (cm)	个体数 (No. of individuals in the plot)	比例 (Proportion) (%)
1～2.5	177	42.0
2.5～5	146	34.7
5～8	22	5.2
8～11	18	4.3
11～15	23	5.5
15～20	16	3.8
≥20	19	4.5

81 云南樱桃 yún nán yīng táo | Fewhairy Cherry

Cerasus yunnanensis，樱属
蔷薇科 | Rosaceae

代码（SpCode）= CERCLA
个体数（Individual number/25 hm^2）= 116
最大胸径（Max DBH）= 30.22 cm
所在林层（Stratum）= 亚乔木层 (Understory)
重要值排序（Importance value rank）= 24/54

落叶灌木或乔木。树皮灰黑色。老枝灰褐色。叶膜质，倒卵形或倒卵状长圆形，先端突渐尖，基部微凹或圆，边缘有近缺刻重锯齿，在叶基部或柄端有2腺体。花伞形着生，花瓣白色或粉红色。核果长圆形，红色。花期4～5月，果期6～8月。

Deciduous shrubs or trees. Bark grayish black. Branchlets grayish brown. Leaf blade membranous, ovate, ovate-oblong, apex acutely acuminate, base slightly concave or rounded, margin nearly incised biserrate, with 2 glands on base or petiole. Umbels, petals white or pink-red. Drupes oblong, red. Fl. Apr.–May, fr. Jun.–Aug..

树干 Trunk
摄影：卢志军 Photo by: Zhijun Lu

小枝和叶片 Branchlet and leaves
摄影：卢志军 Photo by: Zhijun Lu

果枝 Fruiting branches
摄影：卢志军 Photo by: Zhijun Lu

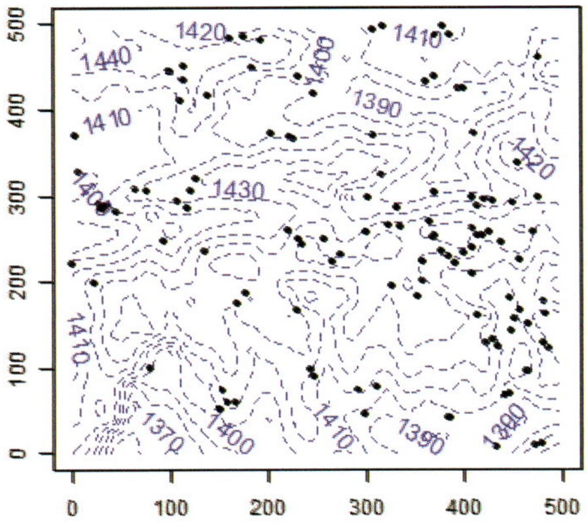

个体分布图 Distribution of individuals

径级分布表 DBH class

胸径区间 (Diameter class) (cm)	个体数 (No. of individuals in the plot)	比例 (Proportion) (%)
1～2.5	25	21.6
2.5～5	19	16.4
5～8	6	5.2
8～11	15	12.9
11～15	10	8.6
15～20	14	12.1
≥20	27	23.3

82 华中樱桃

huá zhōng yīng táo | Conradia Cherry

Cerasus conradinae，樱属
蔷薇科 | Rosaceae

代码（SpCode）= CERCON
个体数（Individual number/25 hm^2）= 2
最大胸径（Max DBH）= 16.52 cm
所在林层（Stratum）= 灌木层 (Shrub)
重要值排序（Importance value rank）= 96/128

落叶乔木。树皮和老枝棕灰色。叶纸质，倒卵形或倒卵状长圆形，少数为圆卵形，先端突狭渐尖，基部圆形或近心形，边缘有锯齿，中部有重锯齿，齿尖有小腺体，腺体2，粗大，在叶柄先端着生。花伞形，或有时为总状伞形，先叶开放；花瓣卵形或倒卵形，白色或淡红色。核果卵形，红色。花期5月，果期7月。

Deciduous trees. Bark and branchlets grayish brown. Leaf blade papery, obovate or obovate-oblong, rarely ovate, apex abruptly acuminate, base rounded or subcordate, margin serrate, biserrate on middle part, teeth with 2 apical glands, thick, on petiole apex. Umbels, or sometime raceme-umbels, appearing before leaves, petals ovate or obovate, white or light red. Drupes ovoid, red. Fl. May, fr. Jul..

树干 Trunk
摄影：卢志军 Photo by: Zhijun Lu

花 Flowers
摄影：江明喜 Photo by: Mingxi Jiang

果枝 Fruiting branch
摄影：江明喜 Photo by: Mingxi Jiang

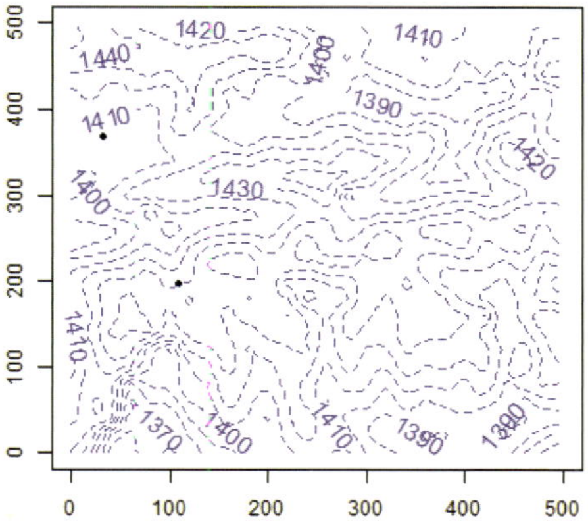

个体分布图 Distribution of individuals

径级分布表 DBH class

胸径区间 (Diameter class) (cm)	个体数 (No. of individuals in the plot)	比例 (Proportion) (%)
1～2	0	0.0
2～3	0	0.0
3～4	1	50.0
4～5	0	0.0
5～7	0	0.0
7～10	0	0.0
⩾10	1	50.0

83 皱叶柳叶栒子 zhòu yè liǔ yè xún zǐ | Rugose Willow-leaf Cotoneaster

Cotoneaster salicifolius var. *rugosus*，栒子属
蔷薇科 | Rosaceae

代码（SpCode）= COTSAL
个体数（Individual number/25 hm^2）= 2
最大胸径（Max DBH）= 3.88 cm
所在林层（Stratum）= 灌木层 (Shrub)
重要值排序（Importance value rank）= 117/128

半常绿或灌木或小乔木。枝条带黑色。叶卵形或椭圆形，两端渐尖，叶上面有皱，边缘反卷，叶脉深陷。聚伞花序；花瓣白色。果近球形，红色。花期6月，果期9月。

Semi-evergreen or evergreen shrubs or small trees. Branchlets black. Leaf blade ovate or elliptic, apex and base acuminate, adaxially deeply rugose, margin revolute, veins deeply impressed. Cymes, petals white. Fruit subglobose, red. Fl. Jun., fr. Sept..

树干 Trunk
摄影：卢志军 Photo by: Zhijun Lu

小枝和叶片 Branchlet and leaves
摄影：卢志军 Photo by: Zhijun Lu

小枝和叶背 Branchlet and leaf abaxial surface
摄影：卢志军 Photo by: Zhijun Lu

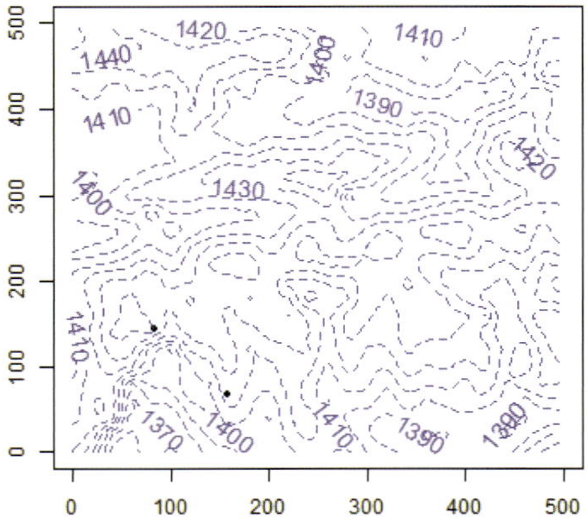

个体分布图 Distribution of individuals

径级分布表 DBH class

胸径区间 (Diameter class) (cm)	个体数 (No. of individuals in the plot)	比例 (Proportion) (%)
1～2	0	0.0
2～3	1	50.0
3～4	1	50.0
4～5	0	0.0
5～7	0	0.0
7～10	0	0.0
⩾10	0	0.0

84 刺叶桂樱

cì yè guì yīng | Spiny Laurocerasus

Laurocerasus spinulosa，桂樱属
蔷薇科 | Rosaceae

代码（SpCode）= LAUSPI
个体数（Individual number/25 hm^2）= 138
最大胸径（Max DBH）= 17.25 cm
所在林层（Stratum）= 亚乔木层 (Understory)
重要值排序（Importance value rank）= 31/54

常绿乔木，稀为灌木。小枝紫褐色或黑褐色。叶片草质至薄革质，长圆形或倒卵状长圆形，先端渐尖至尾尖，基部宽楔形至近圆形，一侧常偏斜，边缘不平而常呈波状，中部以上或近顶端常具少数针状锐锯齿。总状花序生于叶腋，单生；花瓣圆形，白色。果实椭圆形，褐色至黑褐色。花期9～10月，果期11至翌年3月。

Evergreen trees, rarely shrubs. Branchlets purplish brown to blackish brown. Leaf blade herbaceous to thinly leathery, oblong or obovate-oblong, apex acuminate to caudate, base broadly cuneate to subrounded and often unequal, margin undulate and with a few acicular teeth apically from middle to near apex. Racemes axillary, solitary. Petals rounded, white. Drupe ellipsoid, brown to blackish brown. Fl. Sept.–Oct., fr. Nov.–Mar. of following yr..

树干 Trunk
摄影：卢志军 Photo by: Zhijun Lu

小枝和叶片 Branchlet and leaves
摄影：卢志军 Photo by: Zhijun Lu

小枝和叶背 Branchlet and leaf abaxial surface
摄影：卢志军 Photo by: Zhijun Lu

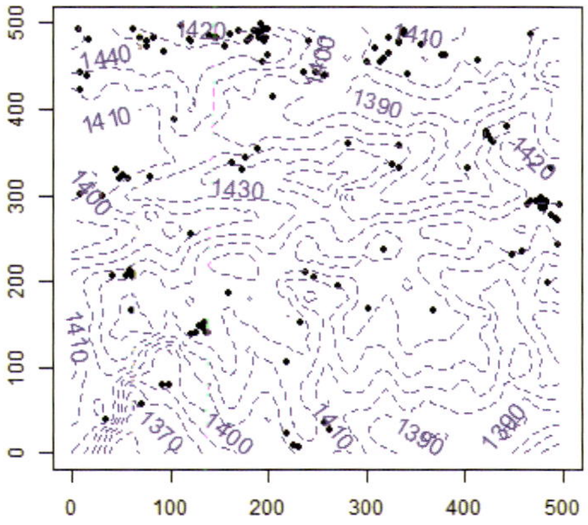

个体分布图 Distribution of individuals

径级分布表 DBH class

胸径区间 (Diameter class) (cm)	个体数 (No. of individuals in the plot)	比例 (Proportion) (%)
1～2.5	70	50.7
2.5～5	44	31.9
5～8	15	10.9
8～11	5	3.6
11～15	3	2.2
15～20	1	0.7
≥20	0	0.0

85 湖北海棠

hú běi hǎi táng | Hupeh Crab-apple

Malus hupehensis，苹果属

蔷薇科 | Rosaceae

代码（SpCode）= MALHUP

个体数（Individual number/25 hm^2）= 1

最大胸径（Max DBH）= 10.73 cm

所在林层（Stratum）= 灌木层 (Shrub)

重要值排序（Importance value rank）= 107/128

树干 Trunk

摄影：卢志军 Photo by: Zhijun Lu

落叶乔木。幼枝初被毛，后脱落。叶卵形至卵状长圆形，先端渐尖，基部宽楔形，少数稍呈圆形，边缘有细锐锯齿。伞状花序，花粉红白色或近于白色，无毛或稍有长柔毛；花柱3或4，较雄蕊稍长。果球形，黄绿色或稍带红色。花期4～5月，果期7～9月。

Deciduous trees. Branchlets initially puberulous, then glabrous. Leaf blade ovate to oblong, apex acuminate, base broadly cuneate, rarely slightly rounded, margin serrulate. Corymb, flowers reddish white or nearly white, glabrous or sparsely villous. Styles 3 or 4, longer than stamens. Fruit globose, yellowish green or slight reddish. Fl. Apr.–May, fr. Jul.–Sept..

花 Flowers

摄影：徐耀粘 Photo by: Yaozhan Xu

果枝 Fruiting branches

摄影：卢志军 Photo by: Zhijun Lu

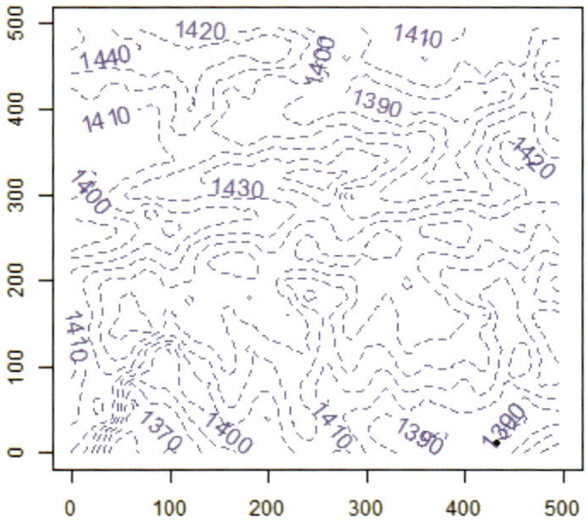

个体分布图 Distribution of individuals

径级分布表 DBH class

胸径区间 (Diameter class) (cm)	个体数 (No. of individuals in the plot)	比例 (Proportion) (%)
1～2	0	0.0
2～3	0	0.0
3～4	0	0.0
4～5	0	0.0
5～7	0	0.0
7～10	0	0.0
≥10	1	100.0

86 灰叶稠李　　huī yè chóu lǐ | Gray Bird Cherry

Padus grayana，稠李属
蔷薇科 | Rosaceae

代码（SpCode）= PADGRA
个体数（Individual number/25 hm^2）= 227
最大胸径（Max DBH）= 73.50 cm
所在林层（Stratum）= 乔木层 (Canopy)
重要值排序（Importance value rank）= 15/50

树干　Trunk
摄影：卢志军　Photo by: Zhijun Lu

落叶乔木。幼枝黑棕色，无毛，有时被微毛。叶膜质，长圆形、倒卵状长圆形或卵形，先端尾状渐尖，基部圆形或近心形，边缘有锐锯齿。总状花序，花白色。核果黑棕色，卵球形，无毛。花期4～6月，果期8～10月。

Deciduous trees. Young Branches blackish brown, glabrous, sometimes pubescent. Leaf blade membranous, oblong, obovate-oblong or ovate, apex caudately acuminate, base rounded to subcordate, margin acutely serrate. Racemes, flowers white. Drupe blackish brown, ovoid-globose, glabrous. Fl. Apr.–Jun., fr. Aug.–Oct..

花枝　Flowering branches
摄影：喻勋林　Photo by: Xunlin Yu

果枝　Fruiting branches
摄影：喻勋林　Photo by: Xunlin Yu

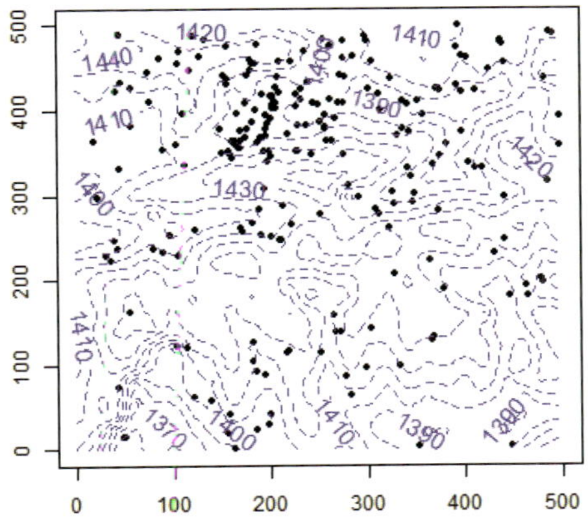

个体分布图 Distribution of individuals

径级分布表 DBH class

胸径区间 (Diameter class) (cm)	个体数 (No. of individuals in the plot)	比例 (Proportion) (%)
1～2.5	22	9.7
2.5～5	9	4.0
5～10	10	4.4
10～25	68	30.0
25～40	74	32.6
40～100	40	17.6
≥100	4	1.8

87 细齿稠李

xì chǐ chóu lǐ | Obtuse-leaf Bird Cherry

Padus obtusata，稠李属

蔷薇科 | Rosaceae

树干 摄影：卢志军

Trunk Photo by: Zhijun Lu

代码（SpCode）= PADOBT

个体数（Individual number/25 hm^2）= 103

最大胸径（Max DBH）= 63.05 cm

所在林层（Stratum）= 亚乔木层 (Understory)

重要值排序（Importance value rank）= 23/54

落叶乔木。老枝紫褐色或暗褐色，无毛，有散生浅色皮孔；小枝幼时红褐色，被短柔毛或无毛。叶片窄长圆形、椭圆形或倒卵形，先端急尖或渐尖，稀圆钝，基部近圆形或宽楔形，稀亚心形，边缘有细密锯齿。总状花序具多花，花瓣白色。核果黑色，卵球形，无毛。花期4～5月，果期6～10月。

Deciduous trees. Branches purplish brown to dark brown, glabrous, with sparse lenticels; branchlets reddish brown when young, pubescent or glabrous. Leaf blade narrowly oblong, elliptic, or obovate, apex acute, acuminate, or rarely obtuse, base subrounded, broadly cuneate, or rarely subcordate, margin crenulate. Racemes, petals white, numerous-flowered. Drupe black, ovoid-globose, glabrous. Fl. Apr.–May, fr. Jun.–Oct..

花枝 摄影：徐耀粘

Flowering branches Photo by: Yaozhan Xu

果枝 摄影：徐耀粘

Fruiting branches Photo by: Yaozhan Xu

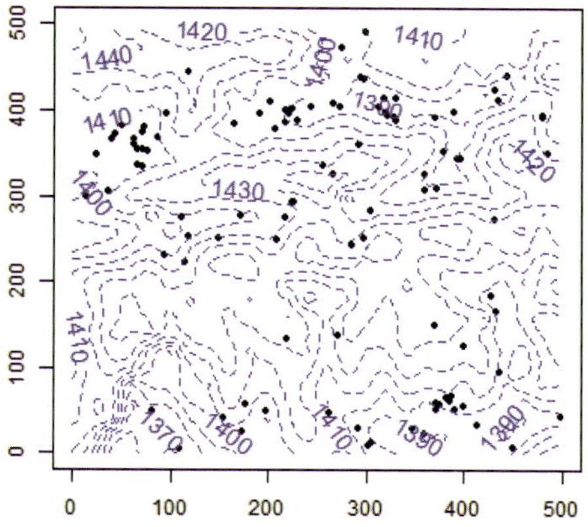

个体分布图 Distribution of individuals

径级分布表 DBH class

胸径区间 (Diameter class) (cm)	个体数 (No. of individuals in the plot)	比例 (Proportion) (%)
1～2.5	26	25.2
2.5～5	29	28.2
5～8	9	8.7
8～11	8	7.8
11～15	9	8.7
15～20	7	6.8
≥20	15	14.6

88 粗梗稠李　cū gěng chóu lǐ | Nepal Bird Cherry

Padus napaulensis，稠李属
蔷薇科 | Rosaceae

代码（SpCode）= PADNAP
个体数（Individual number/25 hm^2）= 5
最大胸径（Max DBH）= 23.73 cm
所在林层（Stratum）= 乔木层 (Canopy)
重要值排序（Importance value rank）= 34/50

树干　Trunk
摄影：卢志军　Photo by: Zhijun Lu

落叶乔木。树皮灰褐色，有圆形皮孔；老枝黑褐色，有明显浅色皮孔，小枝红褐色，无毛。叶片长椭圆形、卵状椭圆形或椭圆状披针形，先端急尖或短渐尖，基部楔形或近圆形，叶边有粗锯齿。总状花序具多花；花瓣白色，倒卵长圆形。核果卵球形，黑色或暗紫色。花期4月，果期7月。

Deciduous trees. Bark grayish brown, round lenticelled. Branches blackish brown, glabrous, with conspicuous pale lenticels; branchlets reddish brown, glabrous. Leaf blade narrowly elliptic, ovate-elliptic, or elliptic-lanceolate, apex acute to shortly acuminate, base cuneate or subrounded, margin coarsely serrate. Racemes, numerous-flowered, petals white, obovate-oblong. Drupe ovoid, black or dark purple. Fl. Apr., fr. Jul..

叶片(背面)　Leaf abaxial surface
摄影：卢志军　Photo by: Zhijun Lu

果枝　Fruiting branch
摄影：鲍大川　Photo by: Dachuan Bao

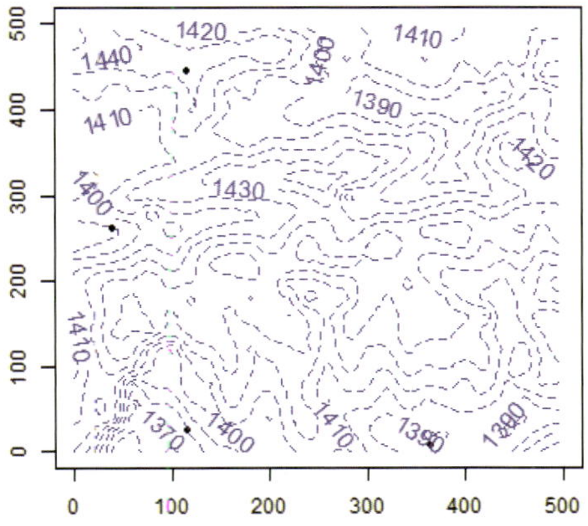

个体分布图　Distribution of individuals

径级分布表　DBH class

胸径区间 (Diameter class) (cm)	个体数 (No. of individuals in the plot)	比例 (Proportion) (%)
1～2.5	1	20.0
2.5～5	3	60.0
5～10	0	0.0
10～25	1	20.0
25～40	0	0.0
40～100	0	0.0
≥100	0	0.0

89 绢毛稠李

juàn máo chóu lǐ | Wilson Padus

Padus wilsonii，稠李属
蔷薇科 | Rosaceae

代码（SpCode）= PADWIL
个体数（Individual number/25 hm^2）= 11
最大胸径（Max DBH）= 26.58 cm
所在林层（Stratum）= 乔木层 (Canopy)
重要值排序（Importance value rank）= 27/50

树干 Trunk
摄影：卢志军 Photo by: Zhijun Lu

落叶乔木。小枝无毛，一年枝棕褐色，老枝紫褐色。叶薄纸质，椭圆形至长圆状倒卵形，先端短渐尖或急尖，基部圆形至楔形，边缘有锐锯齿。总状花序；花瓣白色，宽倒卵形。核果卵形，黑色。花期5月初，果期6～8月。

Deciduous trees. Branchlets glabrous, 1-yr branchlets chestnut brown, old purplish brown. Leaf blade thinly papery, elliptic, oblong to obovate, apex shortly acuminate or acute, base rounded to cuneate, margin acutely serrate. Racemes, petals white, broadly obovate. Drupes ovoid, black. Fl. early May, fr. Jun.–Aug..

叶片 Leaves
摄影：卢志军 Photo by: Zhijun Lu

叶背 Leaf abaxial surface
摄影：卢志军 Photo by: Zhijun Lu

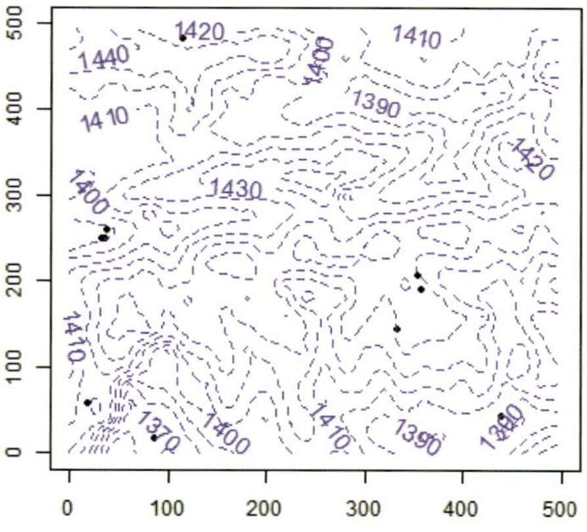

个体分布图 Distribution of individuals

径级分布表 DBH class

胸径区间 (Diameter class) (cm)	个体数 (No. of individuals in the plot)	比例 (Proportion) (%)
1～2.5	3	27.3
2.5～5	2	18.2
5～10	3	27.3
10～25	2	18.2
25～40	1	9.1
40～100	0	0.0
≥100	0	0.0

90 中华石楠 zhōng huá shí nán | Beauverd Photinia

Photinia beauverdiana，石楠属
蔷薇科 | Rosaceae

代码（SpCode）= PHOBEA
个体数（Individual number/25 hm^2）= 2557
最大胸径（Max DBH）= 34.50 cm
所在林层（Stratum）= 灌木层 (Shrub)
重要值排序（Importance value rank）= 6/128

树干 Trunk
摄影：卢志军 Photo by: Zhijun Lu

落叶灌木或小乔木。小枝紫褐色，无毛。叶纸质，长圆形、倒卵状长圆形或卵状披针形，先端突渐尖，基部圆形或楔形，边缘有疏生具腺的锯齿。复伞房花序，花瓣白色。果卵形，紫红色。花期5月，果熟期10月。

Deciduous shrubs or small trees. Branchlets purplish brown, glabrous. Leaf blade papery, oblong, ovate-oblong, or ovate-lanceolate, apex abruptly acuminate, base rounded or cuneate, margin sparsely serrate, teeth with gland. Compound corymbs, petals white. Fruit ovoid, purplish red. Fl. May, fr. maturity Oct..

花枝 Flowering branches
摄影：卢志军 Photo by: Zhijun Lu

果枝 Fruiting branches
摄影：鲍大川 Photo by: Dachuan Bao

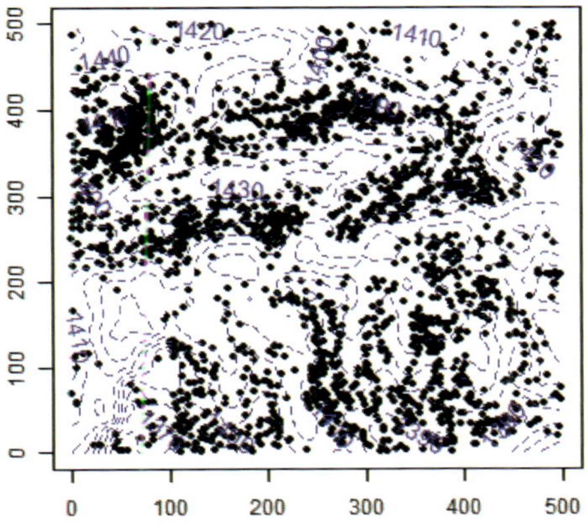

个体分布图 Distribution of individuals

径级分布表 DBH class

胸径区间 (Diameter class) (cm)	个体数 (No. of individuals in the plot)	比例 (Proportion) (%)
1～2	737	28.8
2～3	479	18.7
3～4	282	11.0
4～5	212	8.3
5～7	249	9.7
7～10	240	9.4
≥10	358	14.0

91 绒毛石楠

róng máo shí nán | Schneider Photinia

Photinia schneideriana，石楠属
蔷薇科 | Rosaceae

代码（SpCode）= PHOSCH
个体数（Individual number/25 hm^2）= 87
最大胸径（Max DBH）= 24.49 cm
所在林层（Stratum）= 灌木层 (Shrub)
重要值排序（Importance value rank）= 43/128

落叶灌木或小乔木。小枝紫褐色，老时灰褐色，初被毛，后无毛。叶膜质，长圆状披针形，或长圆状椭圆形，先端长渐尖，基部宽楔形，叶上面初被疏长柔毛，后脱落，下面被疏柔毛，边缘有锐锯齿。聚伞花序顶生；花白色。果卵形，带红色。花期5月，果期10月。

Deciduous shrubs or small trees. Branchlets purplish brown, old grayish brown, initially villous, then glabrous. Leaf blade membranous, oblong-lanceolate or oblong-elliptic, apex acuminate, base broadly cuneate, adaxially initially sparsely villous, then glabrescent, abaxially tomentose, margin sharply serrate. Compound corymbs terminal, flowers white. Fruit ovoid, reddish. Fl. May, fr. Oct..

树干 Trunk
摄影：卢志军 Photo by: Zhijun Lu

花枝 Flowering branches
摄影：刘海波 Photo by: Haibo Liu

小枝和叶背 Branchlets and leaf abaxial surface
摄影：卢志军 Photo by: Zhijun Lu

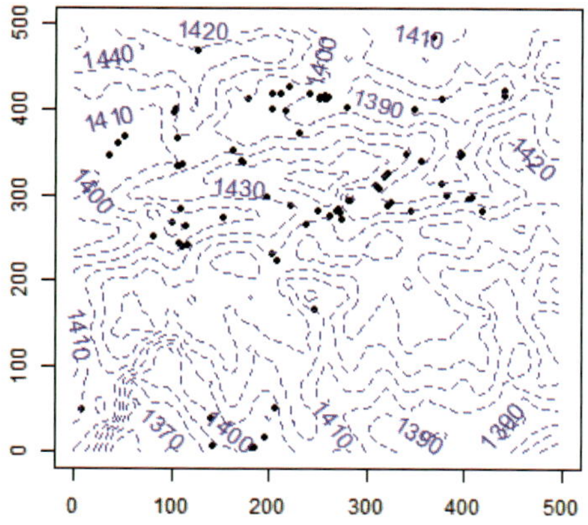

个体分布图 Distribution of individuals

径级分布表 DBH class

胸径区间 (Diameter class) (cm)	个体数 (No. of individuals in the plot)	比例 (Proportion) (%)
1～2	9	10.3
2～3	10	11.5
3～4	11	12.6
4～5	4	4.6
5～7	9	10.3
7～10	14	16.1
⩾10	30	34.5

92 石楠 shí nán | Chinese Photinia

Photinia serratifolia，石楠属
蔷薇科 | Rosaceae

代码（SpCode）= PHOSER
个体数（Individual number/25 hm^2）= 3
最大胸径（Max DBH）= 2.48 cm
所在林层（Stratum）= 灌木层 (Shrub)
重要值排序（Importance value rank）= 113/128

常绿乔木或灌木。无毛。小枝灰褐色。叶革质，长椭圆形、长倒卵形或倒卵状椭圆形，先端渐尖，基部圆形或宽楔形，边缘有疏生带腺体的细锯齿，叶上面暗绿色，叶下面黄绿色。复伞房花序顶生；花白色。果球形，红色。花期5～7月，果期10月。

Evergreen trees or shrubs. Glabrous. Branchlets grayish brown. Leaf blade leathery, oblong, obovate or obovate-elliptic, apex acuminate, base rounded or broadly cuneate, margin sparsely serrulate, teeth glanded, adaxially dark green, abaxially yellowish green. Compound corymbs terminal, flowers white. Fruit obovoid, reddish. Fl. May–Jul., fr. Oct..

树干 Trunk
摄影：卢志军 Photo by: Zhijun Lu

花序 Inflorescence
摄影：江明喜 Photo by: Mingxi Jiang

小枝和叶背 Branchlet and leaf abaxial surface
摄影：卢志军 Photo by: Zhijun Lu

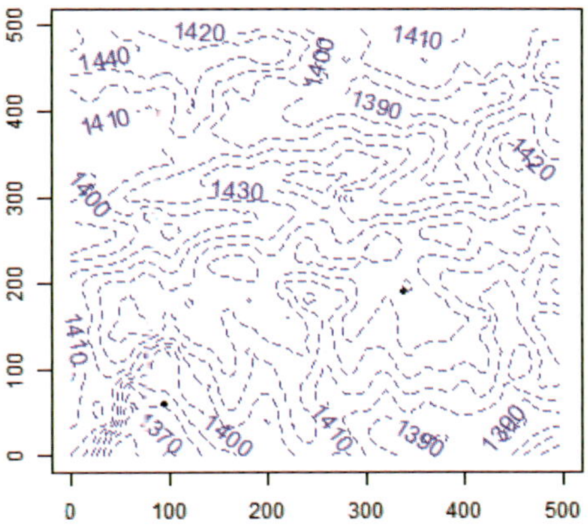

个体分布图 Distribution of individuals

径级分布表 DBH class

胸径区间 (Diameter class) (cm)	个体数 (No. of individuals in the plot)	比例 (Proportion) (%)
1～2	2	66.7
2～3	1	33.3
3～4	0	0.0
4～5	0	0.0
5～7	0	0.0
7～10	0	0.0
⩾10	0	0.0

93 全缘火棘

quán yuán huǒ jí | Entire Fire-thorn

Pyracantha atalantioides，火棘属

蔷薇科 | Rosaceae

代码（SpCode）= PYRATA

个体数（Individual number/25 hm^2）= 5

最大胸径（Max DBH）= 2.90 cm

所在林层（Stratum）= 灌木层 (Shrub)

重要值排序（Importance value rank）= 106/128

常绿灌木或小乔木，通常有枝刺。叶片椭圆形或长圆形，稀长圆状倒卵形，先端微尖或圆钝，有时具尖刺头，基部宽楔形或圆形，全缘或有时具不明显的细锯齿。复伞房花序，花瓣白色。梨果亮红色，扁球形。花期4～5月，果期9～11月。

Evergreen shrubs or small trees, usually with thorny branches. Leaf blade elliptic or oblong, rarely oblong-obovate, apex apiculate or obtuse, sometimes slightly aristate, base broadly cuneate or rounded, margin usually entire, sometimes inconspicuously serrulate. Compound corymb, petals white. Pome bright red, depressed-globose. Fl. Apr.–May, fr. Sept.–Nov..

树干 Trunk

摄影：卢志军 Photo by: Zhijun Lu

小枝和叶片 Branchlet and leaves

摄影：卢志军 Photo by: Zhijun Lu

小枝和叶背 Branchlet and leaf abaxial surface

摄影：卢志军 Photo by: Zhijun Lu

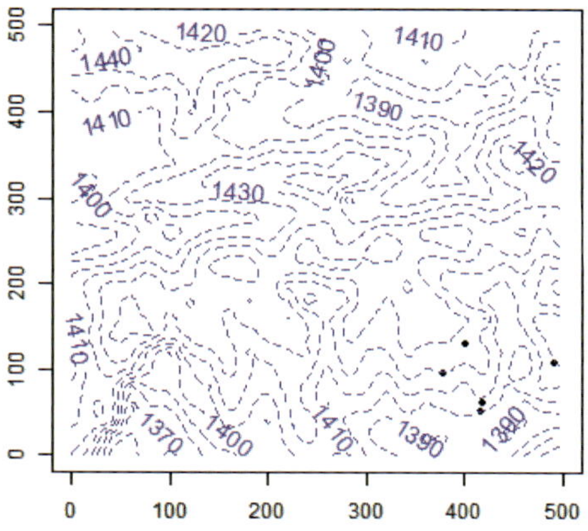

个体分布图 Distribution of individuals

径级分布表 DBH class

胸径区间 (Diameter class) (cm)	个体数 (No. of individuals in the plot)	比例 (Proportion) (%)
1～2	3	60.0
2～3	2	40.0
3～4	0	0.0
4～5	0	0.0
5～7	0	0.0
7～10	0	0.0
≥10	0	0.0

94 卵果蔷薇

luǎn guǒ qiáng wēi | Ovate-fruit Rose

Rosa helenae，蔷薇属
蔷薇科 | Rosaceae

代码（SpCode）= ROSHEL
个体数（Individual number/25 hm^2）= 1
最大胸径（Max DBH）= 1.22 cm
所在林层（Stratum）= 灌木层 (Shrub)
重要值排序（Importance value rank）= 128/128

落叶蔓生灌木。枝条紫褐色，有粗壮下弯的皮刺。叶片长圆状卵形至卵圆状披针形，先端短渐尖，基部圆形或广楔形，边缘有锯齿，叶下面灰绿色，脉上被柔毛。顶生伞房花序，花白色，有芳香。果卵形或长圆状卵形，猩红色。花期6月，果期8月。

Deciduous shrubs diffuse or scandent. Branches purple-brown, prickles stout and curved. Leave blade oblong-ovate to ovate-lanceolate, apex shortly acuminate, base rounded or broadly cuneate, margin serrate, abaxially grayish green, veins villouse. Terminal corgmbs, flowers white, fragrant. Fruit ovoid or oblong-ovate, deep red. Fl. Jun., fr. Aug..

树干 Trunk
摄影：卢志军 Photo by: Zhijun Lu

果枝 Fruiting branches
摄影：徐文斌 Photo by: Wenbin Xu

小枝和叶背 Branchlets and leaf abaxial surface
摄影：卢志军 Photo by: Zhijun Lu

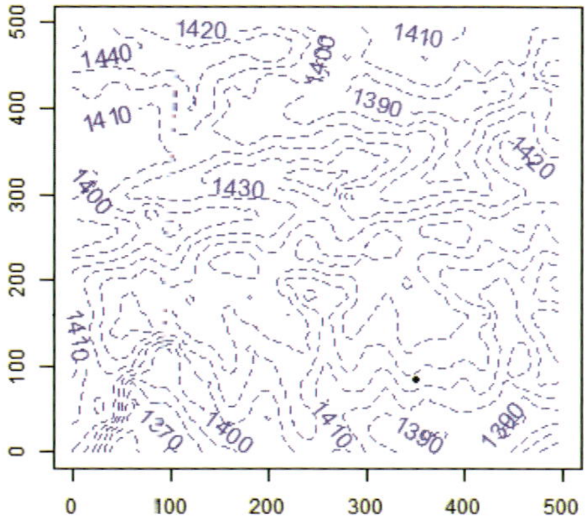

个体分布图 Distribution of individuals

径级分布表 DBH class

胸径区间 (Diameter class) (cm)	个体数 (No. of individuals in the plot)	比例 (Proportion) (%)
1～2	1	100.0
2～3	0	0.0
3～4	0	0.0
4～5	0	0.0
5～7	0	0.0
7～10	0	0.0
≥10	0	0.0

95 棠叶悬钩子 táng yè xuán gōu zǐ | Apple-leaf Rasp-berry

Rubus malifolius，悬钩子属
蔷薇科 | Rosaceae

代码（SpCode）= RUBMAL
个体数（Individual number/25 hm^2）= 23
最大胸径（Max DBH）= 7.59 cm
所在林层（Stratum）= 灌木层 (Shrub)
重要值排序（Importance value rank）= 83/128

落叶攀缘灌木。疏被短弯皮刺。一年枝有不育枝及花枝，不分枝，被柔毛，无刺。叶革质，有短柄，椭圆形或长圆状椭圆形，先端渐尖，边缘有疏齿，齿有短尖。总状花序，顶生，花少。小核果聚为半球形，先端弯曲，稍皱，长花柱宿存。花期5~6月，花期7~8月。

Deciduous shrubs scandent. Sparse prickles short and curved. 1-yr branches, sterile and flowering, nonbranching, pubescent, no prickles. Leaf blade leathery, short petiole, elliptic or oblong-elliptic, apex acuminate, margin sparsely serrate, teeth short tipped. Racemes terminal, few flowered. Aggregate fruit semiglobose with many drupelets, apex curved, crinkled, long style persistent. Fl. May–Jun., fr. Jul.–Aug..

树干 Trunk
摄影：卢志军 Photo by: Zhijun Lu

花枝 Flowering branches
摄影：江明喜 Photo by: Mingxi Jiang

果枝 Fruiting branches
摄影：徐文秀 Photo by: Wenxiu Xu

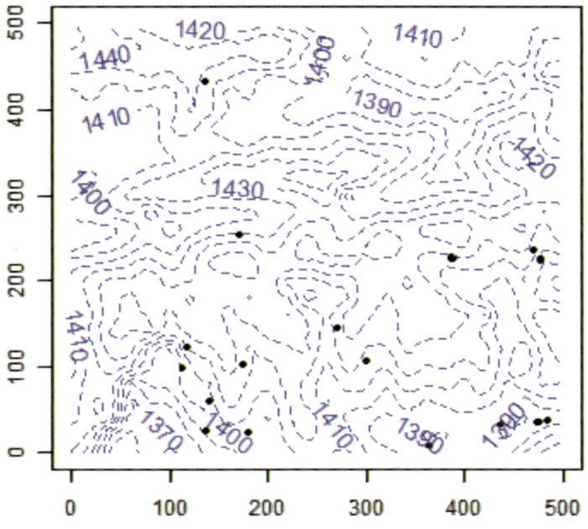

个体分布图 Distribution of individuals

径级分布表 DBH class

胸径区间 (Diameter class) (cm)	个体数 (No. of individuals in the plot)	比例 (Proportion) (%)
1~2	15	65.2
2~3	4	17.4
3~4	3	13.0
4~5	0	0.0
5~7	0	0.0
7~10	1	4.3
≥10	0	0.0

96 木莓 mù méi | Swinhoe Rasp-berry

Rubus swinhoei，悬钩子属
蔷薇科 | Rosaceae

代码（SpCode）= RUBSWI
个体数（Individual number/25 hm^2）= 2
最大胸径（Max DBH）= 2.65 cm
所在林层（Stratum）= 灌木层 (Shrub)
重要值排序（Importance value rank）= 119/128

落叶或半常绿攀缘灌木。茎细弱，圆柱形，疏被绵毛，最后无毛，皮刺疏生下弯。叶长圆状披针形，先端渐尖，基部圆形，有不整齐的锯齿及细锯齿，上面除中脉被毛外无毛，下面被灰色茸毛，或老后无毛。短总状花序顶生，疏花。果黑色，有皱纹。花期6月，果期7月。

Deciduous or semi-evergreen shrubs scandent. Stem delicate and thin, cylindrical, sparsely pubescent, finally glabrous, sparse prickles curved. Leaf blade ovate-lanceolate, apex acuminate, base rounded, margin irregularly serrate and serrulate, adaxially glabrous except midvein pubescent, abaxially grayish pubescent, or glabrous when old. Short raceme terminal, flowers sparse. Aggregate fruit, black, crinkled. Fl. Jun., fr. Jul..

树干 Trunk
摄影：卢志军 Photo by: Zhijun Lu

花枝 Flowering branch
摄影：刘海波 Photo by: Haibo Liu

果枝 Fruiting branch
摄影：卢志军 Photo by: Zhijun Lu

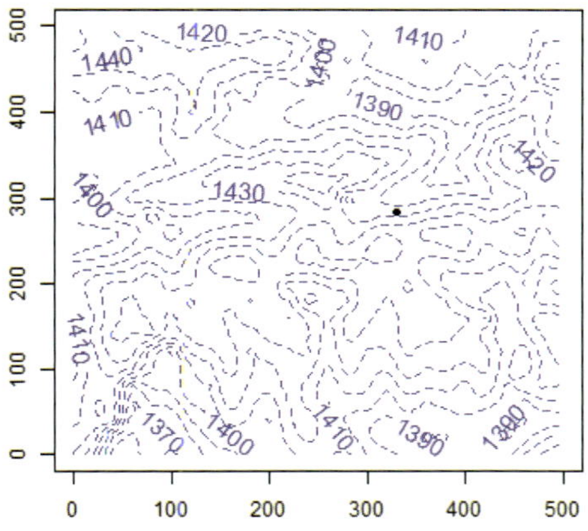

个体分布图 Distribution of individuals

径级分布表 DBH class

胸径区间 (Diameter class) (cm)	个体数 (No. of individuals in the plot)	比例 (Proportion) (%)
1～2	1	50.0
2～3	1	50.0
3～4	0	0.0
4～5	0	0.0
5～7	0	0.0
7～10	0	0.0
⩾10	0	0.0

97 石灰花楸

shí huī huā qiū | Folgner Mountain-ash

Sorbus folgneri，花楸属

蔷薇科 | Rosaceae

代码（SpCode）= SORFOL

个体数（Individual number/25 hm^2）= 856

最大胸径（Max DBH）= 50.68 cm

所在林层（Stratum）= 亚乔木层 (Understory)

重要值排序（Importance value rank）= 11/54

落叶乔木。幼枝被茸毛。叶卵形至椭圆状卵形，先端急尖或短渐尖，基部宽楔形或圆形，边缘有锯齿，有时呈浅裂状，上面深绿色无毛，下面密被灰白茸毛。复伞房花序，花瓣白色。果椭圆形，红色。花期5～6月，果期10月。

Deciduous trees. Young branches pubescent. Leaf blade ovate to elliptic-ovate, apex acute to shortly acuminate, base broadly cuneate or rounded, margin serrate, sometimes shallowly lobed, adaxially dark green, glabrous, abaxially densely grayish white tomentose. Compound corymbs, petals white. Fruit ellipsoid, red. Fl. May–Jun., fr. Oct..

树干 Trunk
摄影：卢志军 Photo by: Zhijun Lu

花枝 Flowering branches
摄影：江明喜 Photo by: Mingxi Jiang

果枝 Fruiting branches
摄影：卢志军 Photo by: Zhijun Lu

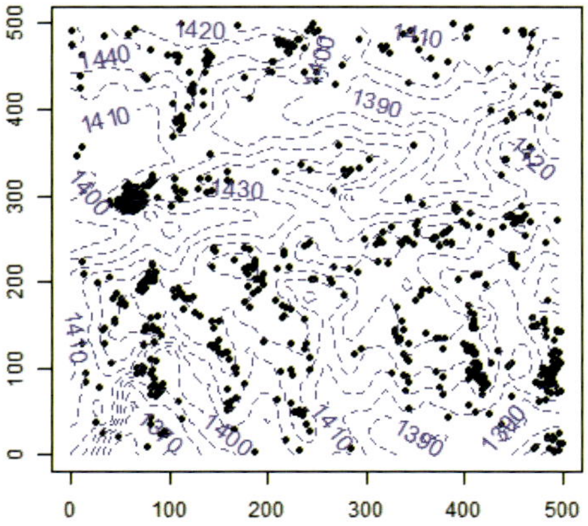

个体分布图 Distribution of individuals

径级分布表 DBH class

胸径区间 (Diameter class) (cm)	个体数 (No. of individuals in the plot)	比例 (Proportion) (%)
1～2.5	71	8.3
2.5～5	127	14.8
5～8	109	12.7
8～11	124	14.5
11～15	130	15.2
15～20	151	17.6
≥20	144	16.8

98 华西花楸 huá xī huā qiū | Wilson Mountain-ash

Sorbus wilsoniana，花楸属
蔷薇科 | Rosaceae

代码（SpCode）= SORWIL
个体数（Individual number/25 hm^2）= 557
最大胸径（Max DBH）= 30.47 cm
所在林层（Stratum）= 亚乔木层 (Understory)
重要值排序（Importance value rank）= 13/54

树干 Trunk
摄影：卢志军 Photo by: Zhijun Lu

落叶乔木。枝条棕色。奇数羽状复叶，小叶6～7对，叶片椭圆形至椭圆状披针形，两面无毛或下面沿中脉有毛，顶端急尖或渐尖，基部宽楔形或圆形，基部全缘，顶部具细锯齿。花序大，疏具柔毛；花小，白色。果橙红色，卵形。花期5～6月，果期7月。

Deciduous trees. Branch brown. Leaves imparipinnate, leaflets 6 or 7 pairs, blades elliptic to elliptic-lanceolate, both surfaces glabrous or abaxially pubescent along midvein, apex acute or acuminate, base broadly cuneate or rounded, margin entire basally, minutely serrate apically. Inflorescences large, sparsely pubescent, flowers small and white. Fruit orangish red, ovoid. Fl. May–Jun., fr. Jul..

花枝 Flowering branch
摄影：刘海波 Photo by: Haibo Liu

果枝 Fruiting branches
摄影：鲍大川 Photo by: Dachuan Bao

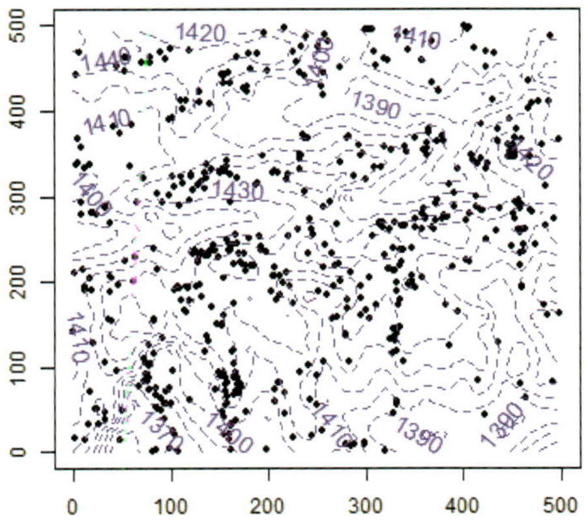

个体分布图 Distribution of individuals

径级分布表 DBH class

胸径区间 (Diameter class) (cm)	个体数 (No. of individuals in the plot)	比例 (Proportion) (%)
1～2.5	30	5.4
2.5～5	76	13.6
5～8	54	9.7
8～11	82	14.7
11～15	128	23.0
15～20	114	20.5
≥20	73	13.1

99 美脉花楸 měi mài huā qiū | Brown-dot Mountain-ash

Sorbus caloneura，花楸属
蔷薇科 | Rosaceae

代码（SpCode）= SORCAL
个体数（Individual number/25 hm^2）= 1
最大胸径（Max DBH）= 11.06 cm
所在林层（Stratum）= 灌木层 (Shrub)
重要值排序（Importance value rank）= 105/128

落叶灌木或乔木。枝条黑栗色，无毛。叶坚纸质，椭圆状长圆形或倒卵状长圆形，先端急尖或渐尖，基部楔形，边缘有重圆齿状锯齿。复伞状花序，花多数，白色。果梨形，有皮孔。花期5～6月，果期9～10月。

Deciduous shrubs or trees. Branchlets dark reddish brown, glabrous. Leaf blade stiff papery, elliptic-oblong or obovate-oblong, apex acute or acuminate, base cuneate, margin obtusely crenate. Compound corymbs, numerous and white flowered. Fruit pyriform, lenticellate. Fl. May–Jun., fr. Sept.–Oct..

树干 Trunk
摄影：卢志军 Photo by: Zhijun Lu

果枝 Fruiting branches
摄影：江明喜 Photo by: Mingxi Jiang

小枝和叶背 Branchlet and leaf abaxial surface
摄影：刘海波 Photo by: Haibo Liu

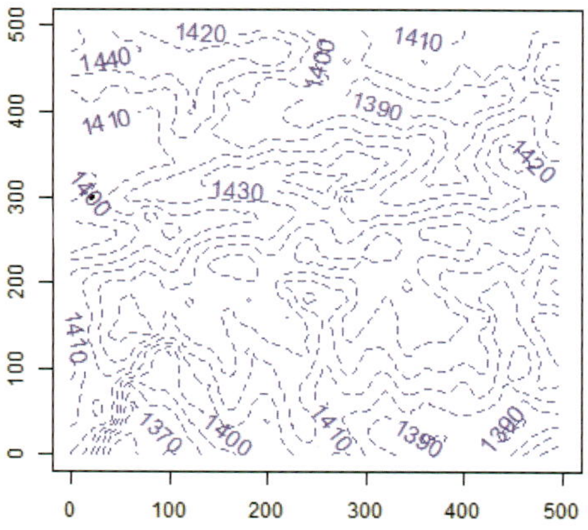

个体分布图 Distribution of individuals

径级分布表 DBH class

胸径区间 (Diameter class) (cm)	个体数 (No. of individuals in the plot)	比例 (Proportion) (%)
1～2	0	0.0
2～3	0	0.0
3～4	0	0.0
4～5	0	0.0
5～7	0	0.0
7～10	0	0.0
≥10	1	100.0

100 毛萼红果树　　máo è hóng guǒ shù | Hairy-calyx Stranvaesia

Stranvaesia amphidoxa，红果树属
蔷薇科 | Rosaceae

代码（SpCode）= STRAMP
个体数（Individual number/25 hm^2）= 1442
最大胸径（Max DBH）= 14.37 cm
所在林层（Stratum）= 灌木层 (Shrub)
重要值排序（Importance value rank）= 24/128

常绿灌木或小乔木。小枝初有黄白色毛，后近无毛，灰棕色。叶倒卵状披针形，先端短渐尖，基部急尖。顶生伞房花序；花瓣白色。果近球形，猩红色。花期5～6月，果期9～10月。

Evergreen shrubs or small trees. Branchlets yellowish white tomentose-villous when young, then glabrous and grayish brown. Leaf blade obovate-lanceolate, apex shortly acuminate, base acute. Corymbs terminal, petals white. Fruit subglobose, dark red. Fl. May–Jun., fr. Sept.–Oct..

树干　Trunk
摄影：卢志军　Photo by: Zhijun Lu

花枝　Flowering branch
摄影：刘海波　Photo by: Haibo Liu

果枝　Fruiting branch
摄影：鲍大川　Photo by: Dachuan Bao

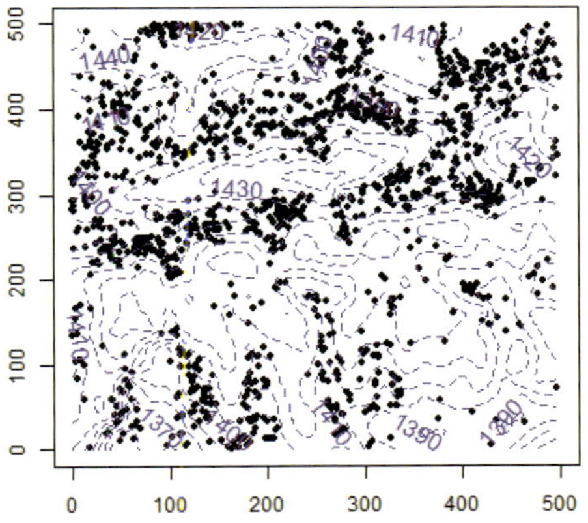

个体分布图 Distribution of individuals

径级分布表 DBH class

胸径区间 (Diameter class) (cm)	个体数 (No. of individuals in the plot)	比例 (Proportion) (%)
1～2	724	50.2
2～3	353	24.5
3～4	149	10.3
4～5	97	6.7
5～7	67	4.7
7～10	35	2.4
≥10	17	1.2

101 波叶红果树 bō yè hóng guǒ shù | Undulate-leaf David Stranvaesia

Stranvaesia davidiana var. *undulata*，红果树属
蔷薇科 | Rosaceae

代码（SpCode）= STRDAV
个体数（Individual number/25 hm^2）= 57
最大胸径（Max DBH）= 7.47 cm
所在林层（Stratum）= 灌木层 (Shrub)
重要值排序（Importance value rank）= 67/128

灌木或小乔木。幼枝有绢毛状长柔毛，后无毛。叶椭圆状长圆形至长圆状披针形，先端急尖，基部宽楔形，全缘，有波状起伏。复伞房花序，顶生有时近无毛。果近球形，橘红色。花期5～6月，果期9～10月。

Shrubs or small trees. Young branches pubescent villous, then glabrous. Leaf blade elliptic-oblong to oblanceolate, apex acute, base broadly cuneate, margin entire, wavy. Compound corymbs terminal, sometimes nearly glabrous. Fruit subglobose, orange red. Fl. May–Jun., fr. Sept.–Oct..

树干 Trunk
摄影：卢志军 Photo by: Zhijun Lu

花枝 Flowering branch
摄影：刘海波 Photo by: Haibo Liu

果枝 Fruiting branch
摄影：卢志军 Photo by: Zhijun Lu

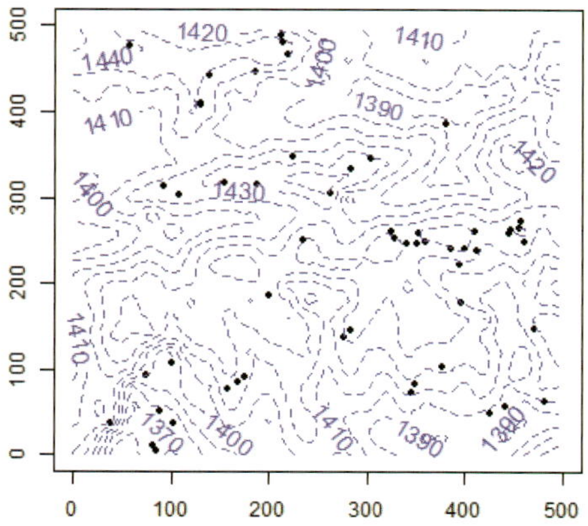

个体分布图 Distribution of individuals

径级分布表 DBH class

胸径区间 (Diameter class) (cm)	个体数 (No. of individuals in the plot)	比例 (Proportion) (%)
1～2	19	33.3
2～3	17	29.8
3～4	7	12.3
4～5	4	7.0
5～7	7	12.3
7～10	3	5.3
≥10	0	0.0

102 湖北紫荆

hú běi zǐ jīng | Hairy-calyx Stranvaesia

Cercis glabra，紫荆属
豆科 | Leguminosae

代码（SpCode）= CERGLA
个体数（Individual number/25 hm^2）= 2
最大胸径（Max DBH）= 45.10 cm
所在林层（Stratum）= 亚乔木层 (Understory)
重要值排序（Importance value rank）= 38/54

落叶乔木。树皮和小枝灰黑色。叶厚纸质或近革质，心脏形或三角状圆形，先端钝或急尖，基部浅心形至深心形。总状花序；花淡紫色或粉红色。荚果狭长圆形，紫红色。花期3～4月，果期9～11月。

Deciduous trees. Bark and branchlets grayish black. Leaf blade thickly papery or subleathery, cordate or triangular-rounded, apex obtuse or acute, base shallowly to deeply cordate. Racemes, flowers light purple or pink. Legume narrowly cylindrical, purplish red. Fl. Mar.–Apr., fr. Sept.–Nov..

树干 Trunk
摄影：鲍大川 Photo by: Dachuan Bao

花枝 Flowering branch
摄影：鲍大川 Photo by: Dachuan Bao

小枝和叶背 Branchlet and leaf abaxial surface
摄影：鲍大川 Photo by: Dachuan Bao

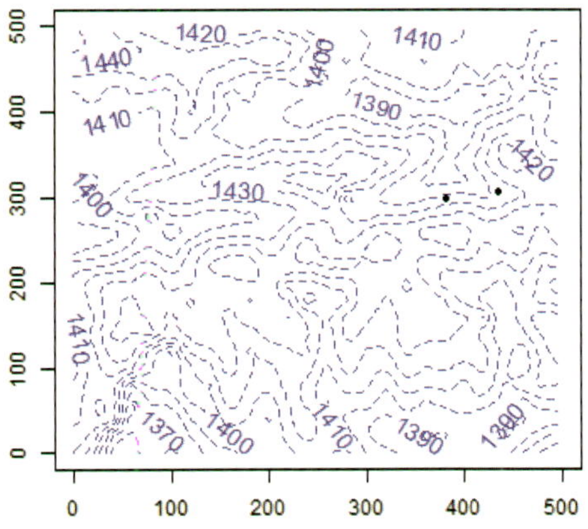

个体分布图 Distribution of individuals

径级分布表 DBH class

胸径区间 (Diameter class) (cm)	个体数 (No. of individuals in the plot)	比例 (Proportion) (%)
1～2.5	0	0.0
2.5～5	0	0.0
5～8	0	0.0
8～11	0	0.0
11～15	0	0.0
15～20	0	0.0
⩾20	2	100.0

103 小花香槐

xiǎo huā xiāng huái | Chinese Yellow-wood

Cladrastis delavayi，香槐属
豆科 | Leguminosae

代码（SpCode）= CLADEL
个体数（Individual number/25 hm^2）= 19
最大胸径（Max DBH）= 41.15 cm
所在林层（Stratum）= 乔木层 (Canopy)
重要值排序（Importance value rank）= 24/50

树干 Trunk
摄影：卢志军 Photo by: Zhijun Lu

落叶乔木。幼枝被灰棕色毛。奇数羽状复叶，小叶9～13，互生，长椭圆状披针形，先端渐尖，基部圆形。圆锥花序顶生；花瓣白色或粉红色。荚果长椭圆至线形，扁平。花果期6月，果期8～10月。

Deciduous trees. Young branches gray-brown hairy. Leaves imparipinnate, leaflets 9-13, alternate, blades long elliptic-lanceolate, apex acuminate, base rounded. Panicles terminal, petals white or pink. Legumes long elliptic to linear, flat. Fl. Jun., fr. Aug.–Oct..

花枝 Flowering branches
摄影：刘海波 Photo by: Haibo Liu

小枝和叶背 Branchlet and leaf abaxial surface
摄影：卢志军 Photo by: Zhijun Lu

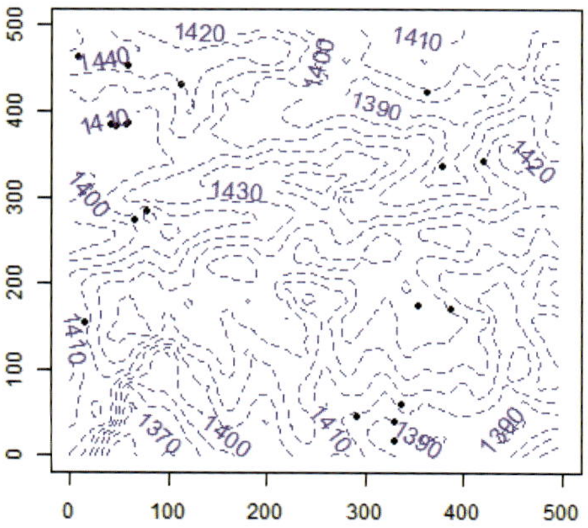

个体分布图 Distribution of individuals

径级分布表 DBH class

胸径区间 (Diameter class) (cm)	个体数 (No. of individuals in the plot)	比例 (Proportion) (%)
1～2.5	0	0.0
2.5～5	0	0.0
5～10	3	15.8
10～25	6	31.6
25～40	9	47.4
40～100	1	5.3
≥100	0	0.0

104 翅荚香槐

chì jiá xiāng huái | Flatfruited Yellowwood

Cladrastis platycarpa，香槐属
豆科 | Leguminosae

代码（SpCode）= CLAPLA
个体数（Individual number/25 hm^2）= 3
最大胸径（Max DBH）= 1.59 cm
所在林层（Stratum）= 乔木层 (Canopy)
重要值排序（Importance value rank）= 42/50

落叶大乔木。树皮暗灰色，多皮孔。一年生枝被褐色柔毛，旋即秃净。奇数羽状复叶；小叶互生或近对生，长椭圆形或卵状长圆形，先端渐尖，基部钝圆或宽楔形。圆锥花序；花萼阔钟状；花冠白色，芳香。荚果扁平，长椭圆形或长圆形，两侧具翅，不开裂；种子长圆形，扁平，深褐色或黑色。花期4～6月，果期7～10月。

Deciduous trees. Bark dark gray, many lenticellate. 1-yr branchlets brown pubescent, soon glabrous. Leaves imparipinnate, alternate or subopposite, blades long elliptic or ovate-oblong, apex acuminate, base obtuse or broadly cuneate. Panicles, calyx campanulate, corolla white and fragrant. Legumes flat, long elliptic or oblong, winged on both sides, indehiscent. Seeds oblong, flat, dark brown or black. Fl. Apr.–Jun., fr. Jul.–Oct..

树干 Trunk
摄影：卢志军 Photo by: Zhijun Lu

复叶(正面) Compound leaves
摄影：卢志军 Photo by: Zhijun Lu

复叶(反面) Compound leaf abaxial surface
摄影：卢志军 Photo by: Zhijun Lu

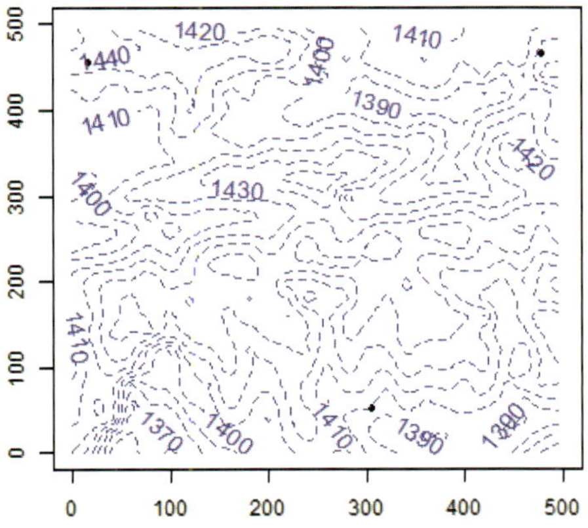

个体分布图 Distribution of individuals

径级分布表 DBH class

胸径区间 (Diameter class) (cm)	个体数 (No. of individuals in the plot)	比例 (Proportion) (%)
1～2.5	3	100.0
2.5～5	0	0.0
5～10	0	0.0
10～25	0	0.0
25～40	0	0.0
40～100	0	0.0
⩾100	0	0.0

105 黄檀

huáng tán | Hupeh Rosewood

Dalbergia hupeana，黄檀属
豆科 | Leguminosae

代码（SpCode）= DALHUP
个体数（Individual number/25 hm^2）= 9
最大胸径（Max DBH）= 15.29 cm
所在林层（Stratum）= 亚乔木层 (Understory)
重要值排序（Importance value rank）= 46/54

落叶乔木。树皮暗灰色。羽状复叶，小叶7～11，叶片长圆形或宽椭圆形，先端钝，有微缺，基部圆形。圆锥花序顶生或上部叶腋生；萼钟状；花冠淡紫色或白色。荚果长圆形，扁平。花期6～8月，果期8～10月。

Deciduous trees. Bark dull gray. Leaves pinnately compound, leaflets 7-11, blades oblong or broadly elliptic, apex obtuse or slightly emarginate, base rounded. Panicles terminal or into axils of uppermost leaves, calyx campanulate, corolla light purple or white. Legume oblong, flat. Fl. Jun.–Aug., fr. Aug.–Oct..

树干 Trunk
摄影：卢志军 Photo by: Zhijun Lu

花枝 Flowering branches
摄影：江明喜 Photo by: Mingxi Jiang

果枝 Fruiting branches
摄影：鲍大川 Photo by: Dachuan Bao

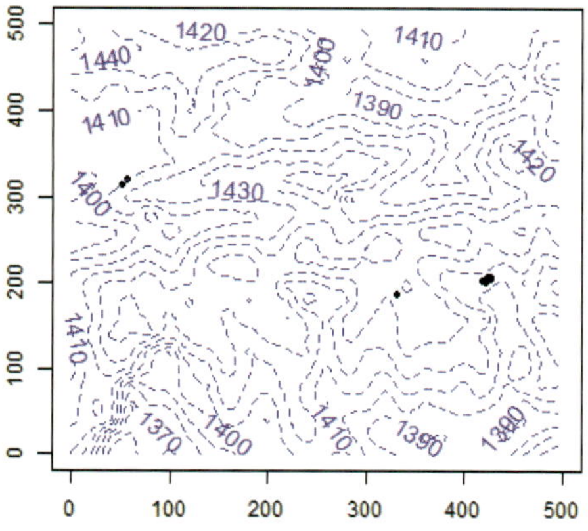

个体分布图 Distribution of individuals

径级分布表 DBH class

胸径区间 (Diameter class) (cm)	个体数 (No. of individuals in the plot)	比例 (Proportion) (%)
1～2.5	1	11.1
2.5～5	2	22.2
5～8	4	44.4
8～11	1	11.1
11～15	0	0.0
15～20	1	11.1
≥20	0	0.0

106 秧青 yāng qīng | Assam Rosewood

Dalbergia assamica，黄檀属
豆科 | Leguminosae

代码（SpCode）= DALASS
个体数（Individual number/25 hm^2）= 2
最大胸径（Max DBH）= 24.34 cm
所在林层（Stratum）= 亚乔木层 (Understory)
重要值排序（Importance value rank）= 49/54

落叶乔木。羽状复叶，小叶13～21对，纸质，长圆形或长圆状椭圆形，先端钝，圆或凹入，基部圆形或楔形。圆锥花序腋生，花冠白色。荚果阔舌状，长圆形至带状。种子肾形，扁平。花期4～7月，果期9～12月。

树干 Trunk
摄影：卢志军 Photo by: Zhijun Lu

Deciduous trees. Pinnately compound leaf, leaflets 13-21, papery, blade oblong or oblong-elliptic, apex obtuse, rounded or emarginated, base rounded or cuneate. Panicles axillary. Corolla white. Legume broadly ligulate or oblong to strap-shaped. Seeds reniform, compressed. Fl. Apr.–Jul., fr. Sept.–Dec..

花枝 Flowering branches
摄影：陈彬 Photo by: Bin Chen

小枝和叶背 Branchlet and leaf abaxial surface
摄影：卢志军 Photo by: Zhijun Lu

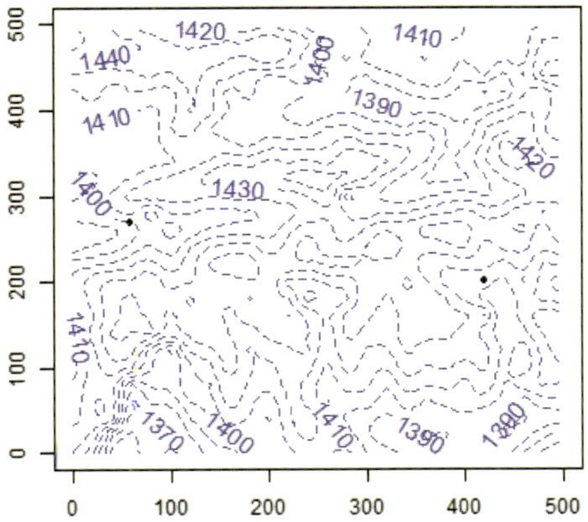

个体分布图 Distribution of individuals

径级分布表 DBH class

胸径区间 (Diameter class) (cm)	个体数 (No. of individuals in the plot)	比例 (Proportion) (%)
1～2.5	0	0.0
2.5～5	0	0.0
5～8	1	50.0
8～11	0	0.0
11～15	0	0.0
15～20	0	0.0
≥20	1	50.0

107 野桐

yě tóng | Thin-leaf Mallotus

Mallotus tenuifolius，野桐属
大戟科 | Euphorbiaceae

代码（SpCode）= MALTEN
个体数（Individual number/25 hm^2）= 101
最大胸径（Max DBH）= 34.44 cm
所在林层（Stratum）= 灌木层 (Shrub)
重要值排序（Importance value rank）= 41/128

树干 Trunk
摄影：卢志军 Photo by: Zhijun Lu

落叶灌木或小乔木。幼枝灰褐色，被星状茸毛，二年生枝带紫黑色，无毛。叶互生，宽卵形或宽三角状圆形，先端急短渐尖，基部截形或心形，全缘或不规则3裂而有钝齿。总状花序顶生，单生。蒴果球形，表面有软刺，被黄褐色星状茸毛；种子近圆形，黑色，有光泽。花期5～6月，果期6～10月。

Deciduous shrubs or small trees. Young branches grayish brown, stellately pubescent, 2-yr branchlets purplish black, glabrous. Leaves alternate, blades broadly ovate or broadly triangular-rounded, apex abruptly and shortly acuminate, base truncate or cordate, margin entire or irregularly 3-lobed, teeth obtuse. Racemes terminal, solitary. Capsules globose, tomentose and densely softly spiny. Seeds subglobose, black, glossy. Fl. May–Jun., fr. Jun.–Oct..

花枝和花序 Flowering branches and inflorescence
摄影：刘海波 Photo by: Haibo Liu

果枝和种子 Fruiting branches and seeds
摄影：鲍大川 Photo by: Dachuan Bao

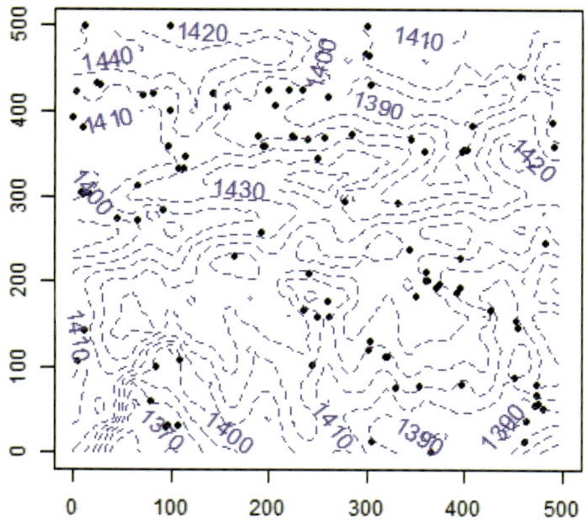

个体分布图 Distribution of individuals

径级分布表 DBH class

胸径区间 (Diameter class) (cm)	个体数 (No. of individuals in the plot)	比例 (Proportion) (%)
1～2	32	31.7
2～3	8	7.9
3～4	4	4.0
4～5	3	3.0
5～7	14	13.9
7～10	9	8.9
≥10	31	30.7

108 交让木 jiāo ràng mù | Macropodous Daphniphyllum

Daphniphyllum macropodum，交让木属
交让木科 | Daphniphyllaceae

代码（SpCode）= DAPMAC
个体数（Individual number/25 hm^2）= 4622
最大胸径（Max DBH）= 21.55 cm
所在林层（Stratum）= 灌木层 (Shrub)
重要值排序（Importance value rank）= 8/128

树干 Trunk
摄影：卢志军 Photo by: Zhijun Lu

常绿乔木。树皮灰棕色，小枝灰黑色，枝条粗。叶互生而丛生于枝端，新叶开放时，老叶常全部凋落，故有“交让”木之称。叶厚革质，长椭圆形至倒卵状披针形，先端渐尖，基部楔形或圆楔形，全缘。花小，淡绿色，雌雄异株，总状花序。核果长椭圆形，蓝黑色，微被白霜。花期4～5月，果期6～11月。

Evergreen trees. Bark grayish brown. Branchlets stout, dark brown. Leaves alternate, clustered on branchlet apex; when new leaves come out, all the old leaves deciduous, the reason for its Chinese name; leaf blade thickly leathery, long elliptic to oblanceolate, apex acuminate, base cuneate or round-cuneate, margin entire. Flowers small, light green, dioecious, racemes. Drupe ellipsoidal, blue black, slightly pruinose. Fl. Apr.–May, fr. Jun.–Nov..

果枝 Fruiting branch
摄影：刘海波 Photo by: Haibo Liu

叶背 Leaf abaxial surface
摄影：卢志军 Photo by: Zhijun Lu

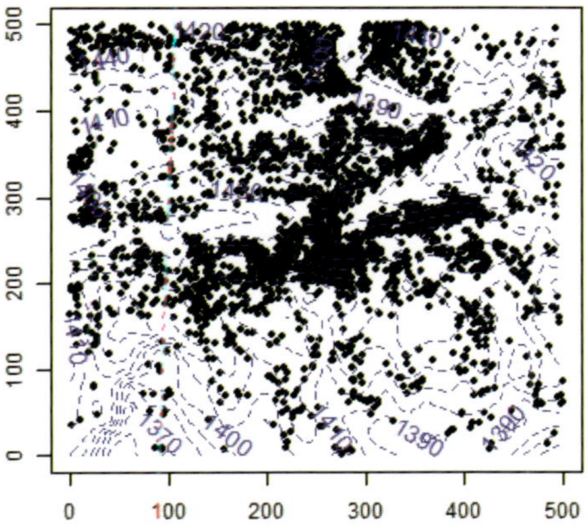

个体分布图 Distribution of individuals

径级分布表 DBH class

胸径区间 (Diameter class) (cm)	个体数 (No. of individuals in the plot)	比例 (Proportion) (%)
1～2	2611	56.5
2～3	1079	23.3
3～4	430	9.3
4～5	182	3.9
5～7	141	3.1
7～10	88	1.9
≥10	91	2.0

109 茵芋 yīn yù | Reeves Skimmia

Skimmia reevesiana，茵芋属
芸香科 | Rutaceae

代码（SpCode）= SKIREE
个体数（Individual number/25 hm^2）= 25
最大胸径（Max DBH）= 2.21 cm
所在林层（Stratum）= 灌木层 (Shrub)
重要值排序（Importance value rank）= 85/128

常绿灌木。树皮苍灰绿色，光滑。枝条髓心空。叶片聚生枝顶，有柑橘属气味；叶狭长圆形或长圆形，两端渐尖。顶生圆锥花序，花芳香，淡黄白色。花序被毛，花密集。果红色，长圆形至卵状长圆形，或倒卵形长圆形。种子扁平卵形。花期3～5月，果期7～11月。

Evergreen shrubs. Bark pale grayish green, smooth. Branches with hollow pith. Leaves clustered at branch apex, with a citrus-leaf odor; leaf blade narrowly oblong or oblong, apex and base acuminate. Panicle terminal, flowers fragrant, light yellowish white. Inflorescences puberulent, with condensed flowers. Fruits red, oblong to ovate-oblong, or obovate-oblong. Seeds compressed ovoid. Fl. Mar.–May, fr. Jul.–Nov..

树干 Trunk
摄影：卢志军 Photo by: Zhijun Lu

花枝 Flowering branch
摄影：徐耀粘 Photo by: Yaozhan Xu

果枝 Fruiting branch
摄影：卢志军 Photo by: Zhijun Lu

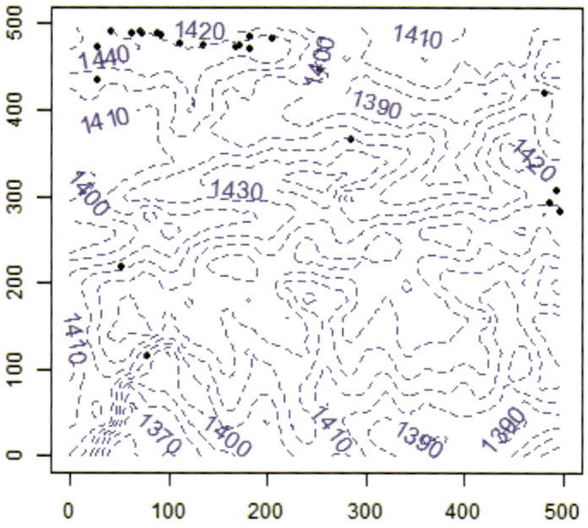

个体分布图 Distribution of individuals

径级分布表 DBH class

胸径区间 (Diameter class) (cm)	个体数 (No. of individuals in the plot)	比例 (Proportion) (%)
1～2	20	80.0
2～3	5	20.0
3～4	0	0.0
4～5	0	0.0
5～7	0	0.0
7～10	0	0.0
≥10	0	0.0

110 楝叶吴萸 liàn yè wú yú | Glabrous-leaf Tetradium

Tetradium glabrifolium，四数花属
芸香科 | Rutaceae

代码（SpCode）= TETGLA
个体数（Individual number/25 hm^2）= 79
最大胸径（Max DBH）= 23.81 cm
所在林层（Stratum）= 亚乔木层 (Canopy)
重要值排序（Importance value rank）= 34/54

常绿灌木或乔木。叶片长14～38 cm，小叶（3或）5～19，叶片宽卵形至披针形或椭圆形或椭圆状长圆形，小叶先端渐尖，侧方小叶基部窄楔形至近心形至近截形，顶端小叶基部窄楔形至楔形，全缘或具细圆齿。花序顶生，白色，长9～19 cm。果实通常5心皮。花期6～9月，果期9～12月。

Evergreen shrubs or trees. Leaves 14-38 cm, (3 or) 5-19-foliolate, leaflet blades broadly ovate to lanceolate or elliptic or elliptic-oblong, apex acuminate, base in lateral leaflets narrowly cuneate to subrounded to subtruncate and in terminal leaflet narrowly cuneate to cuneate, margin entire or ± crenulate. Inflorescences terminal and white, 9-19 cm. Fruit usually 5-carpelled. Fl. Jun.–Sept., fr. Sept.–Dec..

树干 Trunk
摄影：卢志军 Photo by: Zhijun Lu

花枝 Flowering branch
摄影：刘海波 Photo by: Haibo Liu

小枝和叶背 Branchlet and leaf abaxial surface
摄影：卢志军 Photo by: Zhijun Lu

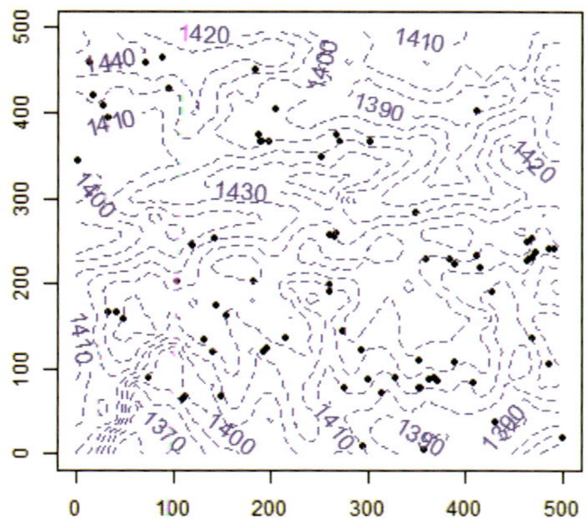

个体分布图 Distribution of individuals

径级分布表 DBH class

胸径区间 (Diameter class) (cm)	个体数 (No. of individuals in the plot)	比例 (Proportion) (%)
1～2.5	56	70.9
2.5～5	13	16.5
5～8	3	3.8
8～11	3	3.8
11～15	2	2.5
15～20	1	1.3
⩾20	1	1.3

111 吴茱萸 wú zhū yú | Medicinal Tetradium

Tetradium ruticarpum，四数花属

芸香科 | Rutaceae

代码（SpCode）= TETRUT

个体数（Individual number/25 hm^2）= 17

最大胸径（Max DBH）= 12.69 cm

所在林层（Stratum）= 灌木层 (Shrub)

重要值排序（Importance value rank）= 90/128

落叶灌木或乔木。叶长15～40 cm，小叶（3或）5～13（或15），叶片椭圆形至卵形，有时披针形，倒披针形，或倒卵形，下面略呈灰绿色，没有凸起，小叶先端渐尖，侧方小叶基部钝圆至窄楔形或有时圆形至楔形或少数渐狭，全缘或具不规则细圆齿。花序顶生，长2.5～18.0 cm。花期4～6月，果期8～11月。

Deciduous shrubs or trees. Leaves 15-40 cm, (3 or) 5-13(or 15)-foliolate; leaflet blades elliptic to ovate or sometimes lanceolate, oblanceolate, or obovate, abaxially slightly glaucous and not papillate, apex acuminate, base in lateral leaflets obtuse to narrowly cuneate or sometimes rounded to cuneate or rarely attenuate, margin entire or irregularly crenulate. Inflorescences terminal 2.5-18.0 cm. Fl. Apr.–Jun., fr. Aug.–Nov..

树干 Trunk

摄影：卢志军 Photo by: Zhijun Lu

复叶 Compound leaves

摄影：卢志军 Photo by: Zhijun Lu

复叶背面 Compound leaf abaxial surface

摄影：卢志军 Photo by: Zhijun Lu

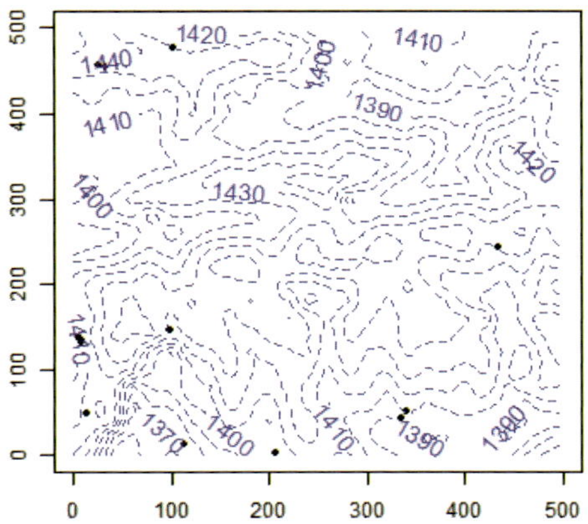

个体分布图 Distribution of individuals

径级分布表 DBH class

胸径区间 (Diameter class) (cm)	个体数 (No. of individuals in the plot)	比例 (Proportion) (%)
1～2	9	52.9
2～3	2	11.8
3～4	0	0.0
4～5	0	0.0
5～7	1	5.9
7～10	0	0.0
≥10	5	29.4

112 野花椒

yě huā jiāo | Flat-spine Prickly-ash

Zanthoxylum simulans，花椒属
芸香科 | Rutaceae

代码（SpCode）= ZANSIM
个体数（Individual number/25 hm^2）= 2
最大胸径（Max DBH）= 4.49 cm
所在林层（Stratum）= 灌木层 (Shrub)
重要值排序（Importance value rank）= 116/128

落叶灌木或小乔木。树干和小枝有刺。复叶，小叶5～15，叶轴有翅，叶片卵形、卵状椭圆形或披针形，顶端急尖至短尖或具微凹尖头，基部略偏斜，边缘具圆齿。花序顶生。果实红棕色。种子小。花期3～5月，果期7～9月。

Deciduous shrubs or small trees. Stems and branchlets with prickles. Leaves compound, 5-15 foliolate; rachis winged; leaflet blades ovate, ovate-elliptic, or lanceolate, apex acute to mucronate or with a retuse tip, base slightly oblique, margin crenate. Inflorescences terminal. Fruit reddish brown. Seeds small. Fl. Mar.–May, fr. Jul.–Sept..

树干 Trunk
摄影：卢志军 Photo by: Zhijun Lu

皮刺和叶片 Prickles and leaves
摄影：陈彬 Photo by: Bin Chen

果枝 Fruiting branch
摄影：陈彬 Photo by: Bin Chen

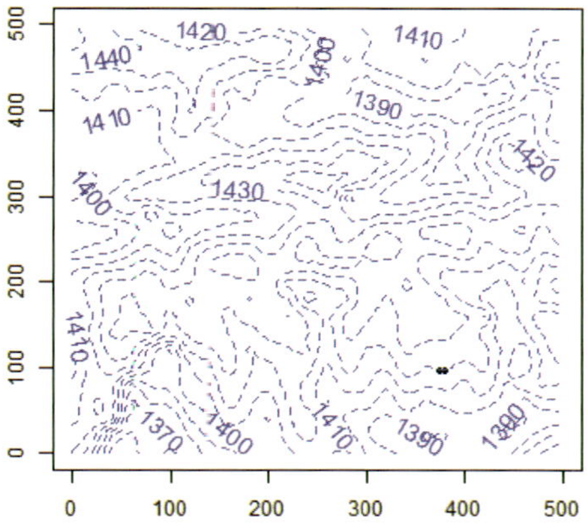

个体分布图 Distribution of individuals

径级分布表 DBH class

胸径区间 (Diameter class) (cm)	个体数 (No. of individuals in the plot)	比例 (Proportion) (%)
1～2	1	50.0
2～3	0	0.0
3～4	0	0.0
4～5	1	50.0
5～7	0	0.0
7～10	0	0.0
⩾10	0	0.0

113 朵花椒

duǒ huā jiāo | Pubescent Prickly-ash

Zanthoxylum molle，花椒属
芸香科 | Rutaceae

树干 Trunk
摄影：卢志军 Photo by: Zhijun Lu

代码（SpCode）= ZANMOL
个体数（Individual number/25 hm^2）= 1
最大胸径（Max DBH）= 29.2 cm
所在林层（Stratum）= 灌木层 (Shrub)
重要值排序（Importance value rank）= 79/128

落叶乔木。树皮棕黑色。幼枝、树干和花序轴有刺。幼枝暗紫红色。复叶，小叶13～19，叶片厚革质，对生，宽卵形、椭圆形，少数为近圆形，顶端急尖，基部对称，少数偏斜圆形或近圆形，全缘或具细圆齿。花序顶生，花多数，白色。果实紫红色，顶端无喙。种子小。花期6～8月，果期10～11月。

Deciduous trees. Bark brownish black. Young branches, trunk and rachis of inflorescences with prickles. Young branches dark purplish red. Leaves compound, 13-19 foliolate; leaflet blades thickly leathery, opposite, broadly ovate, elliptic or rarely suborbicular, apex acute, base symmetrically or rarely obliquely rounded to subcordate, margin entire or crenulate. Inflorescences terminal, many and white flowered. Fruit follicles purplish red, apex not beaked. Seeds small. Fl. Jun.–Aug., fr. Oct.–Nov..

花枝 Flowering branches
摄影：江明喜 Photo by: Mingxi Jiang

叶背 Leaf abaxial surface
摄影：卢志军 Photo by: Zhijun Lu

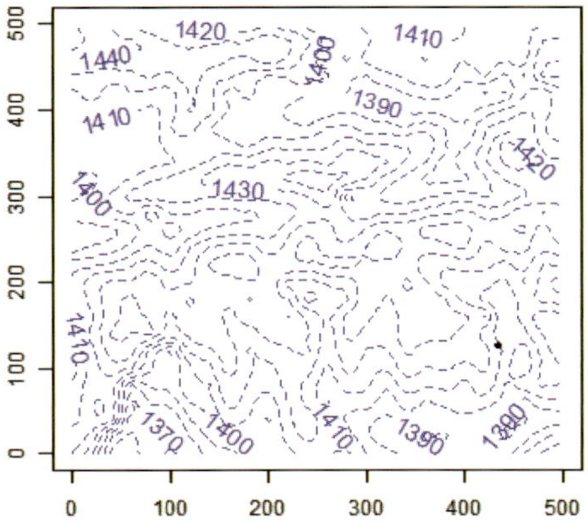

个体分布图 Distribution of individuals

径级分布表 DBH class

胸径区间 (Diameter class) (cm)	个体数 (No. of individuals in the plot)	比例 (Proportion) (%)
1～2	0	0.0
2～3	0	0.0
3～4	0	0.0
4～5	0	0.0
5～7	0	0.0
7～10	0	0.0
≥10	1	100.0

114 苦树 kǔ shù | Indian Quassia-wood

Picrasma quassioides，苦木属
苦木科 I Simaroubaceae

代码（SpCode）= PICQUA
个体数（Individual number/25 hm^2）= 27
最大胸径（Max DBH）= 35.4 cm
所在林层（Stratum）= 灌木层 (Shrub)
重要值排序（Importance value rank）= 49/128

树干 Trunk
摄影：卢志军 Photo by: Zhijun Lu

灌木或乔木。枝条红褐色，有黄色皮孔。叶为奇数羽状复叶，小叶卵形至长圆状卵形，先端渐尖，基部楔形或近圆形，不对称，边有圆齿状锯齿。聚伞花序腋生。果球状倒卵形，蓝色至红色，萼宿存。花期4～6月，果期6～10月。

Shrubs or trees. Branches reddish brown, yellow lenticeled. Leaves imparipinnate, leaflets ovate to obovate, apex acuminate, base cuneate or subrounded, asymmetric, margin crenate. Cymes axillary. Fruits globose-obovate, blue to red, calyx persistent. Fl. Apr.–Jun., fr. Jun.–Oct..

果枝 Fruiting branches
摄影：鲍大川 Photo by: Dachuan Bao

复叶背面 Compound leaf abaxial surface
摄影：卢志军 Photo by: Zhijun Lu

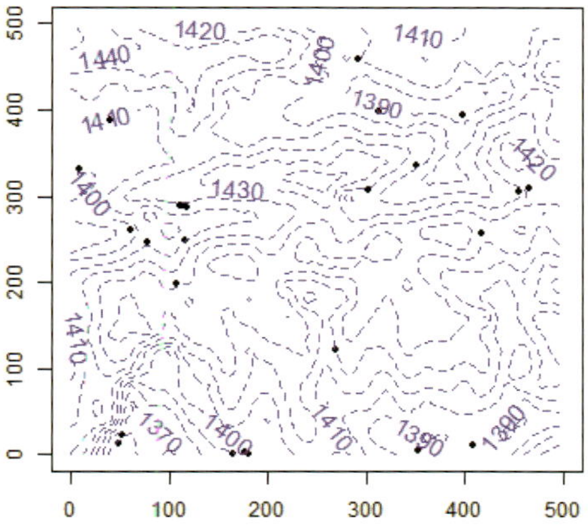

个体分布图 Distribution of individuals

径级分布表 DBH class

胸径区间 (Diameter class) (cm)	个体数 (No. of individuals in the plot)	比例 (Proportion) (%)
1～2	4	14.8
2～3	1	3.7
3～4	2	7.4
4～5	2	7.4
5～7	2	7.4
7～10	2	7.4
≥10	14	51.9

115 香椿

xiāng chūn | Chinese Toona

Toona sinensis，香椿属

楝科 | Meliaceae

树干 Trunk

摄影：卢志军 Photo by: Zhijun Lu

代码（SpCode）= TOOSIN

个体数（Individual number/25 hm^2）= 84

最大胸径（Max DBH）= 73.60 cm

所在林层（Stratum）= 乔木层 (Canopy)

重要值排序（Importance value rank）= 17/50

落叶乔木。树皮灰色至暗棕色，开裂。叶有长柄，偶数羽状复叶，有特殊香气；小叶通常8～20对，对生或互生，卵状披针形至长椭圆形，先端渐尖，基部不对称，边缘具齿、细锯齿，少数全缘。圆锥花序顶生，与叶等长或更长；花白色，有香气，有短柄。果狭椭圆形，先端5瓣裂，褐色；种子上端有长膜质翅。花期6月，果熟期9～10月。

Deciduous trees. Bark gray to dark brown, fissured. Leaves paripinnate with long petiole, specially fragrant; leaflets usually 8-20 pairs, opposite or alternate, blades ovate-lanceolate to long-elliptic, apex acuminate, base asymmetric, margin serrate, serrulate, or rarely entire. Panicles terminal, similar or longer than leaves; flowers white, fragrant, peduncles short. Fruit narrowly ellipsoid, apex 5-lobed dehiscent, brown. Seeds winged at upper end. Fl. Jun., fr. maturity Sept.–Oct..

复叶 Compound leaves

摄影：卢志军 Photo by: Zhijun Lu

果序 Infructescence

摄影：卢志军 Photo by: Zhijun Lu

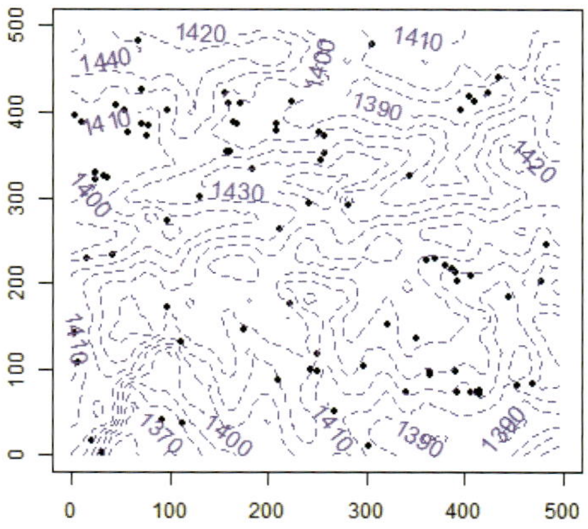

个体分布图 Distribution of individuals

径级分布表 DBH class

胸径区间 (Diameter class) (cm)	个体数 (No. of individuals in the plot)	比例 (Proportion) (%)
1～2.5	25	29.8
2.5～5	8	9.5
5～10	10	11.9
10～25	13	15.5
25～40	10	11.9
40～100	18	21.5
≥100	0	0.0

116 盐麸木

yán fū mù | Chines Sumac

Rhus chinensis，盐肤木属
漆树科 | Anacardiaceae

代码（SpCode）= RHUCHI
个体数（Individual number/25 hm^2）= 25
最大胸径（Max DBH）= 14.07 cm
所在林层（Stratum）= 灌木层 (Shrub)
重要值排序（Importance value rank）= 71/128

落叶灌木或小乔木。小枝、叶轴、花序、果序密被褐色柔毛。奇数羽状复叶，叶轴具宽翅至无翅，小叶7～13对，叶片卵形至卵状长圆形，先端急尖，基部圆楔形，边缘有粗圆齿状锯齿。圆锥花序宽大，顶生，花白色。果近球形，橙红色，密被柔毛。花期8～9月，果期9～11月。

Deciduous shrubs to small trees. Branchlets, rachis, inflorescence, infructescence ferruginous pubescent. Leaves imparipinnate; rachis broadly winged to wingless; leaflets 7-13; leaflet blade ovate to ovate-oblong, apex acute, base rounded-cuneate, margin crenate. Panicles broad and terminal, flowers white. Drupe subglobose, orange red, densely pubescent. Fl. Aug.–Sept., fr. Sept.–Nov..

树干 Trunk
摄影：卢志军 Photo by: Zhijun Lu

花枝和花序 Flowering branch and inflorescence
摄影：刘海波 Photo by: Haibo Liu

果枝 Fruiting branches
摄影：江明喜 Photo by: Mingxi Jiang

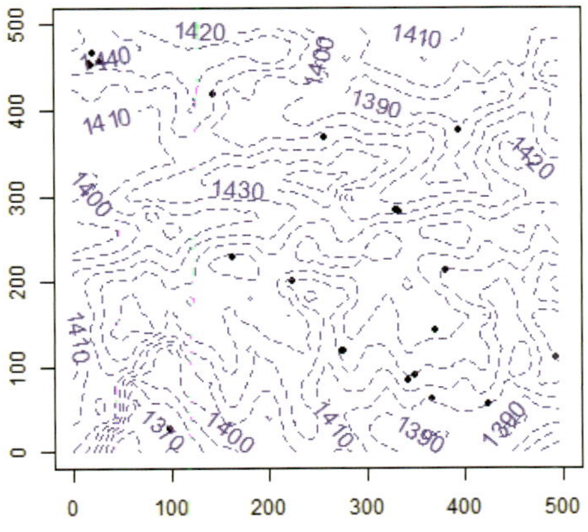

个体分布图 Distribution of individuals

径级分布表 DBH class

胸径区间 (Diameter class) (cm)	个体数 (No. of individuals in the plot)	比例 (Proportion) (%)
1～2	9	36.0
2～3	5	20.0
3～4	3	12.0
4～5	2	8.0
5～7	0	0.0
7～10	1	4.0
≥10	5	20.0

117 野漆 yě qī | Filed Lacquer-tree

Toxicodendron succedaneum，漆树属
漆树科 | Anacardiaceae

树干 Trunk
摄影：卢志军 Photo by: Zhijun Lu

代码（SpCode）= TOXSUC
个体数（Individual number/25 hm^2）= 602
最大胸径（Max DBH）= 40.15 cm
所在林层（Stratum）= 亚乔木层 (Understory)
重要值排序（Importance value rank）= 8/54

落叶乔木。小枝光滑或被茸毛。奇数羽状复叶，小叶5～15，对生或近对生，叶片纸质或薄革质，宽披针形至狭长圆形，先端长渐尖，基部圆或楔形，全缘。圆锥花序，花小，黄绿色。果扁球形，有光泽，无毛，浅黄色。花期5～6月，果期7～10月。

Deciduous trees. Branchlets glabrous to pubescent. Leaves imparipinnate; leaflets 5-15, opposite or subopposite; blade papery or thinly leathery, broad-lanceolate to narrow-oblong, apex long-acuminate, base rounded or cuneate, margin entire. Panicles, flowers small, yellowish green. Drupe compressed globose, glossy, glabrous, light yellow. Fl. May–Jun., fr. Jul.–Oct..

果枝 Fruiting branches
摄影：刘海波 Photo by: Haibo Liu

复叶 Compound leaves
摄影：卢志军 Photo by: Zhijun Lu

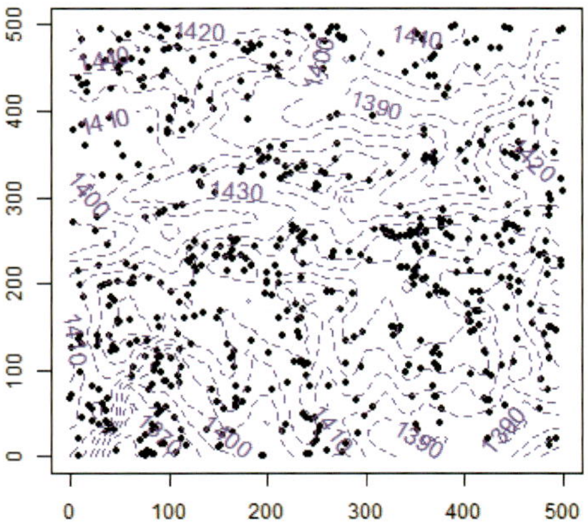

个体分布图 Distribution of individuals

径级分布表 DBH class

胸径区间 (Diameter class) (cm)	个体数 (No. of individuals in the plot)	比例 (Proportion) (%)
1～2.5	38	6.3
2.5～5	29	4.8
5～8	36	6.0
8～11	49	8.1
11～15	85	14.1
15～20	123	20.4
≥20	242	40.2

118 毛漆树

máo qī shù | Hairy-fruit Lacquer-tree

Toxicodendron trichocarpum，漆树属
漆树科 | Anacardiaceae

代码（SpCode）= TOXTRI
个体数（Individual number/25 hm^2）= 55
最大胸径（Max DBH）= 28.52 cm
所在林层（Stratum）= 灌木层 (Shrub)
重要值排序（Importance value rank）= 57/128

落叶乔木。幼枝、叶轴、叶片下面及花序被黄褐色柔毛。奇数羽状复叶，小叶9～15，叶片卵形至长圆形，先端渐尖，基部圆形，不对称，全缘。圆锥花序腋生，花黄绿色，雌雄异株。果扁球形，被带黄色的短刺毛，外皮剥落，现出白色内果皮。花期5～6月，果期7～8月。

Deciduous trees. Young branchlets, rachis, leaflet blade adaxially and inflorescence yellowish brown pubescent. Leaves imparipinnate, leaflets 9-15, leaflet blade ovate or oblong, apex acuminate, base rounded, asymmetric, margin entire. Panicles axillary, flowers yellowish green, dioecious. Drupe oblate, minutely yellow-setaceous, epicarp desquamated, endocarp white. Fl. May–Jun., fr. Jul.–Aug..

树干 Trunk
摄影：卢志军 Photo by: Zhijun Lu

小枝和叶片 Branchlets and leaves
摄影：卢志军 Photo by: Zhijun Lu

花枝 Flowering branches
摄影：鲍大川 Photo by: Dachuan Bao

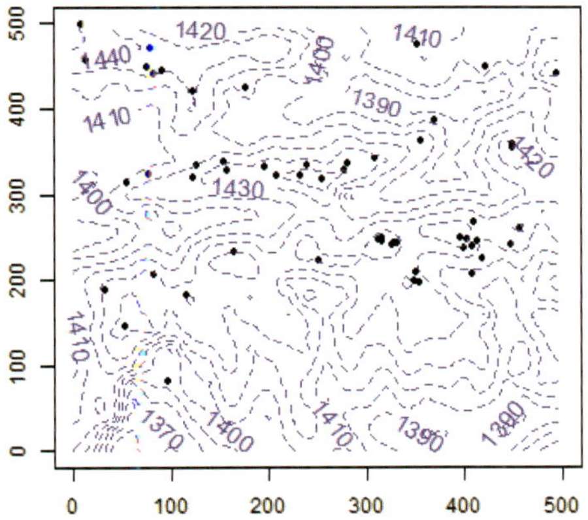

个体分布图 Distribution of individuals

径级分布表 DBH class

胸径区间 (Diameter class) (cm)	个体数 (No. of individuals in the plot)	比例 (Proportion) (%)
1～2	22	40.0
2～3	4	7.3
3～4	5	9.1
4～5	2	3.6
5～7	9	16.4
7～10	8	14.5
≥10	5	9.1

119 五裂枫 wǔ liè fēng | Oliver Maple

Acer oliverianum，槭属
槭树科 | Aceraceae

代码（SpCode）= ACEOLI
个体数（Individual number/25 hm²）= 361
最大胸径（Max DBH）= 30.08 cm
所在林层（Stratum）= 灌木层 (Shrub)
重要值排序（Importance value rank）= 21/128

落叶乔木。树皮淡绿色或灰褐色，平滑，常有白粉。小枝紫绿色，无毛，或微有短柔毛。叶纸质，5裂，裂片三角状卵形或长圆状卵形，先端渐尖或长渐尖，基部近心形或近截形。伞房花序，花白色，顶生无毛，杂性。翅果幼时淡紫色，后变黄褐色；小坚果凸出，脉纹显著，两翅张开近水平。花期5月，果期9月。

树干 Trunk
摄影：卢志军 Photo by: Zhijun Lu

Deciduous trees. Bark light green or grayish brown, smooth, usually pruinose. Branchlets purplish green and glabrous, or slightly pubescent. Leaf blade papery, 5-lobed, lobes triangular-ovate or oblong-ovate, apex acuminate or long-acuminate, base subcordate or truncate. Corymb, flowers white, terminal, glabrous, miscellaneous. Samaras light purple when young, then yellowish brown; nutlets convex, distinctly veined, wings spreading nearly horizontally. Fl. May, fr. Sept..

小枝和叶片 Branchlets and leaves
摄影：陈彬 Photo by: Bin Chen

叶背 Leaf abaxial surface
摄影：徐文斌 Photo by: Wenbin Xu

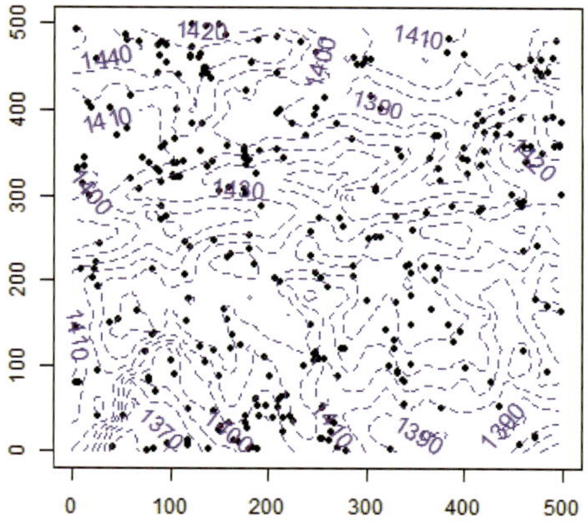

个体分布图 Distribution of individuals

径级分布表 DBH class

胸径区间 (Diameter class) (cm)	个体数 (No. of individuals in the plot)	比例 (Proportion) (%)
1～2	133	36.8
2～3	44	12.2
3～4	22	6.1
4～5	16	4.4
5～7	15	4.2
7～10	27	7.5
≥10	104	28.8

120 青榨枫　　qīng zhà fēng | David Maple

Acer davidii，槭属
槭树科 I Aceraceae

代码（SpCode）= ACEDAV
个体数（Individual number/25 hm^2）= 227
最大胸径（Max DBH）= 39.99 cm
所在林层（Stratum）= 亚乔木层 (Understory)
重要值排序（Importance value rank）= 16/54

落叶乔木。树皮灰褐色或深褐色，纵裂。小枝圆柱状，绿色间褐色，无毛，具稀疏皮孔。叶纸质，卵形，长圆状卵形或近心形，先端渐尖或尾状渐尖，基部近心形或圆形，边缘有不整齐的锯齿。总状花序，顶生下垂，花黄绿色，杂性，与叶同时开放。翅果黄褐色；小坚果扁平，两翅张开成钝角或近水平。花期4～5月，果熟期9月。

树干　　Trunk
摄影：卢志军　　Photo by: Zhijun Lu

Deciduous trees. Bark grayish brown or dark brown, peeling in strips. Branchlets cylindric, brownish green, glabrous, lenticels sparse. Leaf blade papery, ovate, oblong-ovate or subcordate, apex acuminate or caudate-acuminate, base subcordate or rounded, margin irregularly serrate. Racemes, axillary, pendulous, flowers greenish yellow, miscellaneous, appearing with leaves. Samaras yellowish brown; nutlets flat, 2 wings spreading obtusely or horizontally. Fl. Apr–May, fr. maturity Sept..

果枝　　Fruiting branches
摄影：刘海波　　Photo by: Haibo Liu

小枝和叶背　　Branchlet and leaf abaxial surface
摄影：刘海波　　Photo by: Haibo Liu

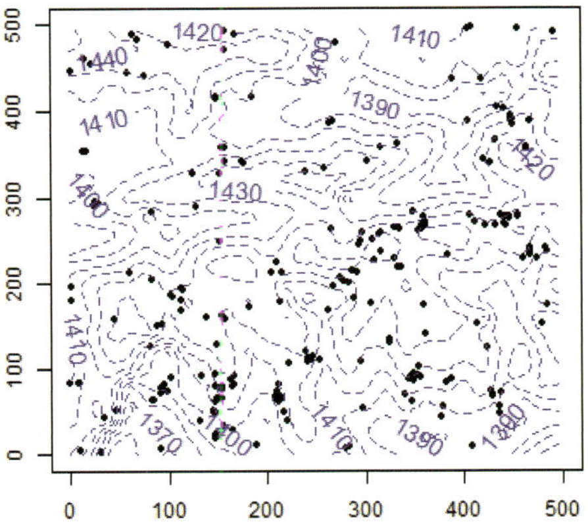

个体分布图 Distribution of individuals

径级分布表 DBH class

胸径区间 (Diameter class) (cm)	个体数 (No. of individuals in the plot)	比例 (Proportion) (%)
1～2.5	31	13.7
2.5～5	22	9.7
5～8	14	6.2
8～11	18	7.9
11～15	48	21.1
15～20	42	18.5
≥20	52	22.9

121 中华枫

zhōng huá fēng | Chinese Maple

Acer sinense，槭属
槭树科 | Aceraceae

代码（SpCode）= ACESIN
个体数（Individual number/25 hm²）= 196
最大胸径（Max DBH）= 47.25 cm
所在林层（Stratum）= 亚乔木层 (Understory)
重要值排序（Importance value rank）= 21/54

树干 Trunk
摄影：卢志军 Photo by: Zhijun Lu

落叶小乔木。树皮黄褐色或深褐色，平滑。小枝紫绿色，无毛。叶近革质，常5裂，裂片长圆状卵形或三角状卵形，先端渐尖，基部心形或近心形，少数为截形，边缘除近基部外疏具紧贴的锯齿。圆锥花序，顶生，下垂，花白色，杂性。翅果淡黄色，无毛；小坚果凸出，两翅张开成钝角或近于水平。花期5月，果期9月。

Deciduous small trees. Bark yellowish brown or dark brown, smooth. Branchlets purplish green, glabrous. Leaves subleathery; leaf blade usually 5-lobed, lobes oblong-ovate or triangular-ovate, apex acuminate, base cordate or subcordate, rarely truncate, margin sparsely serrate except near base. Panicles terminal, pendulous, flowers white, miscellaneous. Samaras light yellow, glabrous, nutlets convex, 2 wings spreading obtusely or horizontally. Fl. May, fr. Sept..

果枝 Fruiting branches
摄影：陈彬 Photo by: Bin Chen

果枝和叶背 Fruiting branch and leaf abaxial surface
摄影：陈彬 Photo by: Bin Chen

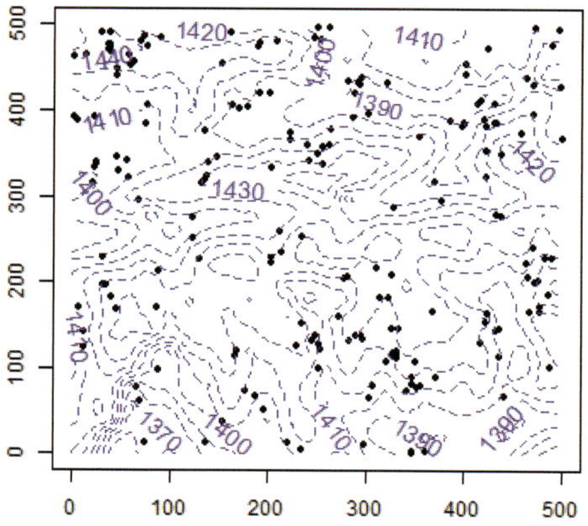

个体分布图 Distribution of individuals

径级分布表 DBH class

胸径区间 (Diameter class) (cm)	个体数 (No. of individuals in the plot)	比例 (Proportion) (%)
1～2.5	60	30.6
2.5～5	42	21.4
5～8	18	9.2
8～11	12	6.1
11～15	16	8.2
15～20	26	13.3
≥20	22	11.2

122 色木枫 sè mù fēng | Colorful Maple

Acer pictum，槭属
槭树科 | Aceraceae

代码（SpCode）= ACEPIC
个体数（Individual number/25 hm^2）= 121
最大胸径（Max DBH）= 39.02 cm
所在林层（Stratum）= 亚乔木层 (Understory)
重要值排序（Importance value rank）= 26/54

落叶乔木。树皮灰色。小枝细瘦，无毛。叶纸质，长圆形，3，5，7或9裂，裂片三角形或卵形，先端锐尖或尾状锐尖，基部截形或近于心形，全缘。圆锥状伞房花序，顶生，花多数，白色，杂性。翅果嫩时紫绿色，成熟时淡黄色；小坚果压扁状；翅长圆形，张开成锐角或近于钝角。花期5月，果期9月。

Deciduous trees. Bark gray. Branchlets slender, glabrous. Leaf blade papery, oblong, 3-, 5-, 7-, or 9-lobed; lobes triangular or ovate, apex acuminate or caudate-acuminate, base truncate or subcordate, margin entire. Inflorescence paniculate-corymbose, terminal, numerous and white flowered, miscellaneous. Samaras purplish green when young, light yellow at maturity, nutlets compressed flat, wings oblong, spreading acutely or obtusely. Fl. May, fr. Sept..

树干 Trunk
摄影：卢志军 Photo by: Zhijun Lu

小枝和叶片 Branchlet and leaves
摄影：卢志军 Photo by: Zhijun Lu

小枝和叶背 Branchlet and leaf abaxial surface
摄影：卢志军 Photo by: Zhijun Lu

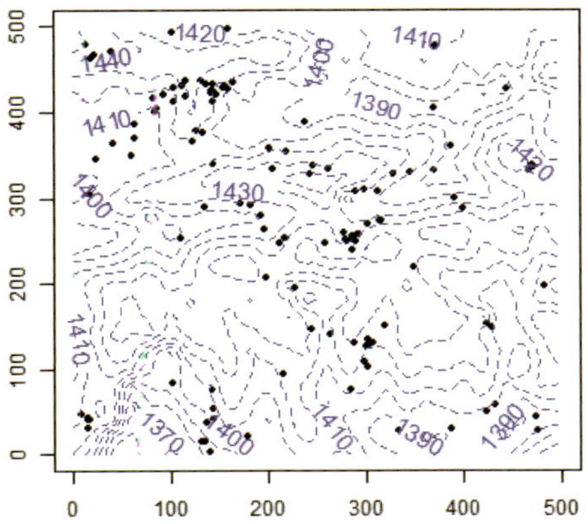

个体分布图 Distribution of individuals

径级分布表 DBH class

胸径区间 (Diameter class) (cm)	个体数 (No. of individuals in the plot)	比例 (Proportion) (%)
1～2.5	43	35.5
2.5～5	30	24.8
5～8	7	5.8
8～11	6	5.0
11～15	8	6.6
15～20	11	9.1
≥20	16	13.2

123 阔叶枫

kuò yè fēng | Broadleaf Maple

Acer amplum，槭属

槭树科 | Aceraceae

代码（SpCode）= ACEAMP

个体数（Individual number/25 hm^2）= 92

最大胸径（Max DBH）= 42.20 cm

所在林层（Stratum）= 亚乔木层 (Understory)

重要值排序（Importance value rank）= 28/54

落叶乔木。树皮褐色或黑褐色，平滑。小枝无毛。叶纸质，通常5裂，少数3裂或不裂，裂片通常卵圆形，先端渐尖或钝尖，基部近心形或截形，全缘。伞房花序，顶生，花多数，黄绿色无毛。翅果扁平，幼时紫色，成熟后变黄褐色，两翅张开成钝角。花期4月，果期8～9月。

Deciduous trees. Bark brownish, dark brown, smooth. Branchlets glabrous. Leaves papery, usually 5-lobed, rarely 3-lobed or unlobed, lobes usually ovate, apex acuminate or obtuse, base subcordate or or truncate, margin entire. Corymbs terminal, numerous flowered, yellowish sreen glabrous. Samaras flat, purple when young, yellowish brown at maturity, wings spreading obtusely. Fl. Apr., fr. Aug.–Sept..

树干 Trunk

摄影：卢志军 Photo by: Zhijun Lu

小枝和叶片 Branchlets and leaves

摄影：陈彬 Photo by: Bin Chen

小枝和叶背 Branchlets and leaf abaxial surface

摄影：卢志军 Photo by: Zhijun Lu

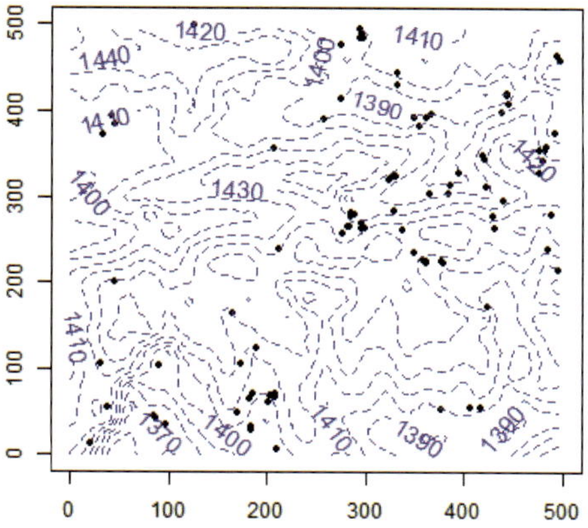

个体分布图 Distribution of individuals

径级分布表 DBH class

胸径区间 (Diameter class) (cm)	个体数 (No. of individuals in the plot)	比例 (Proportion) (%)
1～2.5	24	26.1
2.5～5	26	28.3
5～8	7	7.6
8～11	3	3.3
11～15	11	12.0
15～20	7	7.6
≥20	14	15.2

124 房县枫 fáng xiàn fēng | Franchet Maple

Acer sterculiaceum subsp. *franchetii*，槭属
槭树科 | Aceraceae

代码（SpCode）= ACESTE
个体数（Individual number/25 hm^2）= 38
最大胸径（Max DBH）= 7.46 cm
所在林层（Stratum）= 亚乔木层 (Understory)
重要值排序（Importance value rank）= 42/54

树干 Trunk
摄影：卢志军 Photo by: Zhijun Lu

落叶乔木。树皮黑褐色。小枝紫褐色。叶纸质，近圆形，通常3裂，裂片三角状卵形，先端急尖，基部心形或近心形，少数为圆形，边缘具不整齐的疏锯齿。花黄绿色，单性，雌雄异株。翅果幼时黄绿色，成熟后深黄色；两翅张开成锐角或近直立。花期4～5月，果期9月。

Deciduous trees. Bark black-brown. Branchlets purplish brown. Leaf blade papery, subrounded, usually 3-lobed, lobes triangular-ovate, apex acute, base cordate or subcordate, rarely rounded, margin irregularly sparsely serrate. Flowers yellowish green, unisexual, dioecious. Samaras yellowish green when young, dark yellow at maturity, 2 wings spreading acutely or nearly erectly. Fl. Apr.–May, fr. Sept..

叶片 Leaves
摄影：卢志军 Photo by: Zhijun Lu

果枝 Fruiting branch
摄影：刘海波 Photo by: Haibo Liu

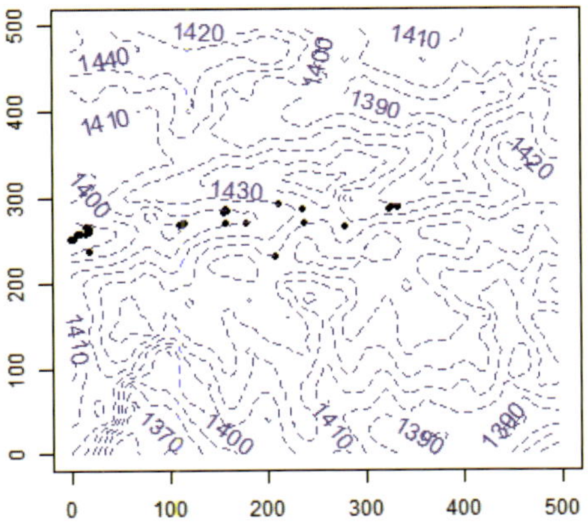

个体分布图 Distribution of individuals

径级分布表 DBH class

胸径区间 (Diameter class) (cm)	个体数 (No. of individuals in the plot)	比例 (Proportion) (%)
1～2.5	25	65.8
2.5～5	7	18.4
5～8	6	15.8
8～11	0	0.0
11～15	0	0.0
15～20	0	0.0
≥20	0	0.0

125 扇叶枫 shàn yè fēng | Flabellateleaf Maple

Acer flabellatum，槭属
槭树科 | Aceraceae

代码（SpCode）= ACEFLA
个体数（Individual number/25 hm^2）= 51
最大胸径（Max DBH）= 30.81 cm
所在林层（Stratum）= 亚乔木层 (Understory)
重要值排序（Importance value rank）= 33/54

树干 Trunk
摄影：卢志军 Photo by: Zhijun Lu

落叶乔木。树皮光滑，褐色或深褐色。小枝紫绿色，无毛。叶薄纸质或膜质，近圆形，常7裂，裂片卵状长圆形或三角状卵形，先端渐尖或尾状渐尖，基部深心形，边缘有不整齐的锯齿。圆锥花序，花黄绿色，杂性，无毛。果序下垂，翅果黄褐色，小坚果凸起，两翅张开近水平。花期6月，果期9～10月。

Deciduous trees. Bark smooth, brown or dark brown. Branchlets purplish green, glabrous. Leaf blade thinly papery or membranous, subrounded, usually 7-lobed, lobes ovate-oblong or triangular-ovate, apex acuminate or caudate acuminate, base deeply cordate, margin irregularly serrate. Panicles, flowers yellowish green, miscellaneous, glabrous. Infructescence pendulous, samaras yellowish brown, nutlets convex, 2 wings spreading nearly horizontally. Fl. Jun., fr. Sept.–Oct..

果枝和叶片 Fruiting branch and leaves
摄影：卢志军 Photo by: Zhijun Lu

小枝和叶背 Branchlet and leaf abaxial surface
摄影：卢志军 Photo by: Zhijun Lu

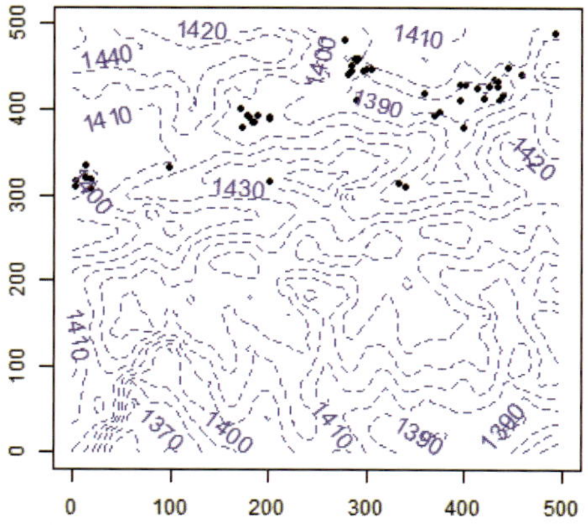

个体分布图 Distribution of individuals

径级分布表 DBH class

胸径区间 (Diameter class) (cm)	个体数 (No. of individuals in the plot)	比例 (Proportion) (%)
1～2.5	15	29.4
2.5～5	14	27.5
5～8	4	7.8
8～11	4	7.8
11～15	5	9.8
15～20	4	7.8
≥20	5	9.8

126 杈叶枫 chā yè fēng | Ceriferous-branch Maple

Acer ceriferum，槭属
槭树科 | Aceraceae

代码（SpCode）= ACECER
个体数（Individual number/25 hm²）= 10
最大胸径（Max DBH）= 12.20 cm
所在林层（Stratum）= 灌木层 (Shrub)
重要值排序（Importance value rank）= 85/128

落叶乔木。小枝淡紫褐色，无毛。叶纸质，掌状7～9裂，裂片卵状长圆形，先端尾状渐尖，基部截形或近心形，边缘具不整齐的锐尖重锯齿。伞房花序顶生，杂性，花淡绿色。翅果黄绿色，小坚果凸出，两翅张开几乎成水平。花期5月，果期9～10月。

Deciduous trees. Branchlets light purplish brown, glabrous. Leaf blade papery, 7-9 palmately lobed, lobes ovate-oblong, apex caudate acuminate, base truncate or subcordate, margin irregularly bi-serrate with sharp teeth. Corymbs terminal, miscellaneous, flowers light green. Samaras yellowish green, nutlets convex, 2 wings spreading nearly horizontally. Fl. May, fr. Sept.–Oct..

树干 Trunk
摄影：卢志军 Photo by: Zhijun Lu

小枝和叶片 Branchlets and leaves
摄影：谷志容 Photo by: Zhirong Gu

叶背 Leaf abaxial surface
摄影：卢志军 Photo by: Zhijun Lu

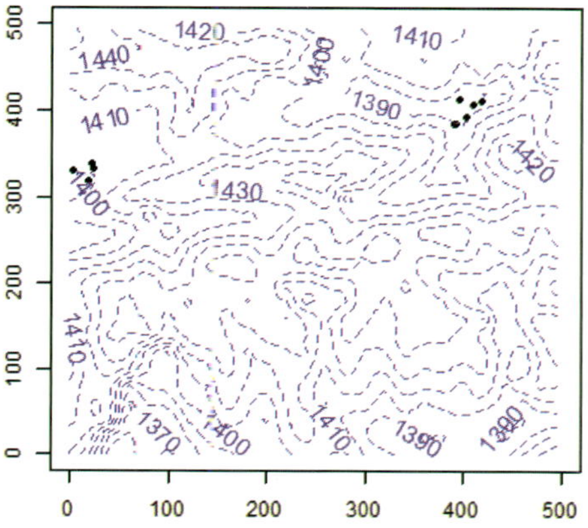

个体分布图 Distribution of individuals

径级分布表 DBH class

胸径区间 (Diameter class) (cm)	个体数 (No. of individuals in the plot)	比例 (Proportion) (%)
1～2	3	30.0
2～3	1	10.0
3～4	0	0.0
4～5	2	20.0
5～7	1	10.0
7～10	2	20.0
≥10	1	10.0

127 三叶枫 sān yè fēng | Henry Maple

Acer henryi，槭属
槭树科 | Aceraceae

代码（SpCode）= ACEHEN
个体数（Individual number/25 hm^2）= 10
最大胸径（Max DBH）= 41.76 cm
所在林层（Stratum）= 亚乔木层 (Understory)
重要值排序（Importance value rank）= 36/54

树干 Trunk
摄影：卢志军 Photo by: Zhijun Lu

落叶乔木。树皮灰褐色。幼枝绿色，被短柔毛。叶纸质，羽状复叶，有3小叶，椭圆形或长圆状椭圆形，先端渐尖，基部楔形或宽楔形，少数近圆形，全缘或有疏锯齿。总状花序，下垂，花绿色，单性，雌雄异株。翅果嫩时淡紫色，成熟后黄褐色；小坚果凸起，脉纹显著，两翅张开成锐角或近直立。花期4月，果期9月。

Deciduous trees. Bark grayish brown. Branchlets green and pubescent. Leaf blades papery, pinnately compound, 3-leafleted, elliptic or oblong-elliptic, apex acuminate, base cuneate or broadly cuneate, rarely subrounded, margin entire or sparsely serrate. Racemes pendulous, flowers green, unisexual, dioecious. Samaras light purple when young, yellowish brown at maturity, nutlets convex, veins conspicuous, 2 wings spreading acutely or nearly erectly. Fl. Apr., fr. Sept..

小枝和叶片 Branchlets and leaves
摄影：卢志军 Photo by: Zhijun Lu

小枝和叶背 Branchlets and leaf abaxial surface
摄影：卢志军 Photo by: Zhijun Lu

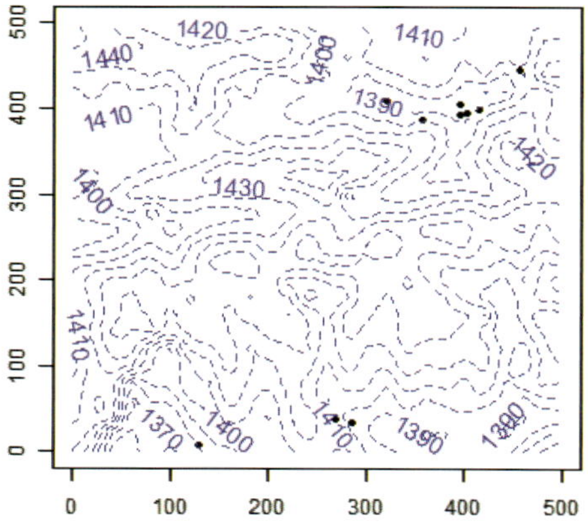

个体分布图 Distribution of individuals

径级分布表 DBH class

胸径区间 (Diameter class) (cm)	个体数 (No. of individuals in the plot)	比例 (Proportion) (%)
1～2.5	1	10.0
2.5～5	1	10.0
5～8	0	0.0
8～11	0	0.0
11～15	3	30.0
15～20	1	10.0
≥20	4	40.0

128 紫果枫 zǐ guǒ fēng | Cordateleaf Maple

Acer cordatum，槭属
槭树科 | Aceraceae

代码（SpCode）= ACECOR
个体数（Individual number/25 hm^2）= 6
最大胸径（Max DBH）= 13.50 cm
所在林层（Stratum）= 灌木层 (Shrub)
重要值排序（Importance value rank）= 88/128

常绿乔木。树皮灰色或黑灰色，平滑。小枝紫色或紫绿色，无毛。叶纸质或近革质，卵状长圆形或卵形，先端短渐尖或渐尖，基部近心形，边缘仅近顶端疏具细锯齿。伞房花序，花淡白色或黄白色，顶生。翅果幼时紫色，成熟后黄褐色；小坚果凸起，两翅张开成钝角或近水平。花期4~5月，果期9月。

树干 Trunk
摄影：卢志军 Photo by: Zhijun Lu

Evergreen trees. Bark gray or black gray, smooth. Branchlets purple or purplish green, glabrous. Leaves papery or subleathery; leaf blade ovate-oblong or ovate, apex shortly acuminate or acuminate, base subcordate, margin only sparsely serrulate near apex. Corymbs terminal, flowers light white or yellowish white. Samaras purple when young, yellowish brown at maturity, nutlets convex, 2 wings spreading obtusely or nearly horizontally. Fl. Apr.–May, fr. Sept..

小枝和叶片 Branchlets and leaves
摄影：刘海波 Photo by: Haibo Liu

小枝和叶背 Branchlets and leaf abaxial surface
摄影：刘海波 Photo by: Haibo Liu

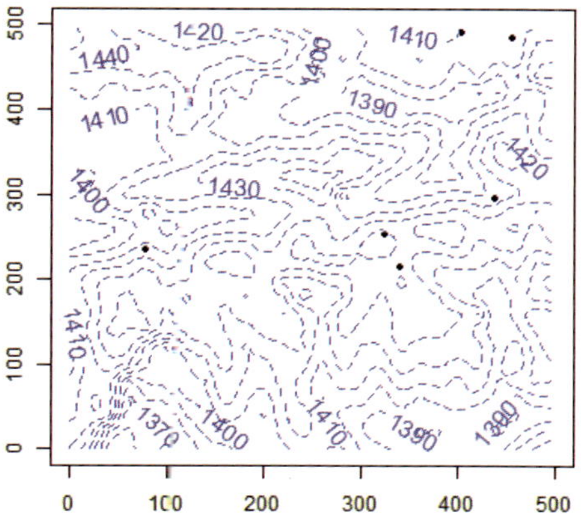

个体分布图 Distribution of individuals

径级分布表 DBH class

胸径区间 (Diameter class) (cm)	个体数 (No. of individuals in the plot)	比例 (Proportion) (%)
1~2	2	33.3
2~3	0	0.0
3~4	0	0.0
4~5	1	16.7
5~7	0	0.0
7~10	1	16.7
≥10	2	33.3

129 鸡爪枫 jī zhǎo fēng | Japanese Maple

Acer palmatum，槭属
槭树科 | Aceraceae

代码（SpCode）= ACEPAL
个体数（Individual number/25 hm^2）= 4
最大胸径（Max DBH）= 7.79 cm
所在林层（Stratum）= 灌木层 (Shrub)
重要值排序（Importance value rank）= 103/128

落叶乔木。树皮深灰色。小枝细瘦，紫色或灰紫色。叶纸质，近圆形，5～9掌状深裂，通常7裂，裂片长卵形或披针形，先端渐尖或长渐尖，基部心形或近心形，少数截形，边缘具紧贴的锐锯齿。伞房圆锥花序，顶生，花紫色，杂性，无毛。翅果幼时紫红色，成熟后变棕黄色，小坚果凸出，两翅张开成钝角。花期5月，果熟期10月。

Deciduous trees. Bark dark gray. Branchlets slender, purple or grayish purple. Leaves papery, blade subrounded, 5-9, usually 7 palmately lobed, lobes long-ovate or lanceolate, apex acuminate or long acuminate, base cordate to subcordate, rarely truncate, margin appressed serrulate. Corymbose-panicles terminal, flowers purple, miscellaneous, glabrous. Samaras purplish red when young, brownish yellow at maturity, nutlets convex, 2 wings spreading obtusely. Fl. May, fr. maturity Oct..

树干 Trunk
摄影：卢志军 Photo by: Zhijun Lu

果枝 Fruiting branches
摄影：江明喜 Photo by: Mingxi Jiang

小枝和叶背 Branchlets and leaf abaxial surface
摄影：卢志军 Photo by: Zhijun Lu

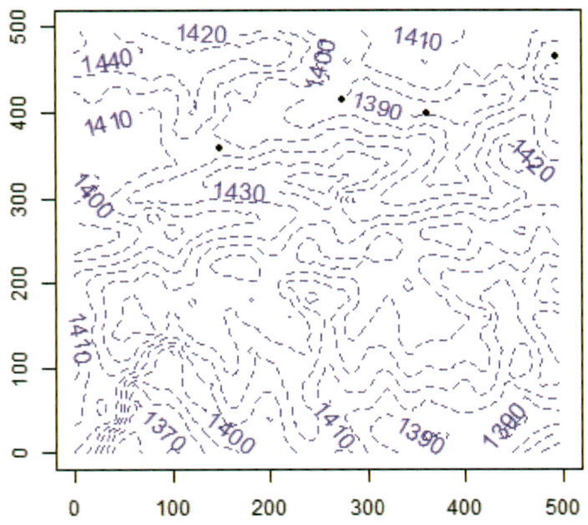

个体分布图 Distribution of individuals

径级分布表 DBH class

胸径区间 (Diameter class) (cm)	个体数 (No. of individuals in the plot)	比例 (Proportion) (%)
1～2	2	50.0
2～3	1	25.0
3～4	0	0.0
4～5	0	0.0
5～7	0	0.0
7～10	1	25.0
≥10	0	0.0

130 三峡枫 sān xiá fēng | Wilson Maple

Acer wilsonii，槭属
槭树科 | Aceraceae

代码（SpCode）= ACEWIL
个体数（Individual number/25 hm^2）= 3
最大胸径（Max DBH）= 35.22 cm
所在林层（Stratum）= 亚乔木层 (Understory)
重要值排序（Importance value rank）= 41/54

树干 Trunk
摄影：卢志军 Photo by: Zhijun Lu

落叶乔木。树皮黑褐色，平滑。小枝绿色或紫绿色，无毛。叶纸质，卵圆形，常3裂，裂片卵状长圆形或披针形，先端尾状渐尖，基部常圆形，少数为截形或近心形，全缘或近顶端有数个小齿。圆锥花序杂性，花白色，无毛。果序下垂，翅果黄褐色；小坚果凸出，脉纹显著，果翅基部狭窄，张开近水平。花期4～5月，果熟期9月。

Deciduous trees. Bark dark brown, smooth. Branchlets green or purplish green, glabrous. Leaf blade papery, ovate, usually 3 lobed, lobes ovate-oblong or lanceolate, apex caudate-acuminate, base usually rounded, rarely truncate or subcordate, margin entire or with a few small serrations apically. Panicles miscellaneous, flowers white, glabrous. Infructescence pendulous, samaras yellowish brown, nutlets convex, veins conspicuous, wing base narrow, spreading nearly horizontally. Fl. Apr.–May, fr. maturity Sept..

花枝 Flowering branch
摄影：江明喜 Photo by: Mingxi Jiang

果枝 Fruiting branches
摄影：江明喜 Photo by: Mingxi Jiang

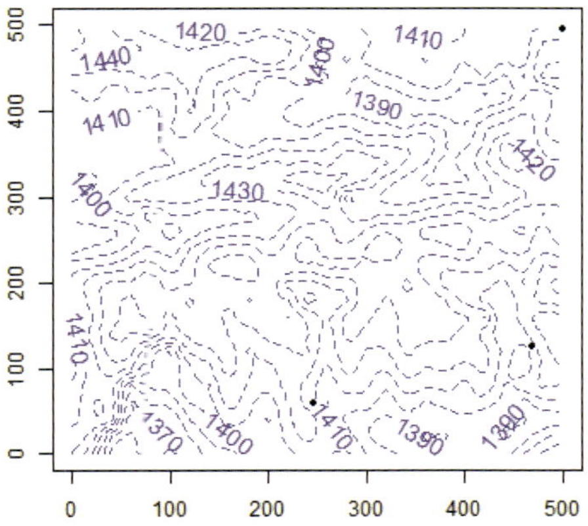

个体分布图 Distribution of individuals

径级分布表 DBH class

胸径区间 (Diameter class) (cm)	个体数 (No. of individuals in the plot)	比例 (Proportion) (%)
1～2.5	0	0.0
2.5～5	0	0.0
5～8	0	0.0
8～11	0	0.0
11～15	1	33.3
15～20	0	0.0
⩾20	2	66.7

131 伯乐树 bó lè shù | Chinese Bretschneidera

Bretschneidera sinensis，伯乐树属
钟萼木科 | Bretschneideraceae

代码（SpCode）= BRESIN
个体数（Individual number/25 hm^2）= 24
最大胸径（Max DBH）= 51.48 cm
所在林层（Stratum）= 乔木层 (Canopy)
重要值排序（Importance value rank）= 23/50

树干 Trunk
摄影：卢志军 Photo by: Zhijun Lu

落叶乔木。树皮灰白色，平滑。奇数羽状复叶，小叶7～15，对生，叶片长圆状披针形，顶端长渐尖，基部圆楔形，有时偏斜，全缘。总状花序顶生，直立，花粉红色，萼钟形。蒴果红色，椭圆形，果皮木质；种子长圆形，带红色。花期5～7月，果期8月至翌年4月。

Deciduous trees. Bark gray-white, smooth. Leaves imparipinnate, leaflets 7-15, opposite, blade oblong-lanceolate, apex long acuminate, base rounded-cuneate, sometimes oblique, margin entire. Racemes terminal, erect, flowers pink, sepals campanulate. Capsules red, ellipsoid, pericarp woody. Seeds oblong, reddish. Fl. May–Jul., fr. Aug.–Apr. of following yr..

果枝 Fruiting branch
摄影：鲍大川 Photo by: Dachuan Bao

复叶背面 Compound leaves abaxial surface
摄影：卢志军 Photo by: Zhijun Lu

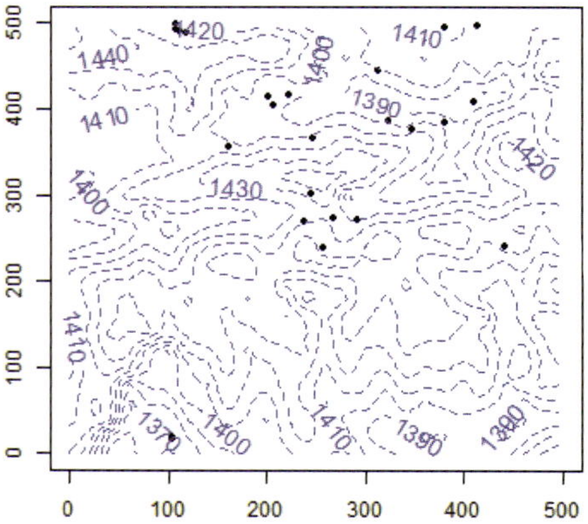

个体分布图 Distribution of individuals

径级分布表 DBH class

胸径区间 (Diameter class) (cm)	个体数 (No. of individuals in the plot)	比例 (Proportion) (%)
1～2.5	16	66.7
2.5～5	2	8.3
5～10	0	0.0
10～25	1	4.2
25～40	4	16.7
40～100	1	4.2
⩾100	0	0.0

132 红柴枝 hóng chái zhī | Oldham Meliosma

Meliosma oldhamii，泡花树属
清风藤科 | Sabiaceae

代码（SpCode）= MELOLD
个体数（Individual number/25 hm^2）= 1477
最大胸径（Max DBH）= 47.22 cm
所在林层（Stratum）= 乔木层 (Canopy)
重要值排序（Importance value rank）= 12/50

落叶乔木。小枝近无毛。奇数羽状复叶，小叶7～15，互生或对生，小叶顶端急尖或渐尖，基部圆形或楔形，叶缘疏具尖齿。圆锥花序顶生，宽大而直立，有毛，花白色，萼片卵状椭圆形。果球形。花期4～5月，果期7～8月。

Deciduous trees. Branchlets glabrescent. Leaves imparipinnate, leaflets 7-15, alternate or opposite, leaflet apex acute or acuminate, base rounded or cuneate, margin sparsely acutely serrate. Panicles terminal, broad, erect, tomentose, flowers white, sepals ovate-elliptic. Fruit globose. Fl. Apr.–May, fr. Jul.–Aug..

树干 Trunk
摄影：卢志军 Photo by: Zhijun Lu

花枝 Flowering branches
摄影：刘海波 Photo by: Haibo Liu

小枝和叶背 Branchlets and leaf abaxial surface
摄影：刘海波 Photo by: Haibo Liu

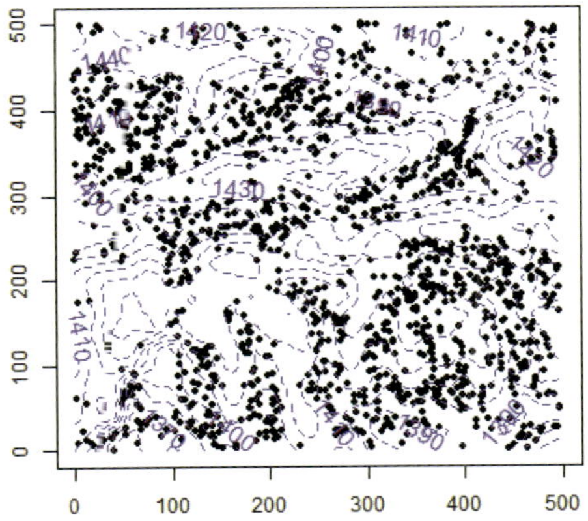

个体分布图 Distribution of individuals

径级分布表 DBH class

胸径区间 (Diameter class) (cm)	个体数 (No. of individuals in the plot)	比例 (Proportion) (%)
1～2.5	460	31.1
2.5～5	248	16.8
5～10	324	21.9
10～25	405	27.4
25～40	39	2.6
40～100	1	0.1
≥100	0	0.0

133 垂枝泡花树

chuí zhī pào huā shù | Flexuous Meliosma

Meliosma flexuosa，泡花树属
清风藤科 I Sabiaceae

代码（SpCode）= MELFLE
个体数（Individual number/25 hm^2）= 129
最大胸径（Max DBH）= 10.28 cm
所在林层（Stratum）= 灌木层 (Shrub)
重要值排序（Importance value rank）= 56/128

落叶灌木或小乔木。小枝褐色，初被长柔毛，后变无毛。叶片倒卵状椭圆形或倒卵状长椭圆形，先端突渐尖，基部渐狭至叶柄，边缘有锐尖的锯齿。圆锥花序顶生，花白色，有香味；萼片不等大，圆卵形，花瓣无毛。核果球形，黑色。花期6～7月，果期10月。

Deciduous shrubs or small trees. Branchlets brown, initially pubescent then glabrous. Leaf blade obovate-elliptic or obovate long elliptic, apex abruptly acuminate, base attenuate to petiole, margin acutely serrate. Panicles terminal, flowers white, fragrant, sepals anisometric, ovate, petals glabrous. Drupes globose, black. Fl. Jun.–Jul., fr. Oct..

树干 Trunk
摄影：卢志军 Photo by: Zhijun Lu

花枝 Flowering branch
摄影：鲍大川 Photo by: Dachuan Bao

小枝和叶背 Branchlet and leaf abaxial surface
摄影：刘海波 Photo by: Haibo Liu

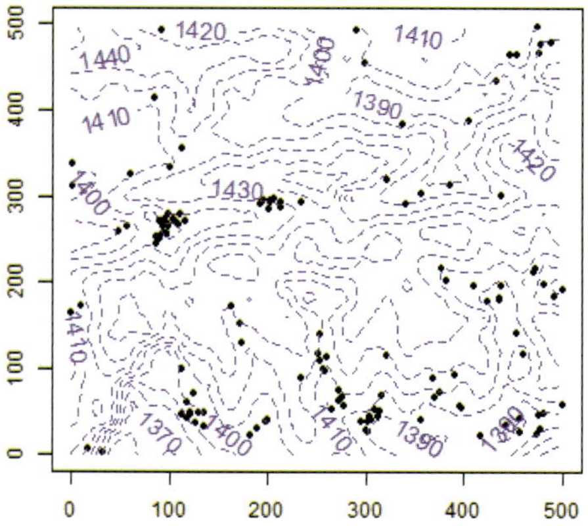

个体分布图 Distribution of individuals

径级分布表 DBH class

胸径区间 (Diameter class) (cm)	个体数 (No. of individuals in the plot)	比例 (Proportion) (%)
1～2	67	51.9
2～3	28	21.7
3～4	18	14.0
4～5	13	10.1
5～7	2	1.6
7～10	0	0.0
≥10	1	0.8

134 暖木 nuǎn mù | Veitch Meliosma

Meliosma veitchiorum，泡花树属
清风藤科 | Sabiaceae

代码（SpCode）= MELVEI
个体数（Individual number/25 hm^2）= 34
最大胸径（Max DBH）= 57.82 cm
所在林层（Stratum）= 亚乔木层 (Understory)
重要值排序（Importance value rank）= 29/54

树干 Trunk
摄影：卢志军 Photo by: Zhijun Lu

落叶乔木。树皮灰色，片状剥落。小枝粗壮，被锈色柔毛，与花轴密被木栓质、凸起的大皮孔。奇数羽状复叶。圆锥花序直立，花白色，芳香，萼片长椭圆形，花瓣倒心形。核果球形，熟时黑色。花期5月，果熟期9月。

Deciduous trees. Bark gray, sheet-like peeling. Branchlets stout, pilose, lenticels subered and convex similar to inflorescence rachis. Leaves imparipinnate. Panicles erect, flowers white, fragrant, sepals long-elliptic, petals obcordate. Drupes globose, black at maturity. Fl. May, fr. maturity Sept..

小枝和叶片 Branchlets and leaves
摄影：刘海波 Photo by: Haibo Liu

复叶背面 Compound leaves abaxial surface
摄影：陈彬 Photo by: Bin Chen

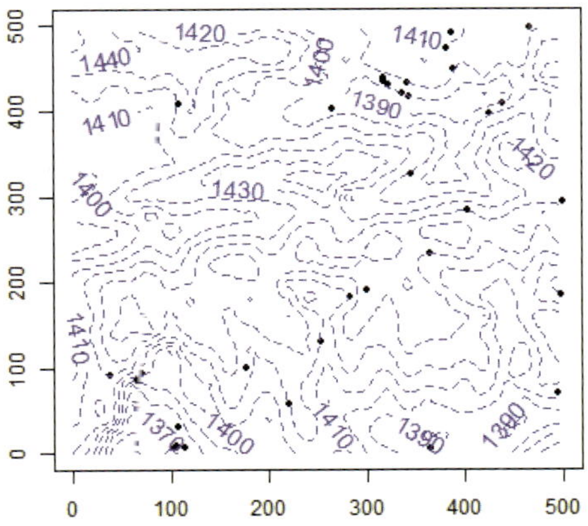

个体分布图 Distribution of individuals

径级分布表 DBH class

胸径区间 (Diameter class) (cm)	个体数 (No. of individuals in the plot)	比例 (Proportion) (%)
1～2.5	8	23.5
2.5～5	5	14.7
5～8	3	8.8
8～11	4	11.8
11～15	1	2.9
15～20	3	8.8
≥20	10	29.4

135 泡花树

pào huā shù | Cuneate-leaf Meliosma

Meliosma cuneifolia，泡花树属
清风藤科 | Sabiaceae

代码（SpCode）= MELCUN
个体数（Individual number/25 hm^2）= 13
最大胸径（Max DBH）= 22.42 cm
所在林层（Stratum）= 灌木层 (Shrub)
重要值排序（Importance value rank）= 68/128

落叶灌木至小乔木。树皮黑棕色。小枝紫褐色，近无毛。叶片倒卵状楔形，先端短渐尖或突尖，基部狭楔形而下延，边缘有锐锯齿。圆锥花序顶生或生于上部叶腋内，花白色，有香味，萼片卵圆形。核果球形，熟时黑色。花期6～7月，果期8～9月。

Deciduous shrubs or small trees. Bark black-brown. Branchlets purplish brown, glabrescent. Leaf blade obovate-cuneate, apex shortly acuminate or acute, base narrowly cuneate and decurrent, margin acutely serrate. Panicles terminal or axillary on apical branchlet, flowers white, fragrant, sepals ovate. Drupes globose, black at maturity. Fl. Jun.–Jul., fr. Aug.–Sept..

树干 Trunk
摄影：卢志军 Photo by: Zhijun Lu

小枝和叶片 Branchlet and leaves
摄影：卢志军 Photo by: Zhijun Lu

小枝和叶背 Branchlet and leaf abaxial surface
摄影：卢志军 Photo by: Zhijun Lu

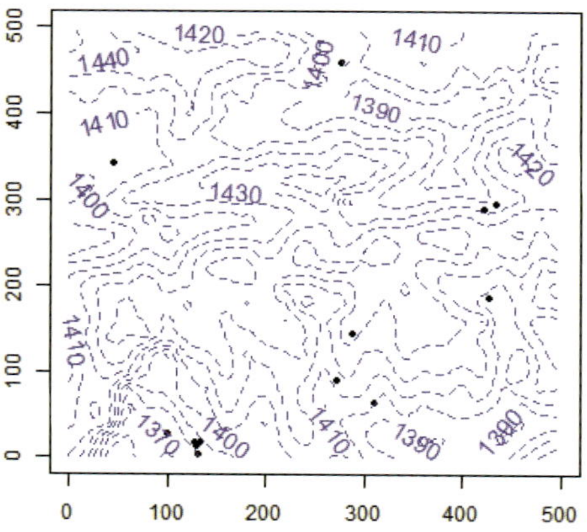

个体分布图 Distribution of individuals

径级分布表 DBH class

胸径区间 (Diameter class) (cm)	个体数 (No. of individuals in the plot)	比例 (Proportion) (%)
1～2	5	38.5
2～3	4	30.8
3～4	3	23.1
4～5	0	0.0
5～7	0	0.0
7～10	0	0.0
⩾10	1	7.7

136 猫儿刺 māo ér cì | Perny Ilex

Ilex pernyi，冬青属
冬青科 I Aquifoliaceae

代码（SpCode）= ILEPER
个体数（Individual number/25 hm^2）= 2314
最大胸径（Max DBH）= 10.96 cm
所在林层（Stratum）= 灌木层 (Shrub)
重要值排序（Importance value rank）= 17/128

常绿灌木或小乔木。树皮银灰色。小枝粗壮，有棱角，有短柔毛。叶革质，四方状卵形或卵形，先端急尖，顶刺状，基部圆形或截形，边缘有大刺齿。聚伞花序，花淡黄色，成密生无梗花簇，腋生于二年生小枝叶腋内。果球形，红色，柱头盘状，果梗短。花期4～5月，果熟期10月。

Evergreen shrubs or small trees. Bark silver-gray. Branchlets terete, angled, pubescent. Leaf blade leathery, square ovate or ovate, apex acute with spinous tip, base rounded or truncate, margin serrate, teeth thorned. Cymes, flowers light yellow, inflorescence dense, peduncles absent, axillary on 2-yr branchlets. Fruit globose, red, stigma discoid, peduncles short. Fl. Apr.–May, fr. maturity Oct..

树干 Trunk
摄影：卢志军 Photo by: Zhijun Lu

花枝 Flowering branch
摄影：刘海波 Photo by: Haibo Liu

果枝 Fruiting branches
摄影：鲍大川 Photo by: Dachuan Bao

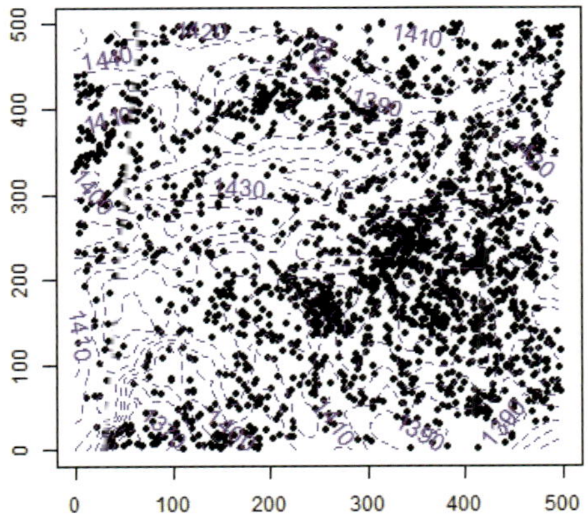

个体分布图 Distribution of individuals

径级分布表 DBH class

胸径区间 (Diameter class) (cm)	个体数 (No. of individuals in the plot)	比例 (Proportion) (%)
1～2	1217	52.6
2～3	532	23.0
3～4	251	10.8
4～5	135	5.8
5～7	112	4.8
7～10	52	2.2
⩾10	15	0.6

137 香冬青 xiāng dōng qīng | Fragrant Ilex

Ilex suaveolens，冬青属
冬青科 | Aquifoliaceae

代码（SpCode）= ILESUA
个体数（Individual number/25 hm^2）= 920
最大胸径（Max DBH）= 32.21 cm
所在林层（Stratum）= 亚乔木层 (Understory)
重要值排序（Importance value rank）= 12/54

常绿乔木。小枝灰褐色，有棱角，皮孔明显，无毛。叶革质，通常椭圆形，有时披针形或卵形，先端渐尖，基部楔形，无毛，边缘有疏锯齿。花序近伞房状，少数为聚伞状。果球形，红色。花期5月，果熟期10月。

树干 Trunk
摄影：卢志军 Photo by: Zhijun Lu

Evergreen trees. Young branchlets grayish brown, angular, lenticellate, glabrous. Leaf blade leathery, usually elliptic, sometimes lanceolate or ovate, apex acuminate, base cuneate, glabrous, margin sparsely serrate. Subcorymbs, rarely cymes. Fruit globose, red. Fl. May, fr. maturity Oct..

花枝 Flowering branches
摄影：刘海波 Photo by: Haibo Liu

果枝 Fruiting branch
摄影：鲍大川 Photo by: Dachuan Bao

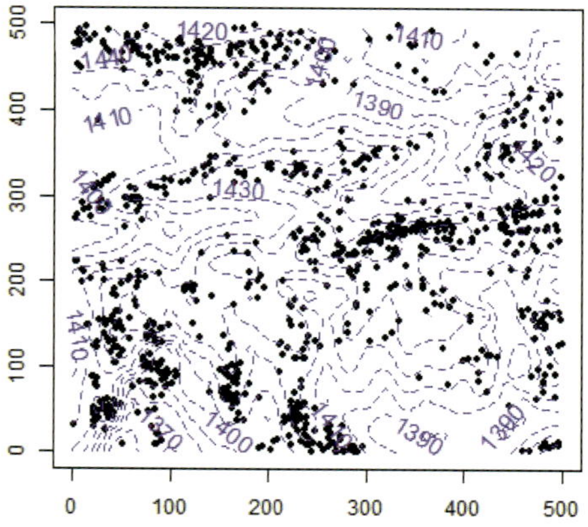

个体分布图 Distribution of individuals

径级分布表 DBH class

胸径区间 (Diameter class) (cm)	个体数 (No. of individuals in the plot)	比例 (Proportion) (%)
1～2.5	261	28.4
2.5～5	252	27.4
5～8	96	10.4
8～11	82	8.9
11～15	91	9.9
15～20	87	9.5
≥20	51	5.5

138 榕叶冬青

róng yè dōng qīng | Fig-like Ilex

Ilex ficoidea，冬青属
冬青科 | Aquifoliaceae

代码（SpCode）= ILEFIC
个体数（Individual number/25 hm^2）= 593
最大胸径（Max DBH）= 35.88 cm
所在林层（Stratum）= 亚乔木层 (Understory)
重要值排序（Importance value rank）= 18/54

常绿乔木。小枝灰色，无毛。叶革质光亮，长圆状椭圆形，先端尾状长尖，基部楔形至圆形，边缘有稀疏的浅圆锯齿，干后略向下反卷。聚伞花序，花白色或绿黄色，有芳香，花序簇生叶腋。果球形，成熟时红色，宿存柱头盘状。花期4～5月，果期8～10月。

Evergreen trees. Branchlets grayish, glabrous. Leaves leathery and shiny, blade oblong-elliptic, apex caudate acuminate, base cuneate to rounded, margin sparsely shallowly crenate, revolute downward after dried. Cymes, flowers white or greenish yellow, fragrant, fascicled and axillary. Fruit globose, red at maturity, stigma discoid and persistent. Fl. Apr.–May, fr. Aug.–Oct..

树干 Trunk
摄影：卢志军 Photo by: Zhijun Lu

果枝 Fruiting branches
摄影：陈彬 Photo by: Bin Chen

小枝和叶背 Branchlet and leaf abaxial surface
摄影：陈彬 Photo by: Bin Chen

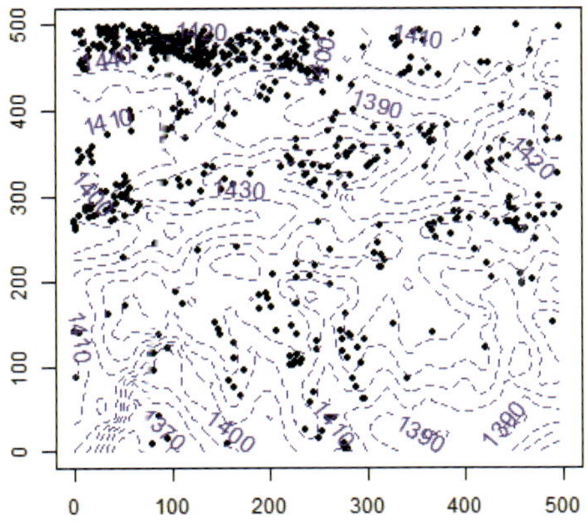

个体分布图 Distribution of individuals

径级分布表 DBH class

胸径区间 (Diameter class) (cm)	个体数 (No. of individuals in the plot)	比例 (Proportion) (%)
1～2.5	216	36.4
2.5～5	235	39.6
5～8	58	9.8
8～11	29	4.9
11～15	30	5.1
15～20	15	2.5
⩾20	10	1.7

139 全缘冬青 quán yuán dōng qīng | Entire-margin Holly

Ilex integra，冬青属
冬青科 I Aquifoliaceae

代码（SpCode）= ILEINT
个体数（Individual number/25 hm^2）= 416
最大胸径（Max DBH）= 10.68 cm
所在林层（Stratum）= 灌木层 (Shrub)
重要值排序（Importance value rank）= 40/128

常绿小乔木。树皮灰白色。小枝粗壮，茶褐色，无毛。叶片厚革质，倒卵形或倒卵状椭圆形，稀倒披针形，先端钝圆，具短的宽钝头，基部楔形，全缘。聚伞花序簇生于当年生枝的叶腋，花瓣长圆状椭圆形，顶端微凹。果球形，成熟时红色。花期4月，果期7～10月。

Evergreen small trees. Bark gray-white. Branchlets stout, brown, glabrous; leaf blade thickly leathery, obovate or obovate-elliptic, rarely oblanceolate, apex obtuse, short-broad-obtuse tipped, base cuneate, margin entire. Cymes fascicled and axillary on current yr branchlets, petals oblong-elliptic, apex emarginate. Fruit globose, red at maturity. Fl. Apr., fr. Jul.–Oct..

树干 Trunk
摄影：卢志军 Photo by: Zhijun Lu

果枝 Fruiting branch
摄影：陈彬 Photo by: Bin Chen

果枝和叶背 Fruiting branch and leaf abaxial surface
摄影：陈彬 Photo by: Bin Chen

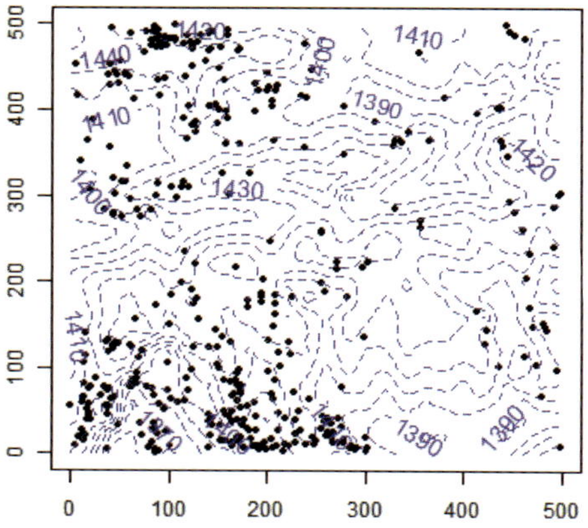

个体分布图 Distribution of individuals

径级分布表 DBH class

胸径区间 (Diameter class) (cm)	个体数 (No. of individuals in the plot)	比例 (Proportion) (%)
1～2	209	50.2
2～3	108	26.0
3～4	53	12.7
4～5	18	4.3
5～7	11	2.6
7～10	9	2.2
≥10	8	1.9

140 大果冬青 dà guǒ dōng qīng | Large-fruit Ilex

Ilex macrocarpa，冬青属
冬青科 | Aquifoliaceae

代码（SpCode）= ILEMAC
个体数（Individual number/25 hm^2）= 153
最大胸径（Max DBH）= 32.58 cm
所在林层（Stratum）= 亚乔木层 (Understory)
重要值排序（Importance value rank）= 22/54

树干 Trunk
摄影：卢志军 Photo by: Zhijun Lu

落叶乔木。枝条无毛，皮孔显著。叶膜质或纸质，互生，卵形或卵状椭圆形，有时为长圆状椭圆形，先端短渐尖，基部宽楔形或圆形，两面无毛，边缘有细锯齿。聚伞花序，有香味，花白色。果球形，熟时黑色；内果皮坚硬。花期5月，果熟期10月。

Trees deciduous. Branchlets glabrous, lenticels conspicuous. Leaf blade membranous or papery, alternate, ovate or ovate-elliptic, sometimes oblong-elliptic, apex shortly acuminate, base broadly cuneate or rounded, glabrous on both surfaces, margin serrulate. Cymes fragrant, flowers white. Fruit globose, black when mature, endocarp stiff. Fl. May, fr. maturity Oct..

果枝 Fruiting branch
摄影：刘海波 Photo by: Haibo Liu

小枝和叶背 Branchlets and leaf abaxial surface
摄影：刘海波 Photo by: Haibo Liu

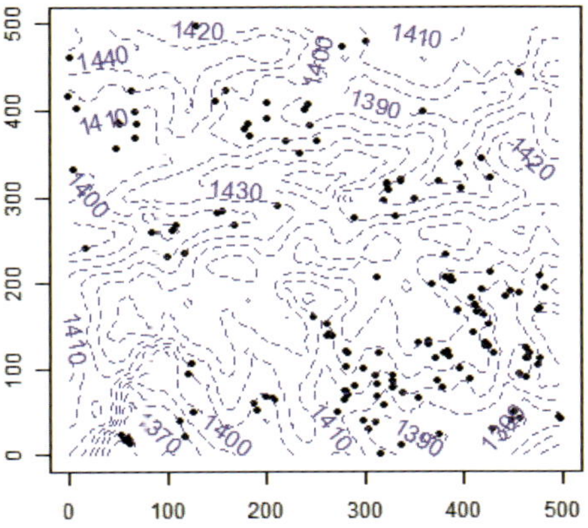

个体分布图 Distribution of individuals

径级分布表 DBH class

胸径区间 (Diameter class) (cm)	个体数 (No. of individuals in the plot)	比例 (Proportion) (%)
1～2.5	28	18.3
2.5～5	35	22.9
5～8	13	8.5
8～11	13	8.5
11～15	14	9.2
15～20	24	15.7
≥20	26	17.0

141 铁冬青

tiě dōng qīng | Ovate-leaf Ilex

Ilex rotunda，冬青属

冬青科 | Aquifoliaceae

代码（SpCode）= ILEROT

个体数（Individual number/25 hm^2）= 130

最大胸径（Max DBH）= 23.71 cm

所在林层（Stratum）= 亚乔木层 (Understory)

重要值排序（Importance value rank）= 32/54

常绿乔木。树皮淡灰色。小枝红褐色，平滑无毛。叶薄革质或纸质，椭圆形或长椭圆形，先端短渐尖，基部圆或钝，全缘。聚伞花序，着生叶腋，花白色，有短梗。果球形，初黄色，熟红色，宿存柱头盘状。花期5～6月，果熟期9～10月。

Evergreen trees. Bark light gray. Branchlets reddish brown, smooth and glabrous. Leaf blade thinly leathery or papery, elliptic or long-elliptic, apex shortly acuminate, base rounded or obtuse, margin entire. Cymes axillary, flowers white, peduncles short. Fruits globose, initially yellow, red at maturity, stigma discoid and persistent. Fl. May–Jun., fr. maturity Sept.–Oct..

树干 Trunk

摄影：卢志军 Photo by: Zhijun Lu

果枝 Fruiting branches

摄影：陈彬 Photo by: Bin Chen

小枝和叶背 Branchlet and leaf abaxial surface

摄影：陈彬 Photo by: Bin Chen

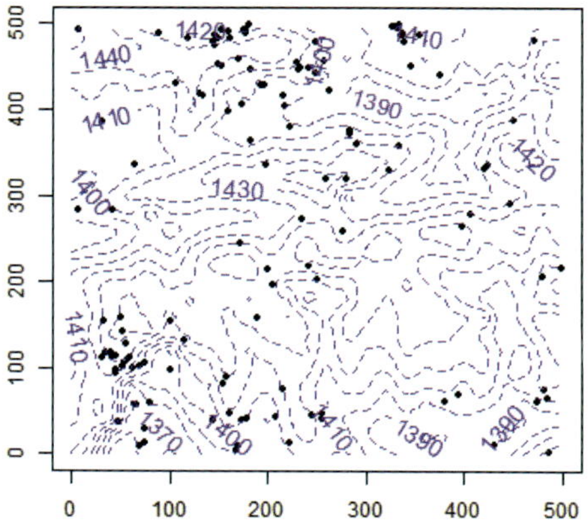

个体分布图 Distribution of individuals

径级分布表 DBH class

胸径区间 (Diameter class) (cm)	个体数 (No. of individuals in the plot)	比例 (Proportion) (%)
1～2.5	60	46.2
2.5～5	58	44.6
5～8	3	2.3
8～11	5	3.8
11～15	2	1.5
15～20	1	0.8
≥20	1	0.8

142 刺叶冬青

cì yè dōng qīng | Spiny-margin Holly

Ilex bioritsensis，冬青属
冬青科 | Aquifoliaceae

代码（SpCode）= ILEBIO
个体数（Individual number/25 hm^2）= 23
最大胸径（Max DBH）= 14.68 cm
所在林层（Stratum）= 灌木层 (Shrub)
重要值排序（Importance value rank）= 75/128

树干 Trunk
摄影：卢志军 Photo by: Zhijun Lu

常绿灌木或小乔木。小枝近圆形，灰褐色，疏被微柔毛或近无毛，平滑，皮孔不明显。叶片革质，卵形至菱形，顶端渐尖，基部圆形或截形，边缘具波状齿，有尖刺。聚伞花序，淡黄绿色，簇生于2年生枝的叶腋内。果椭圆形，成熟时红色。花期4～5月，果期8～10月。

Evergreen shrubs or small trees. Branchlets subterete, gray-brown, sparsely puberulent or glabrescent, smooth, lenticels inconspicuous. Leaf blade leathery, ovate to rhombic, apex acuminate, base rounded or truncate, margin sinuate, with strong spines. Cymes fasciculate, light yellowish green, axillary on 2-yr branchlets. Fruit ellipsoid, red at maturity. Fl. Apr.–May, fr. Aug.–Oct..

小枝和叶片 Branchlet and leaves
摄影：刘海波 Photo by: Haibo Liu

小枝和叶背 Branchlet and leaf abaxial surface
摄影：卢志军 Photo by: Zhijun Lu

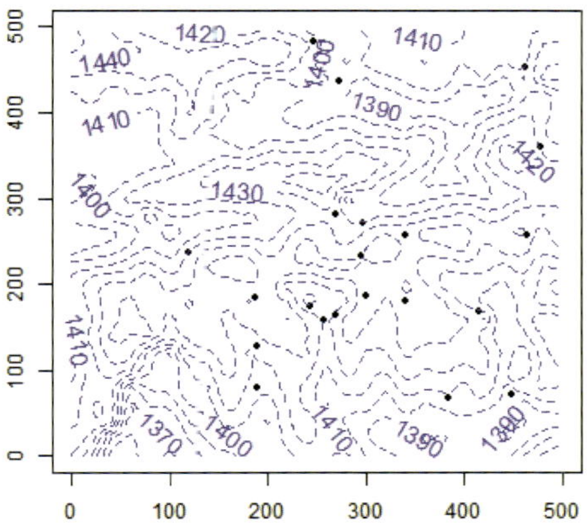

个体分布图 Distribution of individuals

径级分布表 DBH class

胸径区间 (Diameter class) (cm)	个体数 (No. of individuals in the plot)	比例 (Proportion) (%)
1～2	5	21.7
2～3	4	17.4
3～4	5	21.7
4～5	2	8.7
5～7	4	17.4
7～10	1	4.3
≥10	2	8.7

143 尾叶冬青 wěi yè dōng qīng | Wilson Ilex

Ilex wilsonii，冬青属
冬青科 | Aquifoliaceae

代码（SpCode）= ILEWIL
个体数（Individual number/25 hm²）= 5
最大胸径（Max DBH）= 4.81 cm
所在林层（Stratum）= 亚乔木层 (Understory)
重要值排序（Importance value rank）= 53/54

常绿灌木或小乔木。树皮灰白，光滑。小枝灰褐色，有棱角，无毛。叶革质或厚革质，倒卵形或长圆形，先端尾尖，顶部钝，基部楔形，全缘。聚伞花序，簇生二年生小枝的叶腋内。果球形，红色，内果皮革质。花期5～6月，果期8～10月。

Evergreen shrubs or small trees. Bark gray-white, smooth. Branchlets gray-brown, angular, glabrous. Leaf blade leathery or thickly leathery, obovate or oblong, apex caudate with obtuse tip, base cuneate, margin entire. Cymes fasciculate, axillary on 2-yr branchlets. Fruit globose, red, endocarp leathery. Fl. May–Jun., fr. Aug.–Oct..

树干 Trunk
摄影：卢志军 Photo by: Zhijun Lu

小枝和叶片 Branchlet and leaves
摄影：卢志军 Photo by: Zhijun Lu

小枝和叶背 Branchlet and leaf abaxial surface
摄影：卢志军 Photo by: Zhijun Lu

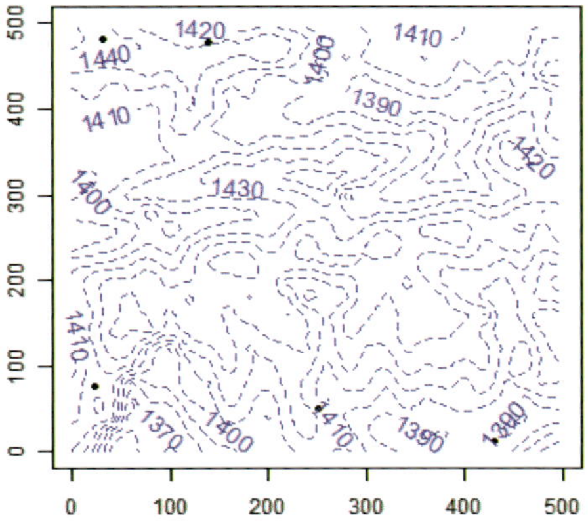

个体分布图 Distribution of individuals

径级分布表 DBH class

胸径区间 (Diameter class) (cm)	个体数 (No. of individuals in the plot)	比例 (Proportion) (%)
1～2.5	2	40.0
2.5～5	3	60.0
5～8	0	0.0
8～11	0	0.0
11～15	0	0.0
15～20	0	0.0
≥20	0	0.0

144 康定冬青 kāng dìng dōng qīng | Franchet Ilex

Ilex franchetiana，冬青属
冬青科 | Aquifoliaceae

代码（SpCode）= ILEFRA
个体数（Individual number/25 hm^2）= 2
最大胸径（Max DBH）= 7.12 cm
所在林层（Stratum）= 灌木层 (Shrub)
重要值排序（Importance value rank）= 111/128

常绿灌木或小乔木。全体无毛。小枝黑褐色，有纵棱。叶近革质或纸质，长圆状椭圆形或倒披针形。聚伞花序，芳香，花白色，花序簇生二年生小枝的叶腋。果球形，红色，内果皮木质。花期5月，果熟期9～10月。

Evergreen shrubs or small trees. Glabrous. Branchlets black-brown, longitudinally angular. Leaf blade subleathery or papery, oblong-elliptic or oblanceolate. Cymes fragrant, flowers white, fasciculate, axillary on 2-yr branchlets. Fruits globose, red, endocarp woody. Fl. May, fr. maturity Sept.–Oct..

树干 Trunk
摄影：卢志军 Photo by: Zhijun Lu

小枝和叶片 Branchlet and leaves
摄影：卢志军 Photo by: Zhijun Lu

小枝和叶背 Branchlet and leaf abaxial surface
摄影：刘海波 Photo by: Haibo Liu

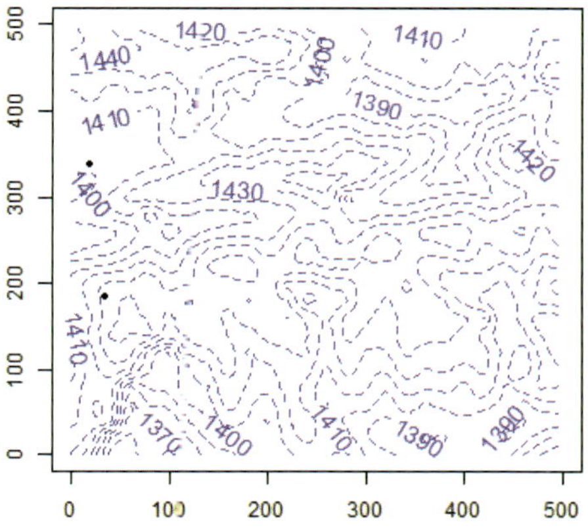

个体分布图 Distribution of individuals

径级分布表 DBH class

胸径区间 (Diameter class) (cm)	个体数 (No. of individuals in the plot)	比例 (Proportion) (%)
1～2	0	0.0
2～3	1	50.0
3～4	0	0.0
4～5	0	0.0
5～7	0	0.0
7～10	1	50.0
≥10	0	0.0

145 大果卫矛

dà guǒ wèi máo | Dense-flower Euonymus

Euonymus myrianthus，卫矛属

卫矛科 | Celastraceae

代码（SpCode）= EUOMYR

个体数（Individual number/25 hm^2）= 1300

最大胸径（Max DBH）= 24.02 cm

所在林层（Stratum）= 灌木层 (Shrub)

重要值排序（Importance value rank）= 14/128

树干 Trunk

摄影：卢志军 Photo by: Zhijun Lu

常绿灌木或小乔木。小枝黄绿色。叶对生，厚革质，倒披针形至长圆形，或倒卵形，先端渐尖或尾状尖，基部楔形，边缘有稀疏圆锯齿。聚伞圆锥花序，近顶生，花黄色。蒴果宽倒卵形或倒心形，成熟时金黄色；种子有橙黄色假种皮。花期4～5月，果熟期9～10月。

Evergreen shrubs or small trees. Branches yellowish green. Leaves opposite, thickly leathery, blades oblanceolate to oblong, or obovate, apex acuminate or caudate, base cuneate, margin sparsely crenate. Cymous panicles, subterminal, flowers yellow. Capsules broadly obovate or obcordate, golden yellow when mature. Seeds aril orange. Fl. Apr.–May, fr. maturity Sept.–Oct..

花枝 Flowering branches

摄影：江明喜 Photo by: Mingxi Jiang

果枝 Fruiting branches

摄影：刘海波 Photo by: Haibo Liu

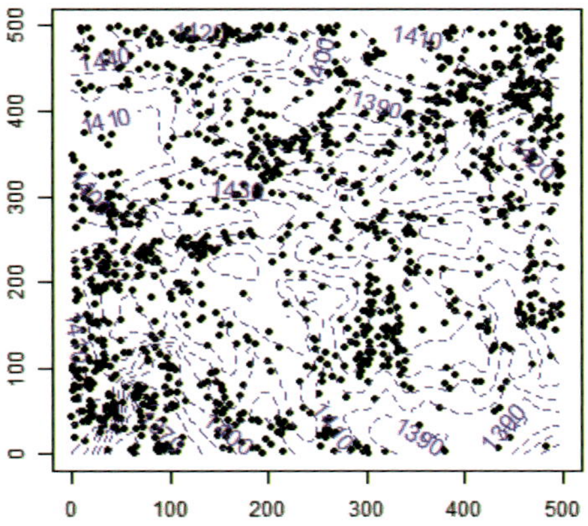

个体分布图 Distribution of individuals

径级分布表 DBH class

胸径区间 (Diameter class) (cm)	个体数 (No. of individuals in the plot)	比例 (Proportion) (%)
1～2	412	31.7
2～3	224	17.2
3～4	109	8.4
4～5	92	7.1
5～7	170	13.1
7～10	135	10.4
≥10	158	12.2

146 卫矛

wèi máo | Winged Euonymus

Euonymus alatus，卫矛属
卫矛科 I Celastraceae

代码（SpCode）= EUOALA
个体数（Individual number/25 hm^2）= 229
最大胸径（Max DBH）= 11.57 cm
所在林层（Stratum）= 灌木层 (Shrub)
重要值排序（Importance value rank）= 46/128

落叶灌木。枝开张，坚硬。叶对生，椭圆形至倒卵形，先端短渐尖，基部楔形。聚伞花序，花淡绿色。蒴果带紫色；种子紫棕色，外有橙红色假种皮。花期5～6月，果熟期9～10。

Deciduous shrubs. Branchlets spreading, stiff. Leaves opposite, blades elliptic to obovate, apex shortly acuminate, base cuneate. Cymes, flowers light green. Capsules purplish, seeds purplish brown, aril orange. Fl. May–Jun., fr. maturity Sept.–Oct..

树干 Trunk
摄影：卢志军 Photo by: Zhijun Lu

花枝 Flowering branch
摄影：徐耀粘 Photo by: Yaozhan Xu

果枝 Fruiting branch
摄影：卢志军 Photo by: Zhijun Lu

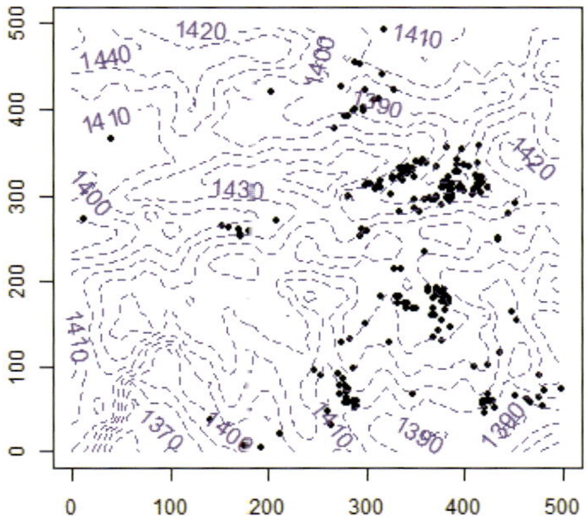

个体分布图 Distribution of individuals

径级分布表 DBH class

胸径区间 (Diameter class) (cm)	个体数 (No. of individuals in the plot)	比例 (Proportion) (%)
1～2	107	46.7
2～3	74	32.3
3～4	31	13.5
4～5	9	3.9
5～7	7	3.1
7～10	0	0.0
≥10	1	0.4

147 裂果卫矛

liè guǒ wèi máo | Diels Euonymus

Euonymus dielsianus，卫矛属

卫矛科 | Celastraceae

代码（SpCode）= EUODIE

个体数（Individual number/25 hm^2）= 19

最大胸径（Max DBH）= 25.42 cm

所在林层（Stratum）= 灌木层 (Shrub)

重要值排序（Importance value rank）= 72/128

常绿灌木或小乔木。叶对生，厚革质，狭倒卵形或狭椭圆形，先端渐尖或短渐尖，基部楔形。聚伞花序，花黄白色，花萼通常深红色，花瓣圆形。蒴果红色，假种皮橙红色。花期6月，果熟期10月。

Evergreen shrubs or small trees. Leaves opposite, thickly leathery, blades narrowly obovate or elliptic, apex acuminate or shortly acuminate, base cuneate. Cymes, flowers yellowish white, calyx usually dark red, petals rounded. Capsules red, aril orange. Fl. Jun., fr. maturity Oct..

树干 Trunk

摄影：卢志军 Photo by: Zhijun Lu

小枝和叶片 Branchlet and leaves

摄影：卢志军 Photo by: Zhijun Lu

小枝和叶背 Branchlet and leaf abaxial surface

摄影：卢志军 Photo by: Zhijun Lu

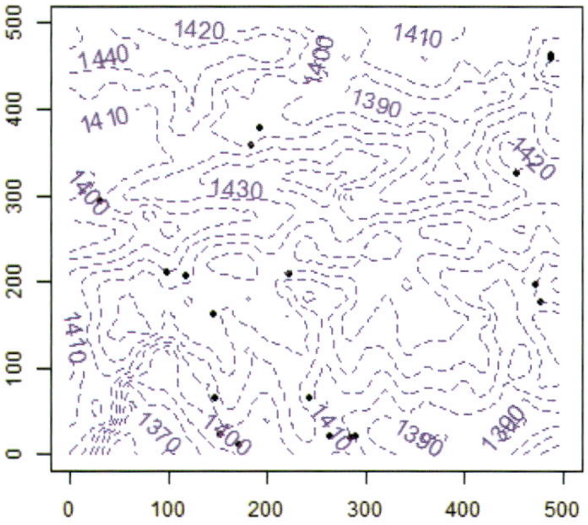

个体分布图 Distribution of individuals

径级分布表 DBH class

胸径区间 (Diameter class) (cm)	个体数 (No. of individuals in the plot)	比例 (Proportion) (%)
1～2	10	52.6
2～3	3	15.8
3～4	2	10.5
4～5	0	0.0
5～7	0	0.0
7～10	1	5.3
≥10	3	15.8

148 角翅卫矛 jiǎo chì wèi máo | Horny Euonymus

Euonymus cornutus，卫矛属
卫矛科 | Celastraceae

代码（SpCode）= EUOCOR
个体数（Individual number/25 hm^2）= 9
最大胸径（Max DBH）= 6.43 cm
所在林层（Stratum）= 灌木层 (Shrub)
重要值排序（Importance value rank）= 100/128

落叶小灌木。小枝纤细，绿色，无毛。叶对生，薄革质或厚纸质，披针形或线状披针形，顶端渐尖，基部近圆形，边缘疏具小圆齿。聚伞花序，花紫红色。蒴果紫红色；种子棕红色，有橙色假种皮。花期5～6月，果期9～10月。

Deciduous shrubs. Branches slender, green, glabrous. Leaves opposite, thinly leathery or thickly papery, lanceolate or linear-lanceolate, apex acuminate, base subrotund, margin sparsely crenulate. Cymes, flowers and capsules purplish red. Seeds brownish red, aril orange. Fl. May–Jun., fr. Sept.–Oct..

树干 Trunk
摄影：卢志军 Photo by: Zhijun Lu

果枝 Fruiting branch
摄影：卢志军 Photo by: Zhijun Lu

小枝和叶背 Branchlet and leaf abaxial surface
摄影：卢志军 Photo by: Zhijun Lu

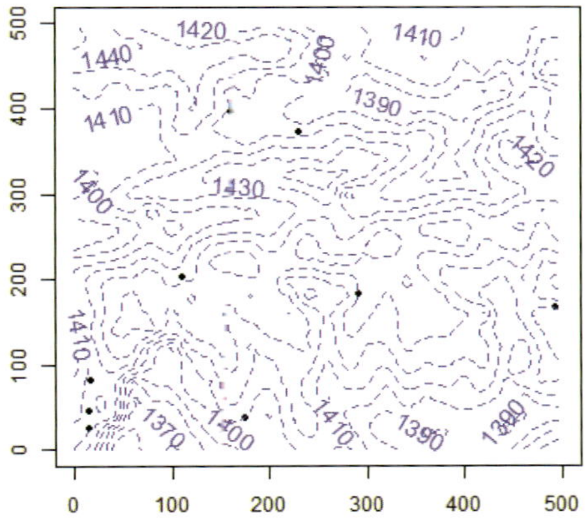

个体分布图 Distribution of individuals

径级分布表 DBH class

胸径区间 (Diameter class) (cm)	个体数 (No. of individuals in the plot)	比例 (Proportion) (%)
1～2	6	66.7
2～3	2	22.2
3～4	0	0.0
4～5	0	0.0
5～7	1	11.1
7～10	0	0.0
⩾10	0	0.0

149 西南卫矛 xī nán wèi máo | Hamilton Euonymus

Euonymus hamiltonianus，卫矛属
卫矛科 | Celastraceae

树干 Trunk
摄影：卢志军 Photo by: Zhijun Lu

代码（SpCode）= EUOHAM
个体数（Individual number/25 hm^2）= 4
最大胸径（Max DBH）= 6.42 cm
所在林层（Stratum）= 灌木层 (Shrub)
重要值排序（Importance value rank）= 104/128

落叶乔木。叶对生，长圆状椭圆形、长圆状卵形或长圆状披针形，先端急尖或短渐尖，基部宽楔形。聚伞花序；花白绿色；花药紫色。蒴果粉红带黄，倒三角形；种子红棕色，有橙红色假种皮。花期5～6月，果期8～10月。

Deciduous trees. Leaves opposite, blades oblong-elliptic, oblong-ovate or oblong-lanceolate, apex acute or shortly acuminate, base broadly cuneate. Cymes, flowers white-green, anther purple. Capsules yellowish pink, inverted triangle. Seeds reddish brown, aril orange-red. Fl. Jun., fr. Aug.–Oct..

花枝 Flowering branches
摄影：卢志军 Photo by: Zhijun Lu

果枝 Fruiting branch
摄影：刘海波 Photo by: Haibo Liu

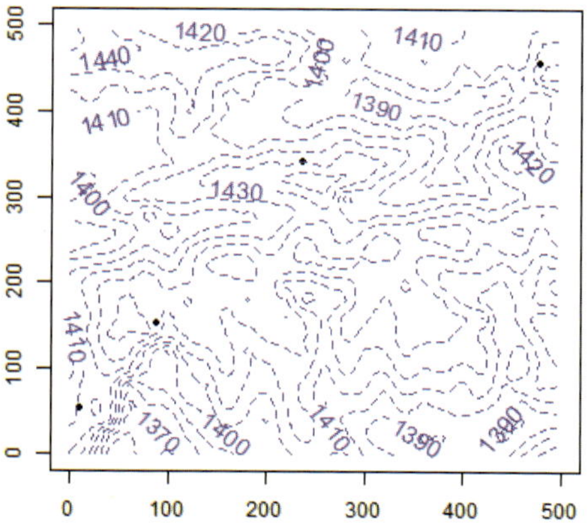

个体分布图 Distribution of individuals

径级分布表 DBH class

胸径区间 (Diameter class) (cm)	个体数 (No. of individuals in the plot)	比例 (Proportion) (%)
1～2	2	50.0
2～3	1	25.0
3～4	0	0.0
4～5	0	0.0
5～7	1	25.0
7～10	0	0.0
≥10	0	0.0

150 野鸦椿

yě yā chūn | Common Euscaphis

Euscaphis japonica，野鸦椿属
省沽油科 | Staphyleaceae

代码（SpCode）= EUSJAP
个体数（Individual number/25 hm^2）= 409
最大胸径（Max DBH）= 46.72 cm
所在林层（Stratum）= 灌木层 (Shrub)
重要值排序（Importance value rank）= 25/128

落叶灌木或小乔木。树皮灰色。小枝红紫色，枝叶揉碎后有恶臭气味。奇数羽状复叶，小叶5～9，叶对生；叶片椭圆形、长圆状卵形；顶端渐尖，基部宽楔形至圆形，叶缘疏具带腺体细锯齿。圆锥花序顶生，花密生，黄白色。蓇葖果紫红色，有纵纹。花期5～6月，果期8～9月。

Deciduous shrubs or small trees. Bark grayish brown. Branchlets reddish purple. Crushed branchlets and leaves foul odored. Leaves imparipinnate, leaflets 5-9, opposite, blade elliptic to oblong-ovate, apex acuminate, base broadly cuneate to rounded. margin sparsely serrulate with glandular teeth. Panicles terminal, flowers dense and yellowish white. Follicle purplish red, longitudinal grained. Fl. May–Jun., fr. Aug.–Sept..

树干 Trunk
摄影：卢志军 Photo by: Zhijun Lu

花枝 Flowering branches
摄影：卢志军 Photo by: Zhijun Lu

果枝 Fruiting branches
摄影：徐耀粘 Photo by: Yaozhan Xu

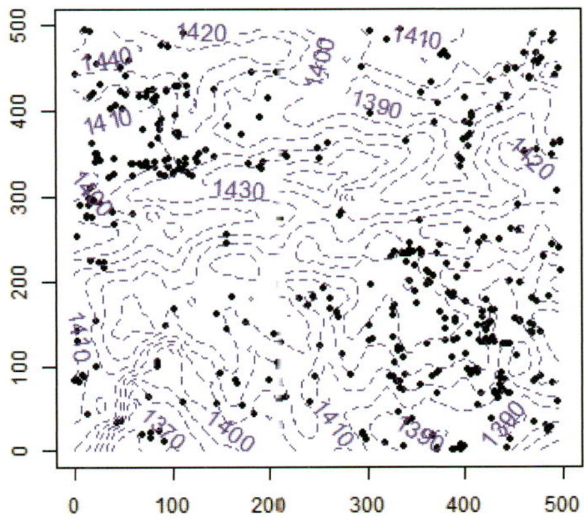

个体分布图 Distribution of individuals

径级分布表 DBH class

胸径区间 (Diameter class) (cm)	个体数 (No. of individuals in the plot)	比例 (Proportion) (%)
1～2	54	13.2
2～3	32	7.8
3～4	27	6.6
4～5	30	7.3
5～7	63	15.4
7～10	91	22.2
≥10	112	27.4

151 瘿椒树 yīng jiāo shù | Chinese Falsepistache

Tapiscia sinensis，野鸦椿属
省沽油科 | Staphyleaceae

代码（SpCode）= TAPSIN
个体数（Individual number/25 hm^2）= 17
最大胸径（Max DBH）= 85.45 cm
所在林层（Stratum）= 乔木层 (Canopy)
重要值排序（Importance value rank）= 25/50

树干 Trunk
摄影：卢志军 Photo by: Zhijun Lu

落叶乔木。树皮灰色至灰黑色。小枝无毛。奇数羽状复叶，小叶5～9，互生；叶片狭卵形至卵形，顶端渐尖，基部心形或近心形，边缘具齿。圆锥花序腋生，花小，有香气，黄色。果近球形。花期6～7月，果期8～9月。

Deciduous trees. Bark gray to grayish black. Branchlets glabrous. Leaves imparipinnate, leaflets 5-9, alternate, blades narrowly ovate to ovate, apex acuminate, base cordate or subcordate, margin serrate. Panicles axillary, flowers small, fragrant, yellow. Fruit subglobose. Fl. Jun.–Jul., fr. Aug.–Sept..

小枝和叶背 Branchlets and leaf abaxial surface
摄影：江明喜 Photo by: Mingxi Jiang

花枝 Flowering branch
摄影：刘海波 Photo by: Haibo Liu

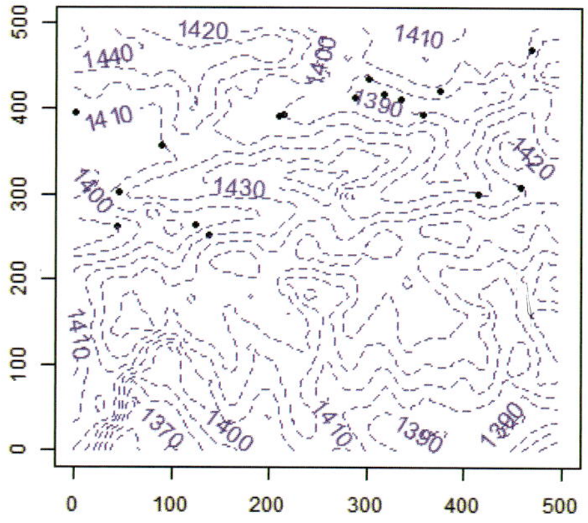

个体分布图 Distribution of individuals

径级分布表 DBH class

胸径区间 (Diameter class) (cm)	个体数 (No. of individuals in the plot)	比例 (Proportion) (%)
1～2.5	3	17.6
2.5～5	2	11.8
5～10	1	5.9
10～25	1	5.9
25～40	3	17.6
40～100	7	41.2
⩾100	0	0.0

152 枳椇

zhǐ jǔ | Raisin Tree

Hovenia acerba，枳椇属
鼠李科 | Rhamnaceae

代码（SpCode）= HOVACE
个体数（Individual number/25 hm^2）= 64
最大胸径（Max DBH）= 77.57 cm
所在林层（Stratum）= 乔木层 (Canopy)
重要值排序（Importance value rank）= 18/50

树干 Trunk
摄影：卢志军 Photo by: Zhijun Lu

落叶乔木。幼枝干后紫褐色，无毛。叶互生，薄纸质，卵形或卵圆形，先端渐尖，基部浅心形或圆形，边缘具细齿。聚伞花序顶生及腋生，花淡黄绿色。果近球形，果梗肥厚，肉质。花期6～7月，果期8～9月。

Deciduous trees. Branchlets purplish brown after dried, glabrous. Leaves alternate, thinly papery, blades ovate or elliptic-ovate, apex acuminate, base shallowly cordate or rounded, margin finely serrulate. Cymes terminal and axillary, flowers yellowish green. Fruits subglobose, peduncles dilated, fleshy. Fl. Jun.–Jul., fr. Aug.–Sept..

花枝 Flowering branches
摄影：江明喜 Photo by: Mingxi Jiang

小枝和叶背 Branchlet and leaf abaxial surface
摄影：刘海波 Photo by: Haibo Liu

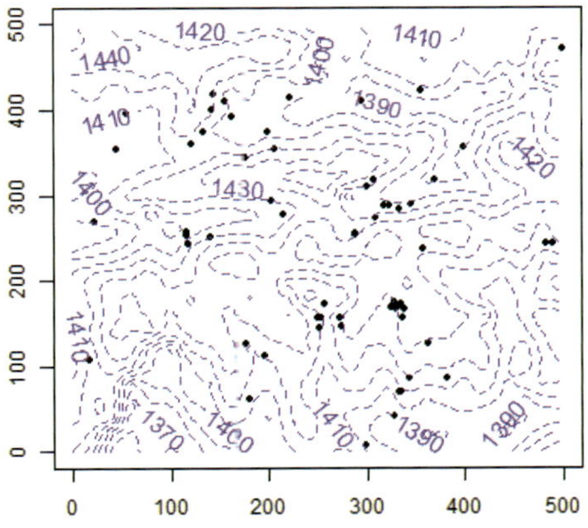

个体分布图 Distribution of individuals

径级分布表 DBH class

胸径区间 (Diameter class) (cm)	个体数 (No. of individuals in the plot)	比例 (Proportion) (%)
1～2.5	13	20.3
2.5～5	7	10.9
5～10	5	7.8
10～25	19	29.7
25～40	11	17.2
40～100	9	14.1
≥100	0	0.0

153 薄叶鼠李

báo yè shǔ lǐ | Thin-leaf Buck-thorn

Rhamnus leptophylla，鼠李属

鼠李科 | Rhamnaceae

代码（SpCode）= RHALEP

个体数（Individual number/25 hm^2）= 143

最大胸径（Max DBH）= 13.4 cm

所在林层（Stratum）= 灌木层 (Shrub)

重要值排序（Importance value rank）= 51/128

落叶灌木。幼枝灰褐色，无毛或有微柔毛，对生或近对生，顶端成针刺状。叶常对生，或互生，薄纸质，倒卵形、椭圆形或长椭圆形；顶端短尖或渐尖，基部楔形，叶缘具圆齿或锯齿。花单性，绿色，聚伞花序。核果球形，成熟后黑色；种子宽倒卵形，背面有纵沟。花期5月，果期8～9月。

Deciduous shrubs. Branchlets grayish brown, glabrous or pubescent, opposite or subopposite, apex spined. Leaves usually opposite, or alternate, thinly papery, blades obovate, elliptic or long-elliptic, apex shortly cuspidate to acuminate, base cuneate, margin crenate to serrate. Flowers unisexual, green, cymes. Drupes globose, black when mature. Seeds broadly obovate, back furrowed. Fl. May, fr. Aug.–Sept..

树干 Trunk

摄影：卢志军 Photo by: Zhijun Lu

果枝 Fruiting branches

摄影：刘海波 Photo by: Haibo Liu

小枝和叶背 Branchlet and leaf abaxial surface

摄影：鲍大川 Photo by: Dachuan Bao

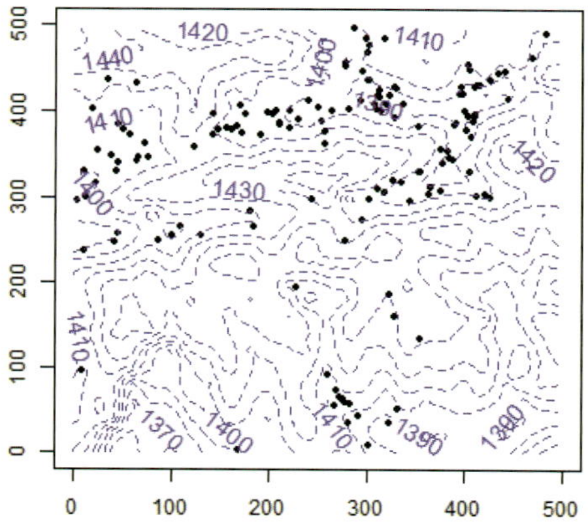

个体分布图 Distribution of individuals

径级分布表 DBH class

胸径区间 (Diameter class) (cm)	个体数 (No. of individuals in the plot)	比例 (Proportion) (%)
1～2	49	34.3
2～3	32	22.4
3～4	26	18.2
4～5	15	10.5
5～7	13	9.1
7～10	7	4.9
≥10	1	0.7

154 长叶冻绿

cháng yè dòng lǜ | Oriental Buck-thorn

Rhamnus crenata，鼠李属
鼠李科 | Rhamnaceae

代码（SpCode）= RHACRE
个体数（Individual number/25 hm^2）= 2
最大胸径（Max DBH）= 6.70 cm
所在林层（Stratum）= 灌木层 (Shrub)
重要值排序（Importance value rank）= 109/128

落叶灌木或小乔木。幼枝红褐色，有锈色短柔毛。叶互生，纸质，长椭圆状披针形或椭圆状倒卵形；顶端渐尖至尾状渐尖，或短急尖，基部楔形或钝，叶缘具细小圆齿。聚伞花序腋生，总花梗短，花淡黄白色至淡黄绿色。核果近球形，成熟后黑色；种子倒卵形。花期5～7月，果期7～10月。

Deciduous shrubs or small trees. Young branchlets reddish brown, ferruginous tomentose. Leaves alternate, papery, blade long-elliptic-lanceolate or elliptic-obovate, apex acuminate to caudate-acuminate, or shortly acute, base cuneate or obtuse, margin finely crenate. Cymes axillary, peduncle short, flowers light yellowish white to yellowish green. Drupes subglobose, black when mature. Seed obovate. Fl. May–Jul., fr. Jul.–Oct..

树干 Trunk
摄影：卢志军 Photo by: Zhijun Lu

花枝 Flowering branch
摄影：江明喜 Photo by: Mingxi Jiang

果枝 Fruiting branch
摄影：卢志军 Photo by: Zhijun Lu

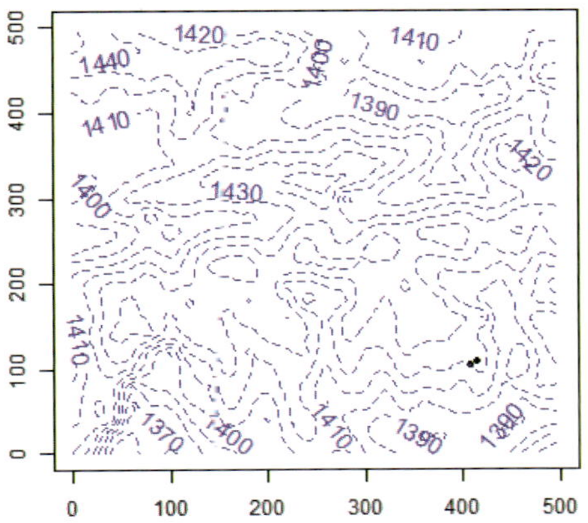

个体分布图 Distribution of individuals

径级分布表 DBH class

胸径区间 (Diameter class) (cm)	个体数 (No. of individuals in the plot)	比例 (Proportion) (%)
1～2	0	0.0
2～3	0	0.0
3～4	0	0.0
4～5	1	50.0
5～7	1	50.0
7～10	0	0.0
≥10	0	0.0

155 山鼠李

shān shǔ lǐ | Wilson Buck-thorn

Rhamnus wilsonii，鼠李属
鼠李科 | Rhamnaceae

代码（SpCode）= RHAWIL
个体数（Individual number/25 hm^2）= 3
最大胸径（Max DBH）= 2.47 cm
所在林层（Stratum）= 灌木层 (Shrub)
重要值排序（Importance value rank）= 112/128

落叶灌木。小枝互生或兼近对生，银灰色或灰褐色，无光泽，枝端有时具钝针刺。叶互生，少数近对生或在当年生枝基部或短枝顶端簇生；叶片椭圆形或宽椭圆形，稀倒卵状披针形或倒卵状椭圆形，顶端渐尖或长渐尖，基部楔形，叶缘具钩状圆齿。花单性，雌雄异株，黄绿色。核果倒卵状球形，成熟时紫黑色或黑色。花期4～5月，果期6～10月。

Deciduous shrubs. Branchlets alternate to subopposite, silvery-gray or gray-brown, dull, apex sometimes obtusely spined. Leaves alternate to rarely subopposite, or fascicled at base of annual branches or on short shoots; blades elliptic or broadly elliptic, rarely obouate-lanceolate or obovate-elliptic, apex acuminate or long acuminate, base cuneate, margin hooked-crenate. Flowers unisexual, dioecious, yellowish green. Drupes obovate-globose, purplish black or black when mature. Fl. Apr.–May, fr. Jun.–Oct..

树干 Trunk
摄影：卢志军 Photo by: Zhijun Lu

小枝和叶片 Branchlet and leaves
摄影：刘海波 Photo by: Haibo Liu

小枝和叶背 Branchlet and leaf abaxial surface
摄影：刘海波 Photo by: Haibo Liu

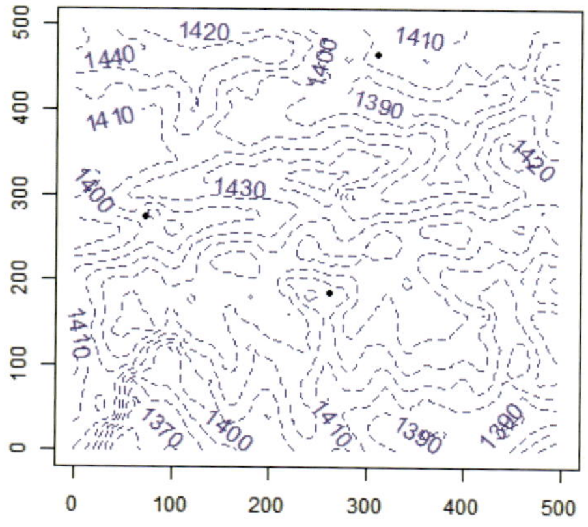

个体分布图 Distribution of individuals

径级分布表 DBH class

胸径区间 (Diameter class) (cm)	个体数 (No. of individuals in the plot)	比例 (Proportion) (%)
1～2	2	66.7
2～3	1	33.3
3～4	0	0.0
4～5	0	0.0
5～7	0	0.0
7～10	0	0.0
≥10	0	0.0

156 鄂椴

è duàn | Oliver Linden

Tilia oliveri，椴树属
椴树科 I Tiliaceae

代码（SpCode）= TILOLI
个体数（Individual number/25 hm^2）= 45
最大胸径（Max DBH）= 55.65 cm
所在林层（Stratum）= 乔木层 (Canopy)
重要值排序（Importance value rank）= 20/50

落叶乔木。树皮浅黑色。小枝无毛。叶宽卵形或卵圆形，先端突尖或渐尖，基部偏斜，截形或心形，叶缘具齿。聚伞花序，花黄色，萼片椭圆状披针形。果椭圆状球形。花期6～8月，果期8～9月。

Deciduous trees. Bark light black. Branchlets glabrous. Leaf blades broadly ovate or elliptic-ovate, apex abruptly acute or acuminate, base obliquely truncate or cordate, margin serrate. Cymes, flowers yellow, sepals elliptic-lanceolate. Fruit ellipsoid-globose. Fl. Jun.–Aug., fr. Aug.–Sept..

树干 Trunk
摄影：卢志军 Photo by: Zhijun Lu

小枝和叶片 Branchlet and leaves
摄影：卢志军 Photo by: Zhijun Lu

小枝和叶背 Branchlet and leaf abaxial surface
摄影：卢志军 Photo by: Zhijun Lu

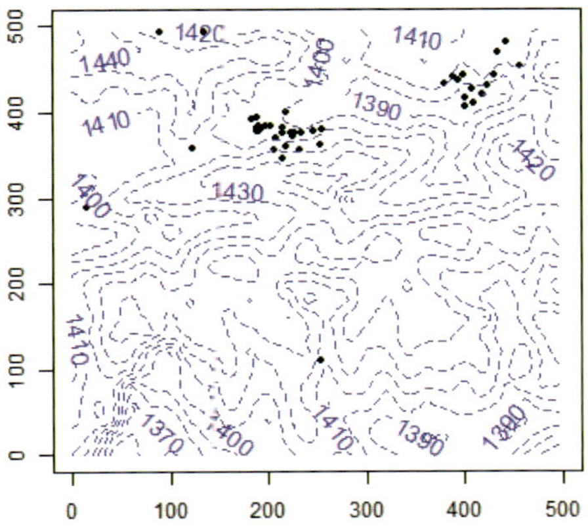

个体分布图 Distribution of individuals

径级分布表 DBH class

胸径区间 (Diameter class) (cm)	个体数 (No. of individuals in the plot)	比例 (Proportion) (%)
1～2.5	11	24.4
2.5～5	11	24.4
5～10	5	11.1
10～25	14	31.1
25～40	1	2.2
40～100	3	6.7
≥100	0	0.0

157 椴树 duàn shù | Tuan Linden

Tilia tuan，椴树属
椴树科 | Tiliaceae

代码（SpCode）= TILTUA
个体数（Individual number/25 hm^2）= 13
最大胸径（Max DBH）= 18.41 cm
所在林层（Stratum）= 乔木层 (Canopy)
重要值排序（Importance value rank）= 26/50

落叶乔木。树皮灰褐色，皮孔凸起，有裂纹。叶斜宽卵形至斜卵形，先端突尖或渐尖，基部极偏斜，截形或心形。聚伞花序，苞片黄绿色，花黄色，萼片卵圆状椭圆形，花瓣卵圆形。果略为球形。花期7月，果熟期9～10月。

树干 Trunk
摄影：卢志军 Photo by: Zhijun Lu

Deciduous trees. Bark gray-brown, lenticels convex, striped. Leaf blade obliquely broad-ovate to oblique-ovate, apex abruptly acute or acuminate, base extremely oblique, truncate or cordate. Cymes, bracts yellowish green, flowers yellow, sepals ovate-elliptic, petals ovate. Fruits slightly globose. Fl. Jul., fr. maturity Sept.–Oct..

花枝 Flowering branches
摄影：江明喜 Photo by: Mingxi Jiang

果枝 Fruiting branch
摄影：江明喜 Photo by: Mingxi Jiang

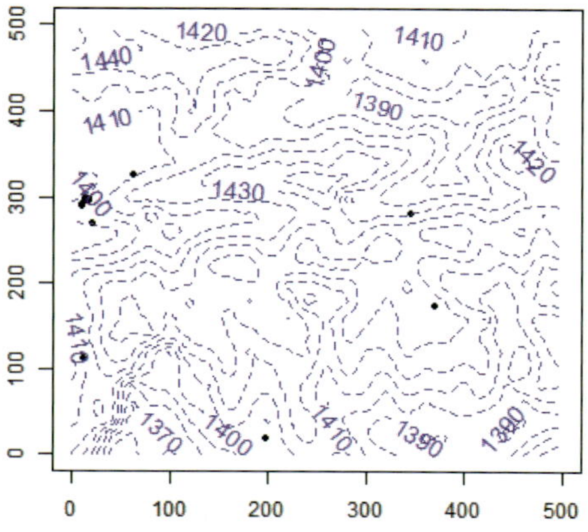

个体分布图 Distribution of individuals

径级分布表 DBH class

胸径区间 (Diameter class) (cm)	个体数 (No. of individuals in the plot)	比例 (Proportion) (%)
1～2.5	6	46.2
2.5～5	3	23.1
5～10	3	23.1
10～25	1	7.7
25～40	0	0.0
40～100	0	0.0
≥100	0	0.0

158 南京椴

nán jīng duàn | Miquel Linden

Tilia miqueliana，椴树属
椴树科 | Tiliaceae

代码（SpCode）= TILMIQ
个体数（Individual number/25 hm²）= 4
最大胸径（Max DBH）= 16.49 cm
所在林层（Stratum）= 乔木层 (Canopy)
重要值排序（Importance value rank）= 36/50

落叶乔木。树皮灰白色；嫩枝有黄褐色茸毛。叶卵圆形，先端急短尖，基部心形，少数稍偏斜，叶缘具齿。聚伞花序。萼片被灰色毛。果期球形，无棱，有小突起。花期7月。

树干 Trunk
摄影：卢志军 Photo by: Zhijun Lu

Deciduous trees. Bark gray-white. Branchlets yellow-brown tomentose. Leaf blade ovate-orbicular, apex acute, base cordate, rarely oblique, margin serrate. Cymes. Sepals gray-white tomentose. Fruit globose, not angled, verrucose. Fl. Jul..

小枝和叶片 Branchlet and leaves
摄影：卢志军 Photo by: Zhijun Lu

小枝和叶背 Branchlet and leaf abaxial surface
摄影：卢志军 Photo by: Zhijun Lu

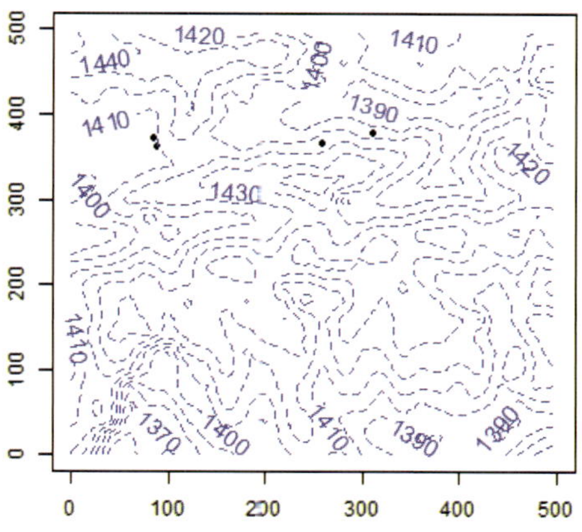

个体分布图 Distribution of individuals

径级分布表 DBH class

胸径区间 (Diameter class) (cm)	个体数 (No. of individuals in the plot)	比例 (Proportion) (%)
1～2.5	0	0.0
2.5～5	2	50.0
5～10	0	0.0
10～25	2	50.0
25～40	0	0.0
40～100	0	0.0
≥100	0	0.0

159 少脉椴

shǎo mài duàn | Few-nerve Linden

Tilia paucicostata，椴树属
椴树科 | Tiliaceae

代码（SpCode）= TILPAU
个体数（Individual number/25 hm^2）= 4
最大胸径（Max DBH）= 18.15 cm
所在林层（Stratum）= 乔木层 (Canopy)
重要值排序（Importance value rank）= 35/50

树干 Trunk
摄影：卢志军 Photo by: Zhijun Lu

落叶大乔木。小枝纤细无毛。叶卵形，薄革质，先端渐尖，基部偏斜，截形或近心形。聚伞花序，通常比桃红色苞片短，花小，带黄色。果球形或梨形，基部尖。花期7月，果熟期9～10月。

Deciduous trees. Branchlets slender, glabrous. Leaf blade ovate, thinly leathery, apex acuminate, base obliquely truncate or subcordate. Cymes, usually shorter than pink bracts, flowers small, yellowish. Fruits globose or pyriform, base peaked. Fl. Jul., fr. maturity Sept.–Oct..

小枝和叶片 Branchlets and leaves
摄影：刘海波 Photo by: Haibo Liu

小枝和叶背 Branchlets and leaf abaxial surface
摄影：刘海波 Photo by: Haibo Liu

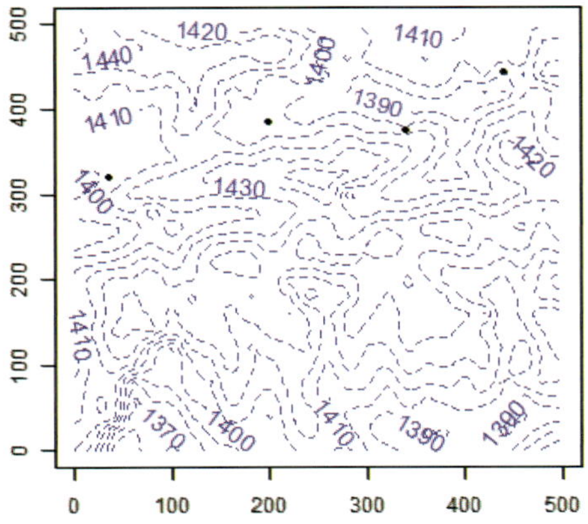

个体分布图 Distribution of individuals

径级分布表 DBH class

胸径区间 (Diameter class) (cm)	个体数 (No. of individuals in the plot)	比例 (Proportion) (%)
1～2.5	1	25.0
2.5～5	1	25.0
5～10	0	0.0
10～25	2	50.0
25～40	0	0.0
40～100	0	0.0
⩾100	0	0.0

160 胡颓子

hú tuí zǐ | Thorny Elaeagnus

Elaeagnus pungens，胡颓子属
胡颓子科 | Elaeagnaceae

代码（SpCode）= ELAPUN
个体数（Individual number/25 hm^2）= 77
最大胸径（Max DBH）= 10.07 cm
所在林层（Stratum）= 灌木层 (Shrub)
重要值排序（Importance value rank）= 66/128

常绿灌木。小枝有刺针，被褐色鳞片。叶革质，椭圆形，先端钝，基部圆形，上面无毛光滑，下面密生淡棕黄色鳞片。花腋生，白色或浅白色，萼片卵圆形。果长椭圆形，被褐色鳞片。花期9～10月，果期11月。

Evergreen shrubs. Branchlets spined, brown scaly. Leaf blade leathery, elliptic, apex obtuse, base rounded, adaxially glabrous and glossy, abaxially with dense light-brown scales. Flowers axillary, white or light white, sepals ovate. Fruits oblong, brown scaly. Fl. Sept.–Oct., fr. Nov..

树干 Trunk
摄影：卢志军 Photo by: Zhijun Lu

花枝 Flowering branch
摄影：徐文斌 Photo by: Wenbin Xu

小枝和叶背 Branchlet and leaf abaxial surface
摄影：徐文斌 Photo by: Wenbin Xu

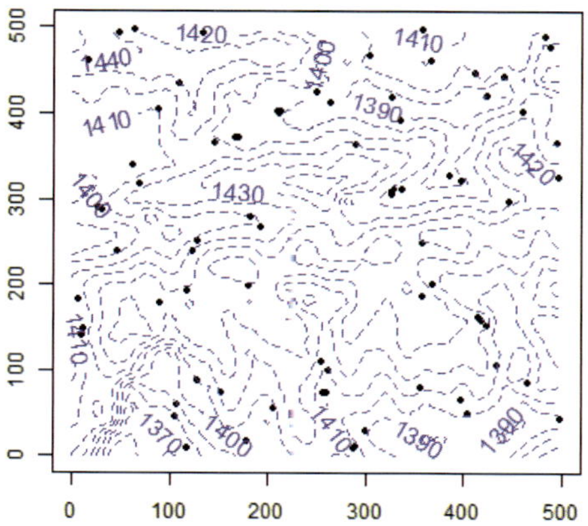

个体分布图 Distribution of individuals

径级分布表 DBH class

胸径区间 (Diameter class) (cm)	个体数 (No. of individuals in the plot)	比例 (Proportion) (%)
1～2	59	76.6
2～3	10	13.0
3～4	3	3.9
4～5	3	3.9
5～7	1	1.3
7～10	0	0.0
≥10	1	1.3

161 长叶胡颓子

cháng yè hú tuí zǐ | Bock Elaeagnus

Elaeagnus bockii，胡颓子属

胡颓子科 | Elaeagnaceae

代码（SpCode）= ELABOC

个体数（Individual number/25 hm^2）= 5

最大胸径（Max DBH）= 7.82 cm

所在林层（Stratum）= 灌木层 (Shrub)

重要值排序（Importance value rank）= 102/128

常绿直立灌木，具长棘刺。小枝密被鳞片。叶纸质或近革质，狭长圆形，两端钝尖或渐尖，边缘略反卷，上面幼时被褐色鳞片，下面灰白色，具鳞片。花银白色；花萼筒钟形。果实长圆形，被银白色鳞片，成熟时红色；果梗下垂。花期10～11月，果期翌年3～4月。

Evergreen erect shrubs, with long spines. Branchlets with dense scales. Leaves papery or subleathery, blades narrowly oblong, apex and base obtuse or acuminate, margin slightly revolute, adaxially brown scaled when young, abaxially gray-white with scales. Flowers silver, calyx tube funnelform. Fruits oblong, silver scaled, red when mature, peduncles pendulous. Fl. Oct.–Nov., fr. May–Apr. of following yr..

树干 Trunk

摄影：卢志军 Photo by: Zhijun Lu

花枝 Flowering branches

摄影：徐文斌 Photo by: Wenbin Xu

小枝和叶背 Branchlet and leaf abaxial surface

摄影：徐文斌 Photo by: Wenbin Xu

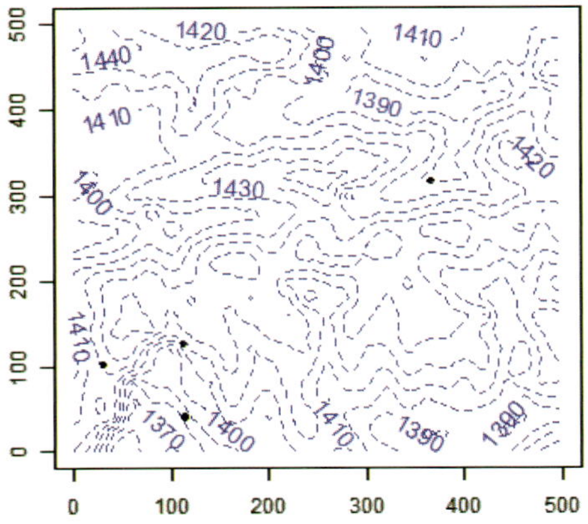

个体分布图 Distribution of individuals

径级分布表 DBH class

胸径区间 (Diameter class) (cm)	个体数 (No. of individuals in the plot)	比例 (Proportion) (%)
1～2	2	40.0
2～3	2	40.0
3～4	0	0.0
4～5	0	0.0
5～7	0	0.0
7～10	1	20.0
≥10	0	0.0

162 宜昌胡颓子

yí chāng hú tuí zǐ | Henry Elaeagnus

Elaeagnus henryi，胡颓子属
胡颓子科 | Elaeagnaceae

代码（SpCode）= ELAHEN
个体数（Individual number/25 hm^2）= 5
最大胸径（Max DBH）= 1.52 cm
所在林层（Stratum）= 灌木层 (Shrub)
重要值排序（Importance value rank）= 108/128

常绿蔓生灌木。枝条黑灰色，常有刺；幼枝被棕色鳞片。叶纸质，宽椭圆形，先端渐尖状急尖，基部钝，上面无毛，下面被鳞片，幼时银色，老时带绿色。花腋生，单生或少花，密被银色鳞片。果被鳞片，长圆形。花期10～11月，果熟期翌年4月。

Shrubs, evergreen, climbing. Branches black-gray, usually spined; young branches with brown scales. Leaf blade papery, broadly elliptic, apex acuminately acute, base obtuse, adaxially glabrous, abaxially scaled, silver when young and greenish when old. Flowers axillary, solitary or a few, densely silver scaled. Drupes with scales, oblong. Fl. Oct.–Nov., fr. Apr. of following yr..

树干 Trunk
摄影：卢志军 Photo by: Zhijun Lu

果枝 Fruiting branch
摄影：徐文斌 Photo by: Wenbin Xu

叶背 Leaf abaxial surface
摄影：徐文斌 Photo by: Wenbin Xu

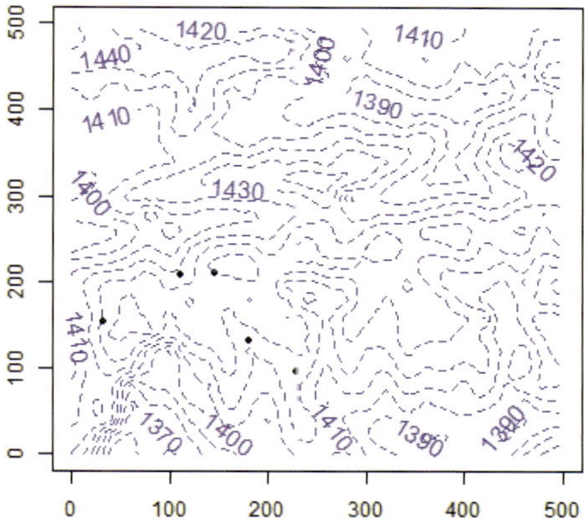

个体分布图 Distribution of individuals

径级分布表 DBH class

胸径区间 (Diameter class) (cm)	个体数 (No. of individuals in the plot)	比例 (Proportion) (%)
1～2	5	100.0
2～3	0	0.0
3～4	0	0.0
4～5	0	0.0
5～7	0	0.0
7～10	0	0.0
≥10	0	0.0

163 披针叶胡颓子

pī zhēn yè hú tuí zǐ | Lanceolate Elaeagnus

Elaeagnus lanceolata，胡颓子属
胡颓子科 | Elaeagnaceae

代码（SpCode）= ELALAN
个体数（Individual number/25 hm^2）= 3
最大胸径（Max DBH）= 2.12 cm
所在林层（Stratum）= 灌木层 (Shrub)
重要值排序（Importance value rank）= 114/128

常绿灌木。枝黑色，小枝被鳞片。叶狭披针形，先端渐尖，基部钝圆，上面无毛，下面有金黄色至银白色鳞片。花序近伞形，腋生，密被银色鳞片，萼片宽三角形，先端钝，里面稍被毛。花期9月，果期翌年5月。

Evergreen shrubs. Branches black; branchlets scaled. Leaf blade narrowly lanceolate, apex acuminate, base obtuse, adaxially glabrous, abaxially with golden to silver scales. Subumbels, flowers axillary, densely silver scaled, sepals broadly triangular, apex obtuse, inside slightly tomentose. Fl. Sept., fr. May of following yr..

树干 Trunk
摄影：卢志军 Photo by: Zhijun Lu

果枝 Fruiting branch
摄影：徐文斌 Photo by: Wenbin Xu

小枝和叶背 Branchlet and leaf abaxial surface
摄影：徐文斌 Photo by: Wenbin Xu

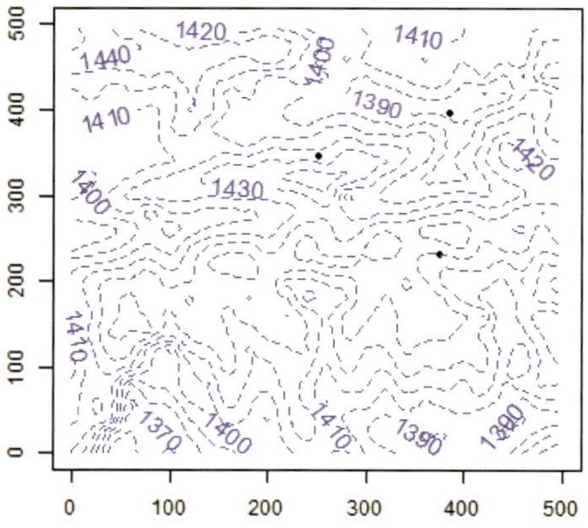

个体分布图 Distribution of individuals

径级分布表 DBH class

胸径区间 (Diameter class) (cm)	个体数 (No. of individuals in the plot)	比例 (Proportion) (%)
1～2	2	66.7
2～3	1	33.3
3～4	0	0.0
4～5	0	0.0
5～7	0	0.0
7～10	0	0.0
⩾10	0	0.0

164 牛奶子

niú nǎi zǐ | Autumn Elaeagnus

Elaeagnus umbellata，胡颓子属
胡颓子科 | Elaeagnaceae

代码（SpCode）= ELAUMB
个体数（Individual number/25 hm^2）= 2
最大胸径（Max DBH）= 2.68 cm
所在林层（Stratum）= 灌木层 (Shrub)
重要值排序（Importance value rank）= 118/128

落叶灌木。小枝初被白色鳞片，后变为灰色。叶纸质，长椭圆状披针形，有时为狭长椭圆形或宽倒披针形，先端稍钝，基部急尖或渐尖，上面初被白色鳞片，下面密被白色鳞片。花腋生，被白色鳞片；花萼筒细，基部渐细，萼片卵形。果稍球形或宽椭圆形。花期4～5月，果期7～8月。

Deciduous shrubs. Branchlets with white scales initially, then gray. Leaves papery, blade long elliptic-lanceolate, rarely narrowly long-elliptic or broadly oblanceolate, apex slightly obtuse, base acute or acuminate, adaxially initially white scaled, abaxially densely white scaled. Flowers axillary, silvery scaled, calyx tube thin, base tapering, sepals ovate. Drupes globose or broadly ellipoid. Fl. Apr.–May, fr. Jul.–Aug..

树干 Trunk
摄影：卢志军 Photo by: Zhijun Lu

小枝和叶片 Branchlets and leaves
摄影：徐文秀 Photo by: Wenxiu Xu

小枝和叶背 Branchlets and leaf abaxial surface
摄影：徐文秀 Photo by: Wenxiu Xu

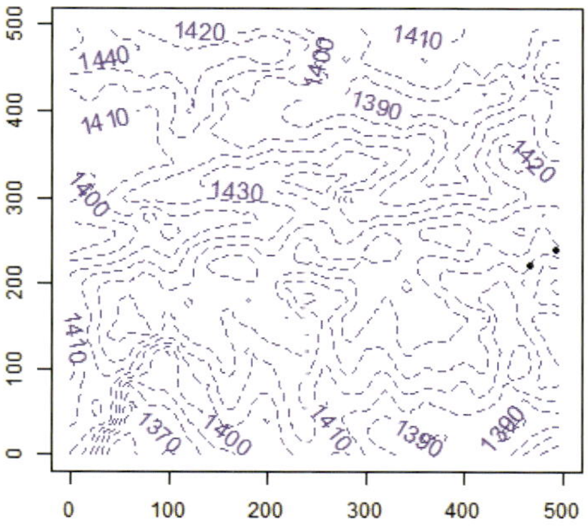

个体分布图 Distribution of individuals

径级分布表 DBH class

胸径区间 (Diameter class) (cm)	个体数 (No. of individuals in the plot)	比例 (Proportion) (%)
1～2	1	50.0
2～3	1	50.0
3～4	0	0.0
4～5	0	0.0
5～7	0	0.0
7～10	0	0.0
≥10	0	0.0

165 巴东胡颓子 bā dōng hú tuí zǐ | Patung Elaeagnus

Elaeagnus difficilis，胡颓子属
胡颓子科 | Elaeagnaceae

代码（SpCode）= ELADIF
个体数（Individual number/25 hm^2）= 1
最大胸径（Max DBH）= 1.23 cm
所在林层（Stratum）= 灌木层 (Shrub)
重要值排序（Importance value rank）= 127/128

常绿蔓生或攀缘灌木，无刺。枝条铁锈色；幼枝被褐色鳞片。叶纸质，披针形，顶端微渐尖，基部圆形或楔形，边缘稍反卷。花短；子房近钟形，红铜色。果实橙红色，圆柱形。花期4月，果期5～7月。

Shrubs, evergreen, spreading or scandent, spines absent. Branches rust brown; young branchlets with brown scales. Leaves papery, blade lanceolate, apex slightly acuminate, base rounded or cuneate, margin narrowly revolute. Flowers short, ovary subfunnelform, purple bronze. Drupes orange-red, cylindric. Fl. Apr., fr. May–Jul..

树干 Trunk
摄影：卢志军 Photo by: Zhijun Lu

小枝和叶片 Branchlet and leaves
摄影：江明喜 Photo by: Mingxi Jiang

小枝和叶背 Branchlet and leaf abaxial surface
摄影：卢志军 Photo by: Zhijun Lu

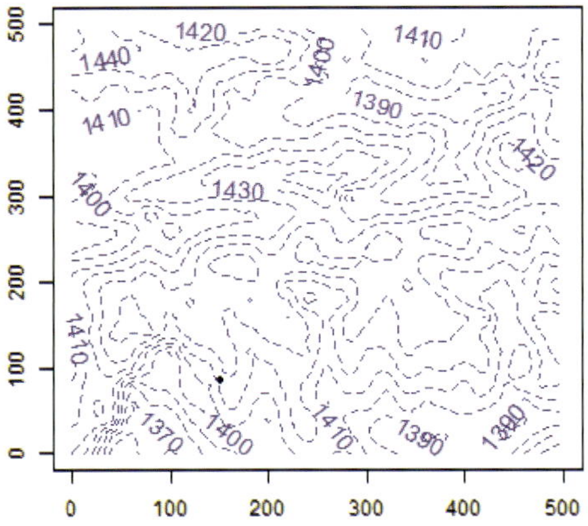

个体分布图 Distribution of individuals

径级分布表 DBH class

胸径区间 (Diameter class) (cm)	个体数 (No. of individuals in the plot)	比例 (Proportion) (%)
1～2	1	100.0
2～3	0	0.0
3～4	0	0.0
4～5	0	0.0
5～7	0	0.0
7～10	0	0.0
≥10	0	0.0

166 山桐子 shān tóng zǐ | Many-fruit Idesia

Idesia polycarpa，山桐子属
大风子科 | Flacourtiaceae

代码（SpCode）= IDEPOL
个体数（Individual number/25 hm^2）= 69
最大胸径（Max DBH）= 29.04 cm
所在林层（Stratum）= 亚乔木层 Shrub
重要值排序（Importance value rank）= 30/54

树干 Trunk
摄影：卢志军 Photo by: Zhijun Lu

落叶乔木。树皮灰白色。叶卵形至卵状心形，先端急尖至短渐尖，基部心形，叶缘具齿。圆锥花序顶生，下垂；花杂性或雌雄异株；无花瓣；萼片黄绿色；子房球形。浆果球形，红色；种子多数。花期5～6月，果期7～8月。

Deciduous trees. Bark grayish white. Leaf blade ovate to ovate-cordate, apex acute to shortly acuminate, base cordate, margin serrate. Panicles terminal, pendulous, miscellaneous or unisexual, petals absent, sepals yellowish green, ovary globose. Berry globose, red. Seeds numerous. Fl. May–Jun., fr. Jul.–Aug..

花枝 Flowering branches
摄影：刘海波 Photo by: Haibo Liu

果枝 Fruiting branches
摄影：鲍大川 Photo by: Dachuan Bao

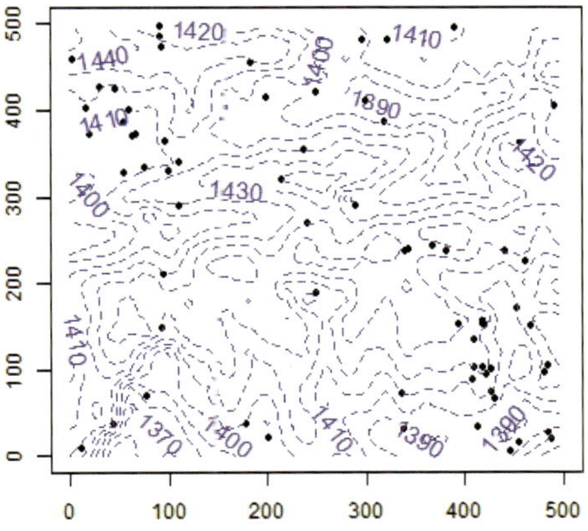

个体分布图 Distribution of individuals

径级分布表 DBH class

胸径区间 (Diameter class) (cm)	个体数 (No. of individuals in the plot)	比例 (Proportion) (%)
1～2.5	6	8.7
2.5～5	13	18.8
5～8	17	24.6
8～11	13	18.8
11～15	8	11.6
15～20	4	5.8
≥20	8	11.6

167 西域旌节花

xī yù jīng jiē huā | Himalayan Stachyurus

Stachyurus himalaicus，旌节花属
旌节花科 I Stachyuraceae

代码（SpCode）= STAHIM
个体数（Individual number/25 hm^2）= 237
最大胸径（Max DBH）= 54.21 cm
所在林层（Stratum）= 灌木层 (Shrub)
重要值排序（Importance value rank）= 34/128

落叶灌木或小乔木。树皮光滑，棕色或深棕色，小枝褐色，具浅色皮孔。叶片坚纸质至薄革质，披针形至长圆状披针形，先端渐尖至长渐尖，基部钝圆，叶缘具细密锯齿。穗状花序腋生，花黄色。果实近球形。花期3～4月，果期5～8月。

Deciduous shrubs or small trees. Bark smooth, brown or dark brown; branchlets brown, lenticels light-colored. Leaves hard papery to thinly leathery, blade lanceolate to oblong-lanceolate, apex acuminate to long acuminate, base obtusely rounded, margin densely serrulate. Spica axillary, flowers yellow. Fruits sub globose. Fl. Mar.–Apr., fr. May–Aug..

树干 Trunk
摄影：卢志军 Photo by: Zhijun Lu

花枝 Flowering branches
摄影：卢志军 Photo by: Zhijun Lu

果枝 Fruiting branches
摄影：刘海波 Photo by: Haibo Liu

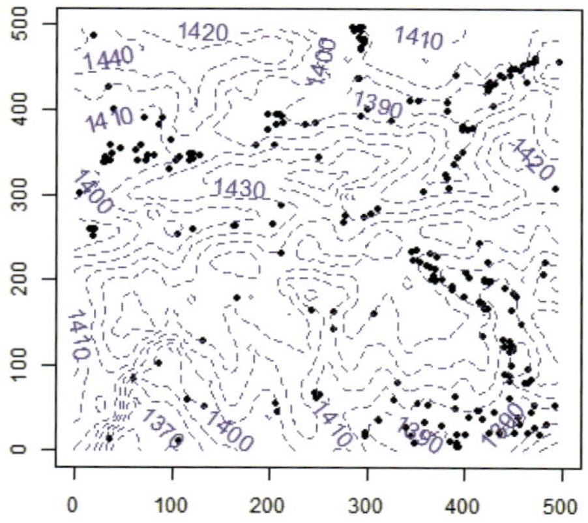

个体分布图 Distribution of individuals

径级分布表 DBH class

胸径区间 (Diameter class) (cm)	个体数 (No. of individuals in the plot)	比例 (Proportion) (%)
1～2	31	13.1
2～3	35	14.8
3～4	28	11.8
4～5	27	11.4
5～7	45	19.0
7～10	39	16.5
≥10	32	13.5

168 毛八角枫

máo bā jiǎo fēng | Kurz Alangium

Alangium kurzii，八角枫属
八角枫科 | Alangiaceae

代码（SpCode）= ALAKUR
个体数（Individual number/25 hm^2）= 55
最大胸径（Max DBH）= 48.36 cm
所在林层（Stratum）= 亚乔木层 (Understory)
重要值排序（Importance value rank）= 27/54

树干 Trunk
摄影：卢志军 Photo by: Zhijun Lu

落叶小乔木，稀灌木。树皮深褐色，平滑。小枝平滑，疏被茸毛。叶互生，纸质，近圆形或阔卵形，顶端长渐尖，基部心形或近心形，稀近圆形，倾斜，两侧不对称，全缘。聚伞花序，花萼漏斗状。核果椭圆形，幼时紫褐色，成熟后黑色。花期5～6月，果期9月。

Deciduous small trees, rarely shrubs. Bark dark brown, smooth. Twigs smooth, sparsely pubescent. Leaves alternate, papery, subrotund or broadly ovate, apex long acuminate, base cordate or subcordate, rarely subrounded, oblique, asymmetric, margin entire. Cymes, calyx funnelform. Drupes ellipoid, purplish brown when young, black at maturity. Fl. May–Jun., fr. Sept..

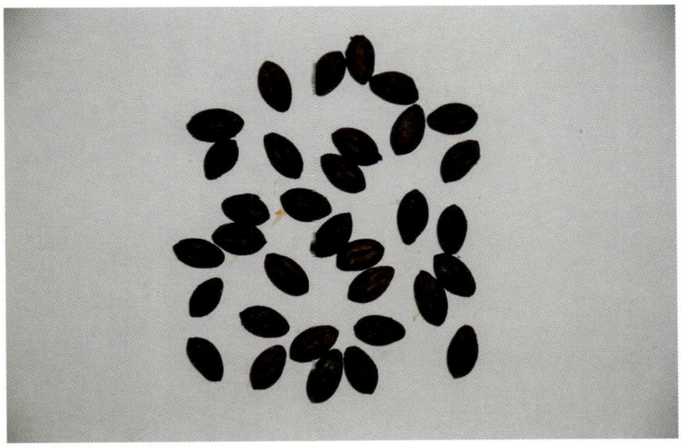

种子 Seeds
摄影：卢志军 Photo by: Zhijun Lu

花枝 Flowering branch
摄影：刘海波 Photo by: Haibo Liu

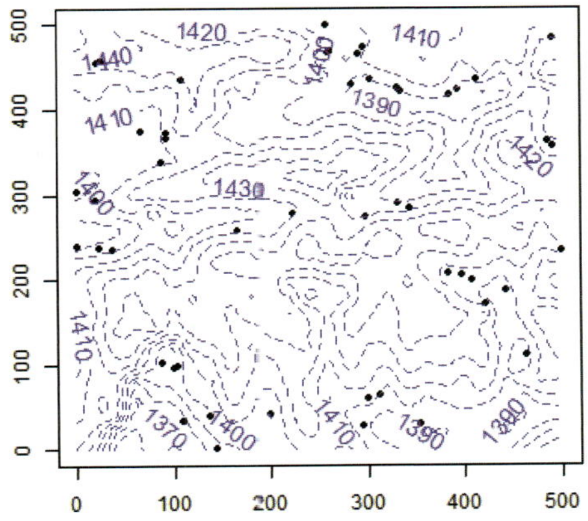

个体分布图 Distribution of individuals

径级分布表 DBH class

胸径区间 (Diameter class) (cm)	个体数 (No. of individuals in the plot)	比例 (Proportion) (%)
1～2.5	12	21.8
2.5～5	8	14.5
5～8	5	9.1
8～11	3	5.5
11～15	3	5.5
15～20	1	1.8
≥20	23	41.8

169 八角枫

bā jiǎo fēng | Chinese Alangium

Alangium chinense，八角枫属
八角枫科 | Alangiaceae

代码（SpCode）= ALACHI
个体数（Individual number/25 hm^2）= 14
最大胸径（Max DBH）= 35.20 cm
所在林层（Stratum）= 亚乔木层 (Understory)
重要值排序（Importance value rank）= 39/54

落叶灌木或小乔木。树皮淡灰色。小枝有黄色疏柔毛。叶互生，卵形或圆形，先端渐尖，基部心形，两侧偏斜，全缘或2～3裂。聚伞花序，腋生，花瓣白色。核果卵圆形，黑色。花期5～7月，果期6～9月。

树干 Trunk
摄影：卢志军 Photo by: Zhijun Lu

Deciduous shrubs or small trees. Bark light gray. Branchlets pubescent, yellow and sparse. Leaves alternate, blade ovate or orbicular, apex acuminate, base cordate, oblique, margin entire or 2-3 lobed. Cymes axillary, petals white. Drupes ovoid, black. Fl. May–Jul., fr. Jun.–Sept..

小枝和叶片 Branchlet and leaves
摄影：卢志军 Photo by: Zhijun Lu

花枝和叶背 Flowering branch and leaf abaxial surface
摄影：刘海波 Photo by: Haibo Liu

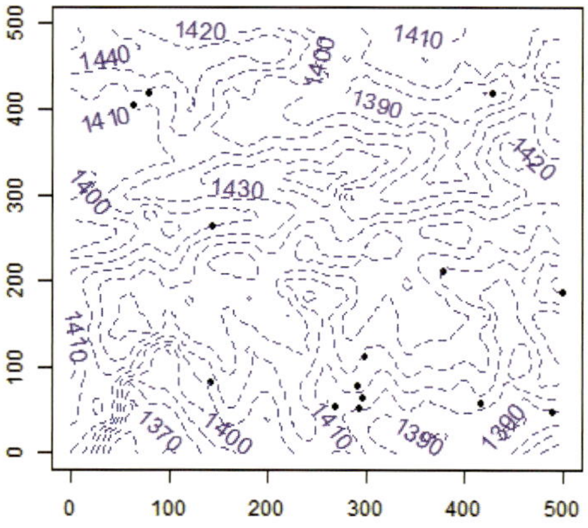

个体分布图 Distribution of individuals

径级分布表 DBH class

胸径区间 (Diameter class) (cm)	个体数 (No. of individuals in the plot)	比例 (Proportion) (%)
1～2.5	3	21.4
2.5～5	2	14.3
5～8	1	7.1
8～11	3	21.4
11～15	2	14.3
15～20	1	7.1
⩾20	2	14.3

170 珙桐 gǒng tóng | Dovetree

Davidia involucrata，珙桐属
蓝果树科 | Nyssaceae

代码（SpCode）= DAVINV
个体数（Individual number/25 hm^2）= 10
最大胸径（Max DBH）= 38.19 cm
所在林层（Stratum）= 乔木层 (Canopy)
重要值排序（Importance value rank）= 29/50

树干 Trunk
摄影：卢志军 Photo by: Zhijun Lu

落叶乔木。树皮深灰褐色，呈不规则薄片脱落。叶互生，宽卵形，先端渐尖，基部心形。花杂性，由多数雄花和1朵两性花组成顶生头状花序；苞片2，长圆形或卵形。核果长卵形，紫绿色，有黄色斑点。花期5～6月，果期6～9月。

Deciduous trees. Bark dark grayish brown, irregularly peeling in thin sheets. Leaves alternate, blade broadly ovate, apex acuminate, base cordate. Flowers miscellaneous; head, terminal, with numerous male flowers and one bisexual flower; bracts 2, oblong or ovate. Drupes long ovoid, purplish green, yellow spotted. Fr. May–Jun., fr. Jun.–Sept..

花枝 Flowering branch
摄影：江明喜 Photo by: Mingxi Jiang

小枝和叶片 Branchlet and leaves
摄影：卢志军 Photo by: Zhijun Lu

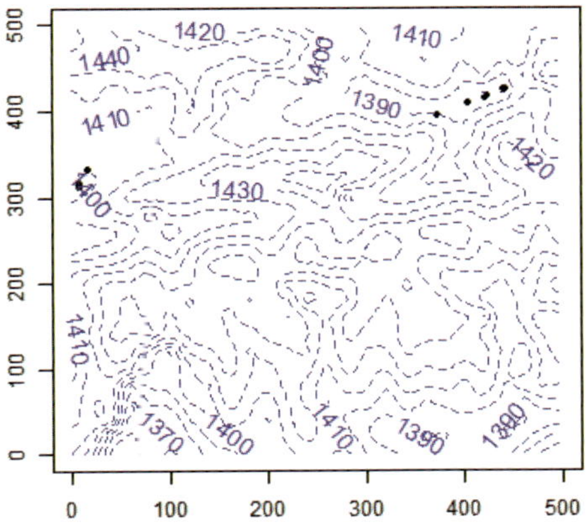

个体分布图 Distribution of individuals

径级分布表 DBH class

胸径区间 (Diameter class) (cm)	个体数 (No. of individuals in the plot)	比例 (Proportion) (%)
1～2.5	1	10.0
2.5～5	1	10.0
5～10	5	50.0
10～25	2	20.0
25～40	1	10.0
40～100	0	0.0
≥100	0	0.0

171 蓝果树

lán guǒ shù | Chinese Tupelo

Nyssa sinensis，蓝果树属
蓝果树科 | Nyssaceae

代码（SpCode）= NYSSIN
个体数（Individual number/25 hm^2）= 804
最大胸径（Max DBH）= 48.79 cm
所在林层（Stratum）= 乔木层 (Canopy)
重要值排序（Importance value rank）= 10/50

落叶乔木。树皮淡褐色或深灰色，粗糙，常裂成薄片状脱落。小枝无毛，皮孔显著。叶纸质或薄革质，互生，椭圆形或长椭圆形，稀卵形或近披针形，顶端急尖，基部近圆形，叶缘略具波状齿。花序伞形或短总状。核果蓝色，长圆形或倒卵形。花期4月下旬，果期9月。

树干 Trunk
摄影：卢志军 Photo by: Zhijun Lu

Deciduous trees. Bark dark light brown or dark gray, coarse, usually fissured and peeling in thin sheets. Branchlets glabrous and lenticellate. Leaves papery or thinly leathery, alternate, blade elliptic or long-elliptic, rarely ovate or sublanceolate, apex acute, base subrounded, margin slightly sinuate. Umbels or short racemes. Drupe bluish, oblong or obovoid. Fl. the last ten-day of Apr., fr. Sept..

花枝 Flowering branch
摄影：江明喜 Photo by: Mingxi Jiang

小枝和叶片 Branchlets and leaves
摄影：卢志军 Photo by: Zhijun Lu

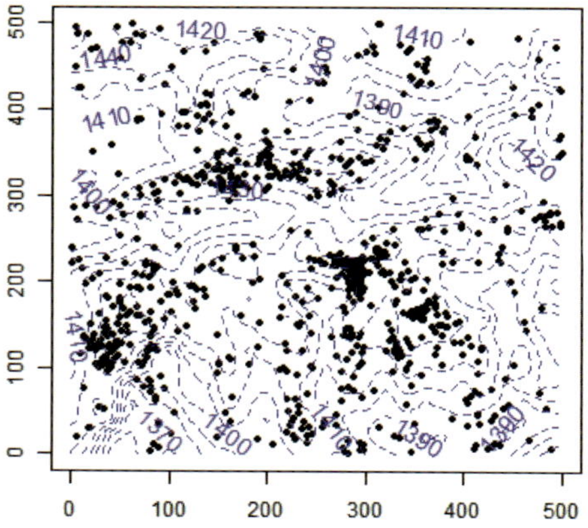

个体分布图 Distribution of individuals

径级分布表 DBH class

胸径区间 (Diameter class) (cm)	个体数 (No. of individuals in the plot)	比例 (Proportion) (%)
1～2.5	36	4.5
2.5～5	74	9.2
5～10	106	13.2
10～25	400	49.8
25～40	165	20.5
40～100	23	2.9
≥100	0	0.0

172 桃叶珊瑚

táo yè shān hú | Chinese Aucuba

Aucuba chinensis，桃叶珊瑚属
山茱萸科 | Cornaceae

代码（SpCode）= AUCCHI
个体数（Individual number/25 hm^2）= 29
最大胸径（Max DBH）= 3.03 cm
所在林层（Stratum）= 灌木层 (Shrub)
重要值排序（Importance value rank）= 80/128

常绿灌木。老枝具白色皮孔。叶对生，革质或厚革质，长圆状披针形至倒卵状长圆形，先端渐尖或尾状渐尖，基部钝圆，边缘具疏锯齿。雄花序为大型圆锥花序，雌花序为短圆锥花序，花黄绿色或带黄色。浆果长圆形，成熟时深红色。花期5～6月，果期9～10月。

Evergreen shrubs. Old branches lenticels white. Leaves alternate, leathery or thickly leathery, blade oblong-lanceolate to obovate-oblong, apex acuminate or caudately acuminate, base obtusely rounded, margin sparsely serrate. Staminate inflorescences largely paniculate; carpellate inflorescence shortly paniculate, flowers yellow green or yellowish. Berry oblong, dark red when mature. Fl. May–Jun., fr. Sept.–Oct..

树干 Trunk
摄影：卢志军 Photo by: Zhijun Lu

果枝和叶片 Fruiting branch and leaves
摄影：卢志军 Photo by: Zhijun Lu

小枝和叶背 Branchlet and leaf abaxial surface
摄影：卢志军 Photo by: Zhijun Lu

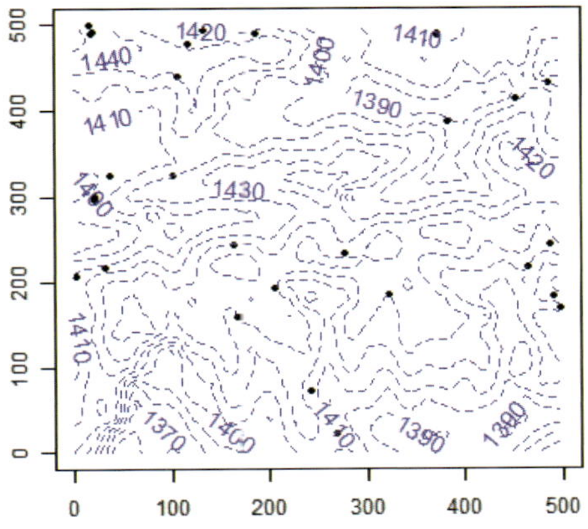

个体分布图 Distribution of individuals

径级分布表 DBH class

胸径区间 (Diameter class) (cm)	个体数 (No. of individuals in the plot)	比例 (Proportion) (%)
1～2	22	75.9
2～3	6	20.7
3～4	1	3.4
4～5	0	0.0
5～7	0	0.0
7～10	0	0.0
≥10	0	0.0

173 倒披针叶珊瑚 dǎo pī zhēn yè shān hú | Oblanceolate Himalayan Aucuba

Aucuba himalaica var. *oblanceolata*，桃叶珊瑚属
山茱萸科 | Cornaceae

代码（SpCode）= AUCOBL
个体数（Individual number/25 hm^2）= 7
最大胸径（Max DBH）= 6.78 cm
所在林层（Stratum）= 灌木层 (Shrub)
重要值排序（Importance value rank）= 101/128

常绿叶片较厚，常为倒披针形，叶下面密被短柔毛，顶端具尖，基部楔形，叶缘具小齿。雌花序长。雄花序为总状圆锥花序，花红色；雌花序紫红色。花期4～5月，果期9～10月。

Evergreen leaf blade thick, usually oblanceolate, abaxially densely pubescent with short, soft trichomes, apex cuspidate, base cuneate, margin sparsely denticulate. Carpellate inflorescence long. Staminate inflorescences racemose panicles, flowers red; female inflorescences reddish reddish purple. Fl. Apr.–May, fr. Sept.–Oct..

树干 Trunk
摄影：卢志军 Photo by: Zhijun Lu

叶片 Leaves
摄影：卢志军 Photo by: Zhijun Lu

叶背 Leaf abaxial surface
摄影：卢志军 Photo by: Zhijun Lu

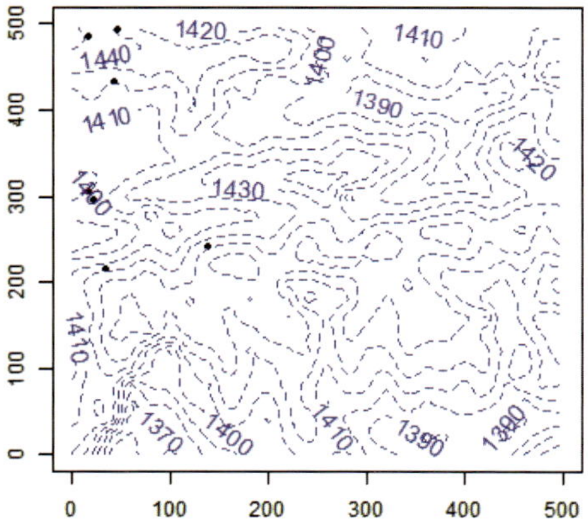

个体分布图 Distribution of individuals

径级分布表 DBH class

胸径区间 (Diameter class) (cm)	个体数 (No. of individuals in the plot)	比例 (Proportion) (%)
1～2	5	71.4
2～3	0	0.0
3～4	0	0.0
4～5	1	14.3
5～7	1	14.3
7～10	0	0.0
≥10	0	0.0

174 喜马拉雅珊瑚 xǐ mǎ lā yǎ shān hú | Himalayan Aucuba

Aucuba himalaica，桃叶珊瑚属
山茱萸科 | Cornaceae

树干 Trunk
摄影：卢志军 Photo by: Zhijun Lu

代码（SpCode）= AUCHIM
个体数（Individual number/25 hm^2）= 1
最大胸径（Max DBH）= 2.05 cm
所在林层（Stratum）= 灌木层 (Shrub)
重要值排序（Importance value rank）= 125/128

常绿灌木或小乔木。树皮灰绿色，幼枝疏生伏毛，二年生枝条无毛。叶对生，厚纸质，卵状椭圆形或卵状披针形，顶端渐尖，基部楔形，边缘反卷，有锯齿。雌雄异株；雄花序为总状圆锥花序，花红色；雌花序紫红色。核果浆果状，红色，长卵圆形。花期4～5月，果期9～10月。

Evergreen shrubs or small trees. Bark grayish green. Young branches pubescent; 2-yr branches glabrous. Leaves alternate, thickly papery, blade ovate-elliptic or ovate-lanceolate, apex acuminate, base cuneate, margin revolute and serrate. Dioecious; staminate inflorescences racemose panicles, flowers red; carpellate inflorescences reddish purple. Berry-like drupe red, ovoid-oblong. Fl. Apr–May, fr. Sept.–Oct..

花枝 Flowering branches
摄影：徐文斌 Photo by: Wenbin Xu

果枝 Fruiting branch
摄影：徐文斌 Photo by: Wenbin Xu

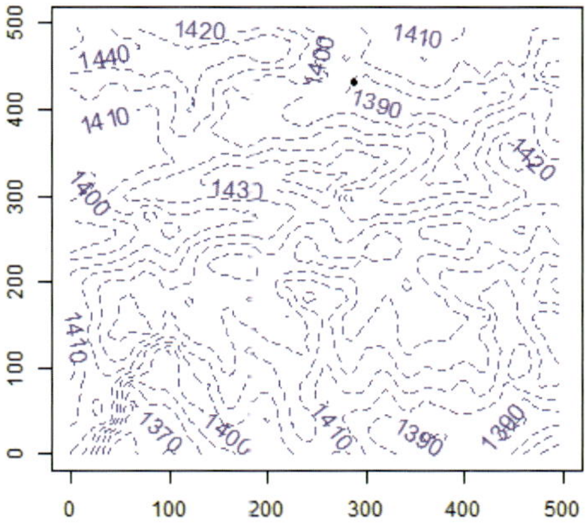

个体分布图 Distribution of individuals

径级分布表 DBH class

胸径区间 (Diameter class) (cm)	个体数 (No. of individuals in the plot)	比例 (Proportion) (%)
1～2	0	0.0
2～3	1	100.0
3～4	0	0.0
4～5	0	0.0
5～7	0	0.0
7～10	0	0.0
≥10	0	0.0

175 四照花

sì zhào huā | Kousa Dogwood

Cornus kousa subsp. *chinensis*，山茱萸属
山茱萸科 | Cornaceae

代码（SpCode）= CORKOU
个体数（Individual number/25 hm^2）= 4436
最大胸径（Max DBH）= 32.59 cm
所在林层（Stratum）= 亚乔木层 (Understory)
重要值排序（Importance value rank）= 3/54

落叶小乔木。枝条灰褐色，无毛；嫩枝被白色柔毛。叶对生，纸质，卵形或卵状椭圆形，先端渐尖，基部宽楔形或圆形，常稍偏斜。头状花序，苞片白色，椭圆形至长圆状卵形。花期5～6月，果期8月。

Deciduous trees. Branches grayish brown, glabrous; white pubescent when young. Leaves opposite, papery, ovate or ovate-elliptic, apex acuminate, base broadly cuneate or rounded, usually oblique. Head, bracts white, elliptic to oblong-ovate. Fl. May–Jun., fr. Aug..

树干 Trunk
摄影：卢志军 Photo by: Zhijun Lu

花 Flowers
摄影：刘海波 Photo by: Haibo Liu

果枝 Fruiting branches
摄影：卢志军 Photo by: Zhijun Lu

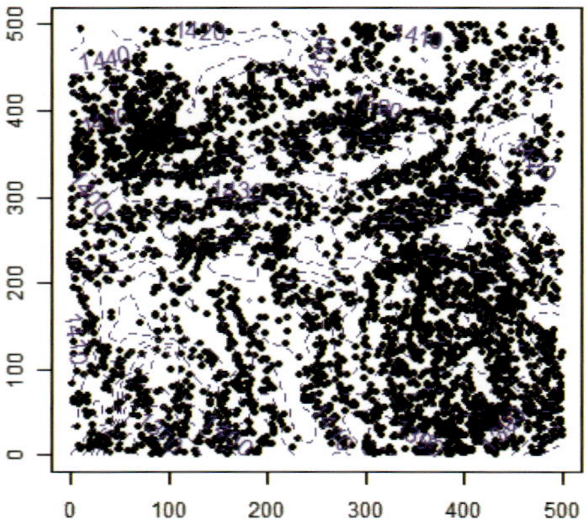

个体分布图 Distribution of individuals

径级分布表 DBH class

胸径区间 (Diameter class) (cm)	个体数 (No. of individuals in the plot)	比例 (Proportion) (%)
1～2.5	450	10.1
2.5～5	1098	24.8
5～8	927	20.9
8～11	770	17.4
11～15	694	15.6
15～20	385	8.7
≥20	112	2.5

176 尖叶四照花 jiān yè sì zhào huā | Elliptic-leaf Dogwood

Cornus elliptica，山茱萸属
山茱萸科 | Cornaceae

代码（SpCode）= CORELL
个体数（Individual number/25 hm^2）= 1135
最大胸径（Max DBH）= 33.08 cm
所在林层（Stratum）= 灌木层 (Shrub)
重要值排序（Importance value rank）= 13/128

常绿灌木或小乔木。树皮灰或灰棕色。嫩枝被白色粗毛。叶对生，革质，长椭圆形或椭圆状卵形，先端渐尖，基部楔形或宽楔形。头状花序生于小枝顶端，近球形；总苞4片，卵形，白色。果实紫红色。花期5～6月，果期9月。

Evergreen shrubs or small trees. Bark gray or grayish brown. Young branches pubescent with white coarse trichomes. Leaves opposite, leathery, long-elliptic or elliptic-ovate, apex acuminate, base cuneate or broadly cuneate. Head on branchlet apex, globose, bracts 4, ovate, white. Fruit purplish red. Fl. May–Jun., fr. Sept..

树干 Trunk
摄影：卢志军 Photo by: Zhijun Lu

花枝 Flowering branches
摄影：刘海波 Photo by: Haibo Liu

果枝 Fruiting branch
摄影：徐耀粘 Photo by: Yaozhan Xu

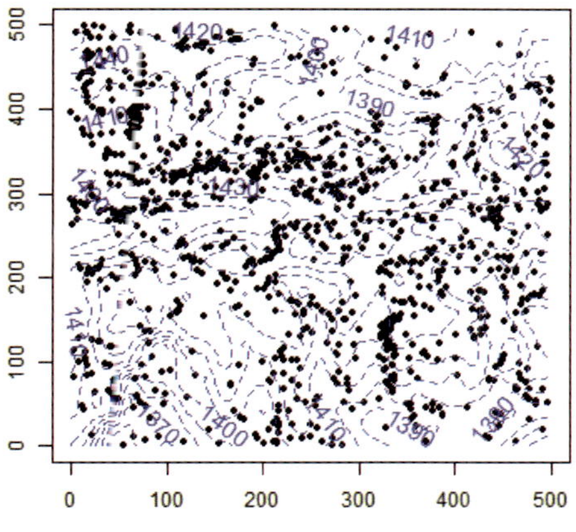

个体分布图 Distribution of individuals

径级分布表 DBH class

胸径区间 (Diameter class) (cm)	个体数 (No. of individuals in the plot)	比例 (Proportion) (%)
1～2	272	24.0
2～3	182	16.0
3～4	119	10.5
4～5	85	7.5
5～7	119	10.5
7～10	135	11.9
⩾10	223	19.6

177 灯台树

dēng tái shù | Giant Dogwood

Cornus controversa，山茱萸属
山茱萸科 | Cornaceae

代码（SpCode）= CORCON
个体数（Individual number/25 hm^2）= 316
最大胸径（Max DBH）= 40.17 cm
所在林层（Stratum）= 亚乔木层 (Understory)
重要值排序（Importance value rank）= 15/54

落叶乔木。树皮紫褐色或暗灰色；枝条紫红色，无毛。叶互生，有长柄，常簇生于枝端，宽卵形或宽椭圆形，先端渐尖或短尾状渐尖，基部圆形或近楔形。伞房状聚伞花序，顶生，花白色，多数。核果球形。花期5月，果期7～8月。

Deciduous trees. Bark purplish brown or dark gray. Branches purplish red, glabrous. Leaves alternate, petiole long, fascicled on branch apex, blade broadly ovate or broadly elliptic, apex acuminate or short-caudate acuminate, base rounded or subcuneate. Corymbose cymes terminal, flowers white, numerous. Drupes globose. Fl. May, fr. Jul.–Aug..

树干 Trunk
摄影：卢志军 Photo by: Zhijun Lu

花枝 Flowering branch
摄影：江明喜 Photo by: Mingxi Jiang

果枝 Fruiting branches
摄影：鲍大川 Photo by: Dachuan Bao

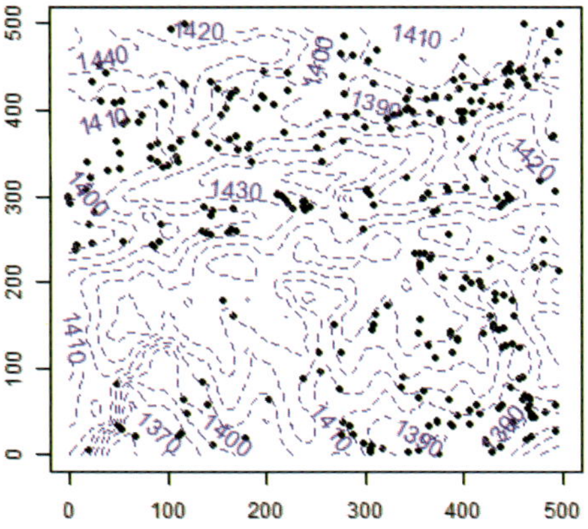

个体分布图 Distribution of individuals

径级分布表 DBH class

胸径区间 (Diameter class) (cm)	个体数 (No. of individuals in the plot)	比例 (Proportion) (%)
1～2.5	46	14.5
2.5～5	41	13.0
5～8	28	8.9
8～11	26	8.2
11～15	34	10.8
15～20	67	21.2
⩾20	74	23.4

178 川鄂山茱萸 chuān è shān zhū yú | Chinese Dogwood

Cornus chinensis，山茱萸属
山茱萸科 | Cornaceae

代码（SpCode）= CORCHI
个体数（Individual number/25 hm^2）= 8
最大胸径（Max DBH）= 11.43 cm
所在林层（Stratum）= 灌木层 (Shrub)
重要值排序（Importance value rank）= 69/128

落叶乔木。嫩枝紫褐色。叶对生，纸质，卵状椭圆形，先端长渐尖，基部楔形，或近圆形。伞形花序腋生，先叶开放，下具4枚宽卵圆形先端急尖小型苞片；花黄色；花梗与花萼管均密被细毛；萼片披针形；花瓣披针形或卵形。核果长椭圆形，成熟时紫黑色。花期4～5月，果期7月。

树干 Trunk
摄影：卢志军 Photo by: Zhijun Lu

Deciduous trees. Yong branches purplish brown. Leaves opposite, papery, ovate-elliptic, apex long acuminate, base cuneate, or subrounded. Umbels axillary, appearing before leaves, with 4 broadly ovate small bracts with acute apex, flowers yellow, peduncle and calyx tube densely pubescent, sepal lanceolate, petals lanceolate or ovate. Fruits long-ellipoid, purplish black when mature. Fl. Apr.–May, fr. Jul..

小枝和叶片 Branchlet and leaves
摄影：刘海波 Photo by: Haibo Liu

小枝和叶背 Branchlet and leaf abaxial surface
摄影：刘海波 Photo by: Haibo Liu

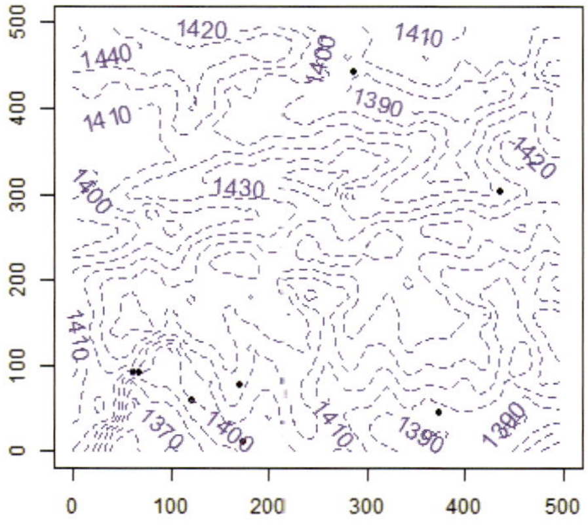

个体分布图 Distribution of individuals

径级分布表 DBH class

胸径区间 (Diameter class) (cm)	个体数 (No. of individuals in the plot)	比例 (Proportion) (%)
1～2	1	12.5
2～3	1	12.5
3～4	0	0.0
4～5	0	0.0
5～7	0	0.0
7～10	1	12.5
≥10	5	62.5

179 毛梾

máo lái | Walter Dogwood

Cornus walteri，山茱萸属
山茱萸科 | Cornaceae

代码（SpCode）= CORWAL
个体数（Individual number/25 hm^2）= 1
最大胸径（Max DBH）= 20.75 cm
所在林层（Stratum）= 亚乔木层 (Understory)
重要值排序（Importance value rank）= 51/54

落叶乔木。树皮黑褐色，常纵裂成条；幼枝疏被伏毛，后无毛。叶膜质，对生，椭圆形至长椭圆形。伞房状聚伞花序，花白色，有香味，萼片三角形，花瓣长披针状，背面疏被伏毛。核果球形，黑色。花期5～6月，果期9～10月。

树干 Trunk
摄影：卢志军 Photo by: Zhijun Lu

Deciduous trees. Bark dark gray, rectangularly splitting. Young branches sparsely pubescent with trichomes, then glabrous. Leaves membranous, opposite, blade elliptic to long elliptic. Corymbose cymes, flowers white and fragrant sepals triangular, petals oblong-lanceolate, abaxially sparsely pubescent. Drupes globose, black. Fl. May–Jun., fr. Sept.–Oct..

小枝和叶片 Branchlet and leaves
摄影：卢志军 Photo by: Zhijun Lu

小枝和叶背 Branchlet and leaf abaxial surface
摄影：卢志军 Photo by: Zhijun Lu

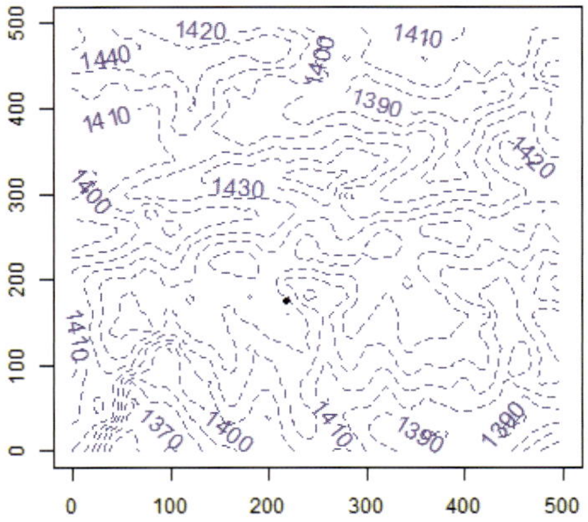

个体分布图 Distribution of individuals

径级分布表 DBH class

胸径区间 (Diameter class) (cm)	个体数 (No. of individuals in the plot)	比例 (Proportion) (%)
1～2.5	0	0.0
2.5～5	0	0.0
5～8	0	0.0
8～11	0	0.0
11～15	0	0.0
15～20	0	0.0
⩾20	1	100.0

180 青荚叶

qīng jiá yè | Japanese Helwingia

Helwingia japonica，青荚叶属
山茱萸科 | Cornaceae

代码（SpCode）= HELJAP
个体数（Individual number/25 hm^2）= 33
最大胸径（Max DBH）= 4.18 cm
所在林层（Stratum）= 灌木层 (Shrub)
重要值排序（Importance value rank）= 78/128

落叶灌木。树皮深棕色或灰黑色，光滑。嫩枝绿色或紫绿色，具条纹，叶痕突起。叶互生，纸质，卵形、卵状椭圆形，稀为卵状披针形，顶端渐尖，稀尾状渐尖，基部宽楔形，叶缘具针刺锯齿或圆齿。伞形花序，雌雄异株，花淡绿色。核果近球形，成熟后黑色。花期5月，果期7月。

Deciduous shrubs. Bark dark brown or grayish black, smooth. Young branches green or purplish green, striped, leaf scars convex. Leaves alternate, papery, blade ovate or ovate-elliptic, rarely ovate-lanceolate, apex acuminate, rarely caudate-acuminate, base broadly cuneate, margin spiculate serrate to spiculate crenate. Umbels, dioecious, petals light green. Drupes subglobose, black when mature. Fl. May, fr. Jul..

树干 Trunk
摄影：卢志军 Photo by: Zhijun Lu

花枝 Flowering branches
摄影：江明喜 Photo by: Mingxi Jiang

果枝 Fruiting branches
摄影：江明喜 Photo by: Mingxi Jiang

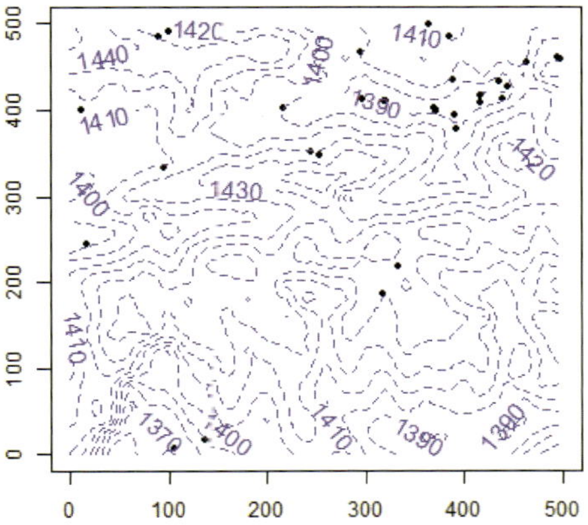

个体分布图 Distribution of individuals

径级分布表 DBH class

胸径区间 (Diameter class) (cm)	个体数 (No. of individuals in the plot)	比例 (Proportion) (%)
1～2	26	78.8
2～3	5	15.2
3～4	1	3.0
4～5	1	3.0
5～7	0	0.0
7～10	0	0.0
≥10	0	0.0

181 中华青荚叶

zhōng huá qīng jiá yè | Chinese Helwingia

Helwingia chinensis，青荚叶属

山茱萸科 | Cornaceae

代码（SpCode）= HELCHI

个体数（Individual number/25 hm^2）= 2

最大胸径（Max DBH）= 2.35 cm

所在林层（Stratum）= 灌木层 (Shrub)

重要值排序（Importance value rank）= 122/128

常绿灌木。树皮深灰或淡灰棕色。嫩枝紫绿色。叶互生，革质或近革质，线状披针形或卵状披针形，顶端长渐尖，基部楔形，叶缘疏具带腺体尖刺锯齿。伞形花序生于叶片上面中脉近中部或幼枝的上部。核果近球形，干后黑色。花期4～5月，果期8～10月。

Evergreen shrubs. Bark dark gray or light grayish brown. Young branches purplish green. Leaves alternate, leathery or subleathery, blade linear-lanceolate or ovate-lanceolate, apex long acuminate, base cuneate, margin sparsely glandular spiculate-serrate. Umbels near mid point of midvein on adaxial surface of blade or on upper part of young branches. Drupes subglobose, black after dried. Fl. Apr.–May, fr. Aug.–Oct..

树干 Trunk

摄影：卢志军 Photo by: Zhijun Lu

花枝 Flowering branch

摄影：江明喜 Photo by: Mingxi Jiang

花 Flower

摄影：江明喜 Photo by: Mingxi Jiang

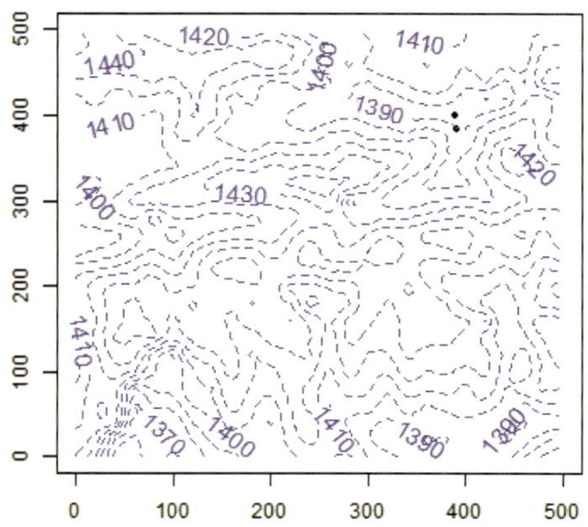

个体分布图 Distribution of individuals

径级分布表 DBH class

胸径区间 (Diameter class) (cm)	个体数 (No. of individuals in the plot)	比例 (Proportion) (%)
1～2	1	50.0
2～3	1	50.0
3～4	0	0.0
4～5	0	0.0
5～7	0	0.0
7～10	0	0.0
≥10	0	0.0

182 楤木

cōng mù | Japanese Aralia

Aralia elata，楤木属
五加科 | Araliaceae

代码（SpCode）= ARAELA
个体数（Individual number/25 hm^2）= 64
最大胸径（Max DBH）= 10.62 cm
所在林层（Stratum）= 灌木层 (Shrub)
重要值排序（Importance value rank）= 61/128

落叶灌木或乔木。树皮灰色，疏生粗壮直刺。小枝通常呈淡灰棕色，有黄棕色茸毛，疏生细刺。叶为二回或三回羽状复叶，小叶5～11，叶片宽卵形至椭圆卵形或者狭卵形，顶端渐尖，基部心形至近心形或圆形，叶缘具细锯齿。由伞形花序组成的圆锥花序，单生枝顶。果黑色，有5棱。花期7～8月，果期8～9月。

Deciduous shrubs or small trees. Bark gray, with sparse stout prickles. Branchlets usually light grayish brown, yellowish brown tomentose, with sparse spines. Leaves 2 or 3-pinnately compound, leaflets 5-11, blade broadly ovate to elliptic-ovate or narrowly ovate, apex acuminate, base cordate to subcordate or rounded, margin serrulate. Panicles consisting of umbels, solitary, terminal. Fruits black, 5-edged. Fl. Jul.–Aug., fr. Aug.–Sept..

树干 Trunk
摄影：卢志军 Photo by: Zhijun Lu

复叶 Compound leaves
摄影：卢志军 Photo by: Zhijun Lu

复叶背面 Compound leaves abaxial surface
摄影：卢志军 Photo by: Zhijun Lu

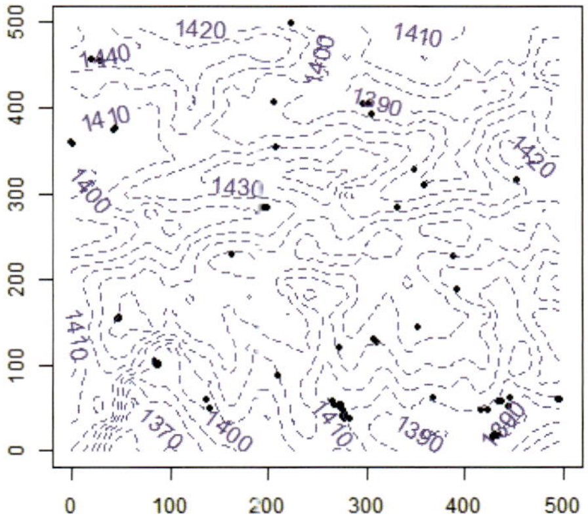

个体分布图 Distribution of individuals

径级分布表 DBH class

胸径区间 (Diameter class) (cm)	个体数 (No. of individuals in the plot)	比例 (Proportion) (%)
1～2	25	39.1
2～3	14	21.9
3～4	9	14.1
4～5	6	9.4
5～7	3	4.7
7～10	5	7.8
≥10	2	3.1

183 棘茎楤木 jí jīng cōng mù | Spinystem Aralia

Aralia echinocaulis，楤木属
五加科 | Araliaceae

代码（SpCode）= ARAECH
个体数（Individual number/25 hm^2）= 8
最大胸径（Max DBH）= 5.20 cm
所在林层（Stratum）= 灌木层 (Shrub)
重要值排序（Importance value rank）= 99/128

落叶小乔木。小枝密生细长直刺。叶为二回羽状复叶，小叶5~9，叶片长圆卵形至披针形，顶端渐尖，基部圆形至钝形，叶缘具齿。伞形花序组成圆锥花序，大，顶生，淡褐色，花白色。果球形，5棱。花期6~8月，果期9~10月。

Deciduous small trees. Branchlets with dense slender prickles. Leaves 2-pinnately compound, leaflets 5-9, blade oblong-ovate to lanceolate, apex acuminate, base rounded to obtuse, margin serrate. Panicles consisting of umbels, large, terminal, light brown, flowers white. Fruit globose, 5-edged. Fl. Jun.–Aug., fr. Sept.–Oct..

树干 Trunk
摄影：卢志军 Photo by: Zhijun Lu

复叶 Compound leaves
摄影：卢志军 Photo by: Zhijun Lu

复叶背面 Compound leaves abaxial surface
摄影：卢志军 Photo by: Zhijun Lu

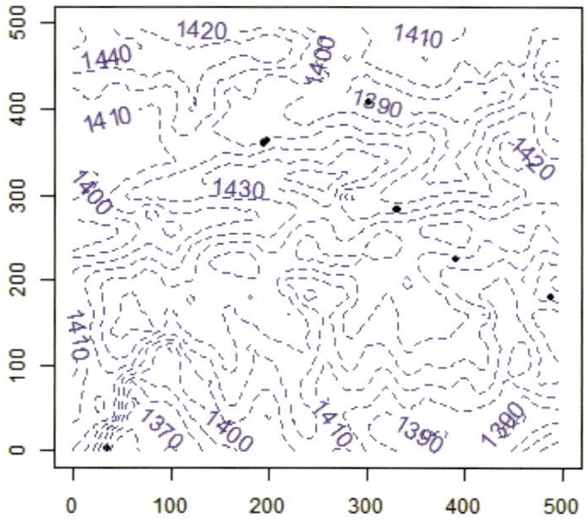

个体分布图 Distribution of individuals

径级分布表 DBH class

胸径区间 (Diameter class) (cm)	个体数 (No. of individuals in the plot)	比例 (Proportion) (%)
1~2	1	12.5
2~3	3	37.5
3~4	1	12.5
4~5	2	25.0
5~7	1	12.5
7~10	0	0.0
≥10	0	0.0

184 人参木

rén shēn mù | Farges Chengiopanax

Chengiopanax fargesii，人参木属
五加科 | Araliaceae

代码（SpCode）= CHEFAR
个体数（Individual number/25 hm^2）= 709
最大胸径（Max DBH）= 54.05 cm
所在林层（Stratum）= 灌木层 (Shrub)
重要值排序（Importance value rank）= 20/128

落叶乔木。枝、叶和花序幼时密被短锈色星状毛。掌状复叶，小叶5～7，叶片椭圆形至长圆形或披针形，顶端急尖至渐尖，基部宽楔形，叶缘具微小锯齿。伞房状圆锥花序顶生。果实扁球形。花期9月，果期11～12月。

Deciduous trees. Branches, leaves, and inflorescences densely shortly ferruginous stellate tomentose when young. Leaves palmately compound, leaflets 5-7, blade elliptic to oblong or lanceolate, apex acute to acuminate, base broadly cuneate, margin minutely serrulate. Corymbose panicles terminal. Fruits flat globose. Fl. Sept., fr. Nov.–Dec..

树干 Trunk
摄影：卢志军 Photo by: Zhijun Lu

小枝和叶片 Branchlet and leaves
摄影：刘海波 Photo by: Haibo Liu

小枝和叶背 Branchlet and leaf abaxial surface
摄影：卢志军 Photo by: Zhijun Lu

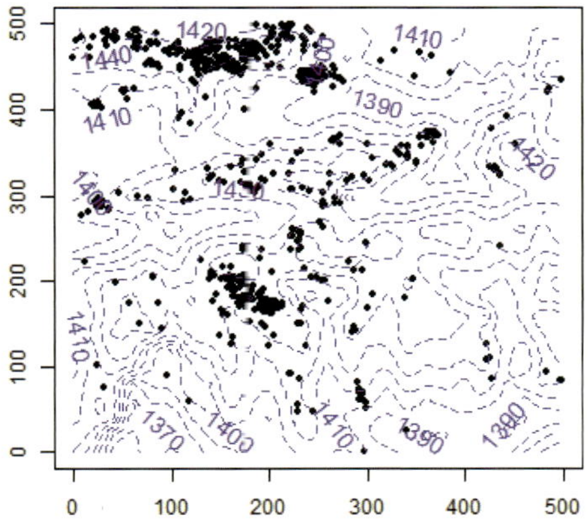

个体分布图 Distribution of individuals

径级分布表 DBH class

胸径区间 (Diameter class) (cm)	个体数 (No. of individuals in the plot)	比例 (Proportion) (%)
1～2	388	54.7
2～3	135	19.0
3～4	63	8.9
4～5	21	3.0
5～7	21	3.0
7～10	12	1.7
≥10	69	9.7

185 树参

shù shēn | Dentigerous Dendropanax

Dendropanax dentiger, 树参属
五加科 | Araliaceae

代码（SpCode）= DENDEN
个体数（Individual number/25 hm^2）= 225
最大胸径（Max DBH）= 26.95 cm
所在林层（Stratum）= 灌木层 (Shrub)
重要值排序（Importance value rank）= 37/128

常绿灌木或小乔木。叶厚纸质或革质，叶形变异很大，不裂叶通常椭圆形，很少为长椭圆形、椭圆状披针形至线状披针形；裂片叶2～3掌状裂，裂片狭三角卵形或卵状披针形。伞形花序单个或2～5组成复伞形花序，顶生。果长圆状球形，很少近球形。花期8～10月，果期10～12月。

Evergreen shrubs or small trees. Leaves thickly papery or leathery, blade with great variation, unlobed blades usually elliptic, rarely long-elliptic, elliptic-lanceolate to linear lanceolate; lobed blades palmately 2-or 3-lobed, lobes narrowly triangular-ovate or ovate-lanceolate. Umbels solitary or compound with 2-5 flowers, terminal. Fruit ellipsoid, rarely subglobose. Fl. Aug.–Oct., fr. Oct.–Dec..

树干 Trunk
摄影：卢志军 Photo by: Zhijun Lu

小枝和叶片 Branchlet and leaves
摄影：卢志军 Photo by: Zhijun Lu

小枝和叶背 Branchlet and leaf abaxial surface
摄影：卢志军 Photo by: Zhijun Lu

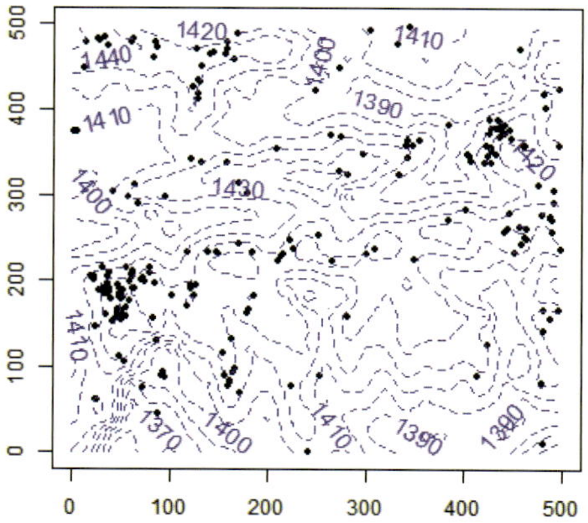

个体分布图 Distribution of individuals

径级分布表 DBH class

胸径区间 (Diameter class) (cm)	个体数 (No. of individuals in the plot)	比例 (Proportion) (%)
1～2	103	45.8
2～3	20	8.9
3～4	9	4.0
4～5	10	4.4
5～7	15	6.7
7～10	25	11.1
≥10	43	19.1

186 吴茱萸五加

wú zhū yú wǔ jiā | Ciliate Gamblea

Gamblea ciliata var. *evodiifolia*，萸叶五加属
五加科 | Araliaceae

代码（SpCode）= GAMCIL
个体数（Individual number/25 hm^2）= 579
最大胸径（Max DBH）= 22.42 cm
所在林层（Stratum）= 亚乔木层 (Understory)
重要值排序（Importance value rank）= 20/54

树干 Trunk
摄影：卢志军 Photo by: Zhijun Lu

落叶灌木或乔木。老枝暗色；新枝红棕色，光滑。掌状复叶，小叶3～5，叶片椭圆形至卵形，顶端短渐尖至渐尖，基部楔形，叶缘近全缘或具细锯齿。复伞形花序或伞形花序组成的圆锥花序。果球形，黑色。花期5～7月，果期6～9月。

Deciduous shrubs or trees. Old branches dull, new branches reddish brown, smooth. Leaves palmately compound, leaflets 3-5, blade elliptic to ovate, apex shortly acuminate to acuminate, base cuneate, margin subentire or minutely serrulate. Compound umbel or panicles of umbels. Fruit globose, black. Fl. May–Jul., fr. Jun.–Sept..

花枝 Flowering branch
摄影：刘海波 Photo by: Haibo Liu

果枝 Fruiting branch
摄影：刘海波 Photo by: Haibo Liu

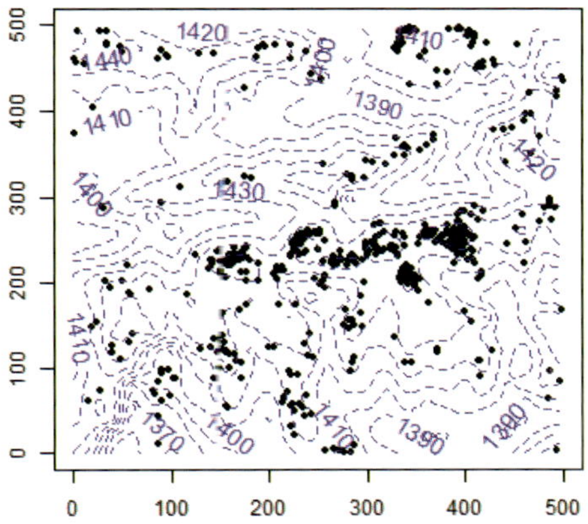

个体分布图 Distribution of individuals

径级分布表 DBH class

胸径区间 (Diameter class) (cm)	个体数 (No. of individuals in the plot)	比例 (Proportion) (%)
1～2.5	358	61.8
2.5～5	142	24.5
5～8	36	6.2
8～11	17	2.9
11～15	9	1.6
15～20	7	1.2
≥20	10	1.7

187 刺楸

cì qiū | Septemlobate Kalopanax

Kalopanax septemlobus，刺楸属
五加科 | Araliaceae

代码（SpCode）= KALSEP
个体数（Individual number/25 hm^2）= 2
最大胸径（Max DBH）= 37.70 cm
所在林层（Stratum）= 乔木层 (Canopy)
重要值排序（Importance value rank）= 44/50

树干 Trunk
摄影：卢志军 Photo by: Zhijun Lu

落叶乔木。树皮暗灰棕色。小枝散生粗刺，刺基部宽阔扁平。叶纸质，圆形或近圆形，5～7裂，裂片宽三角卵形至长圆卵形，顶端渐尖，基部心形或圆形至近截形，叶缘具齿。伞形花序，花冠白色或淡绿黄色。果蓝熟时黑色。花期7～8月，果期8～10月。

Deciduous trees. Bark dark grayish brown. Branchlets with sparse and stout prickles, prickle base broad and flat. Leaf blade papery, orbicular or suborbicular, 5-7-lobed, lobes broadly triangular-ovate to oblong-ovate, apex acuminate, base cordate or rounded to nearly truncate, margin serrate. Umbels, corolla white or light yellowish green. Fruit dark blue at maturity. Fl. Jul.–Aug., fr. Aug.–Oct..

小枝和叶片 Branchlet and leaves
摄影：卢志军 Photo by: Zhijun Lu

小枝和叶背 Branchlet and leaf abaxial surface
摄影：卢志军 Photo by: Zhijun Lu

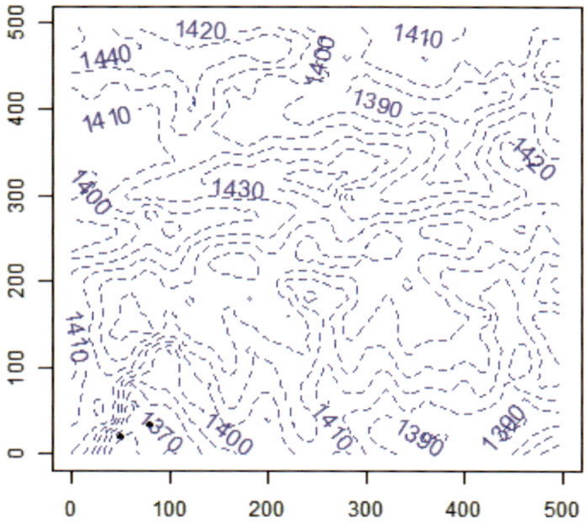

个体分布图 Distribution of individuals

径级分布表 DBH class

胸径区间 (Diameter class) (cm)	个体数 (No. of individuals in the plot)	比例 (Proportion) (%)
1～2.5	0	0.0
2.5～5	0	0.0
5～10	1	50.0
10～25	0	0.0
25～40	1	50.0
40～100	0	0.0
⩾100	0	0.0

188 异叶梁王茶 yì yè liáng wáng chá | David Metapanax

Metapanax davidii，梁王茶属
五加科 I Araliaceae

代码（SpCode）= METDAV
个体数（Individual number/25 hm²）= 9
最大胸径（Max DBH）= 12.10 cm
所在林层（Stratum）= 灌木层 (Shrub)
重要值排序（Importance value rank）= 89/128

常绿灌木或小乔木。树皮灰黑色或褐色。小枝绿色或灰绿色，皮孔明显。单叶，不裂或2~3浅裂至深裂。圆锥花序，花白色或淡黄色。果实黑色。花期7~9月，果期8~11月。

树干 Trunk
摄影：卢志军 Photo by: Zhijun Lu

Evergreen shrubs or small trees. Bark grayish black or brown. Branchlets green or grayish green, lenticels conspicuous. Leaves simple, unlobed, 2-3 shallowly to deeply lobed. Panicles, flowers white or light yellow. Fruits black. Fl. Jul.–Sept., fr. Aug.–Nov..

小枝 Branchlet
摄影：谷志容 Photo by: Zhirong Gu

花序 Inflorescence
摄影：江明喜 Photo by: Mingxi Jiang

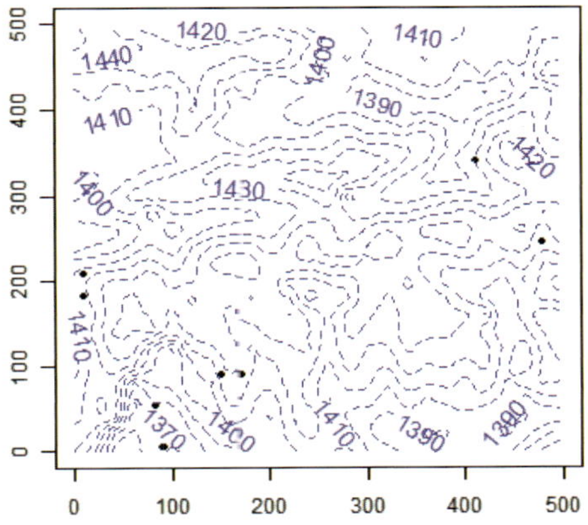

个体分布图 Distribution of individuals

径级分布表 DBH class

胸径区间 (Diameter class) (cm)	个体数 (No. of individuals in the plot)	比例 (Proportion) (%)
1~2	3	33.3
2~3	1	11.1
3~4	0	0.0
4~5	1	11.1
5~7	2	22.2
7~10	1	11.1
≥10	1	11.1

189 城口桤叶树 chéng kǒu qī yè shù | Farges Clethra

Clethra fargesii，桤叶树属
山柳科 | Clethraceae

代码（SpCode）= CLEFAR
个体数（Individual number/25 hm^2）= 562
最大胸径（Max DBH）= 37.65 cm
所在林层（Stratum）= 亚乔木层 (Understory)
重要值排序（Importance value rank）= 17/54

落叶灌木或乔木。幼枝密生星状柔毛。叶纸质，互生，披针状椭圆形，先端渐尖，基部楔形或钝，叶缘锯齿。总状花序，3～7枝排列成近伞形圆锥花序，花瓣5，白色。蒴果近圆球形。花期7～8月，果期9～10月。

树干 Trunk
摄影：卢志军 Photo by: Zhijun Lu

Deciduous shrubs or trees. Young branches pubescent with stellate pilose. Leaves papery, alternate, lanceolate-elliptic, apex acuminate, base cuneate or obtuse, margin serrate. Racemes, 3-7 forming sub-umbellate-panicle, petals 5 and white. Capsules subglobose. Fl. Jul.–Aug., fr. Sept.–Oct..

花枝 Flowering branch
摄影：卢志军 Photo by: Zhijun Lu

果枝 Fruiting branches
摄影：鲍大川 Photo by: Dachuan Bao

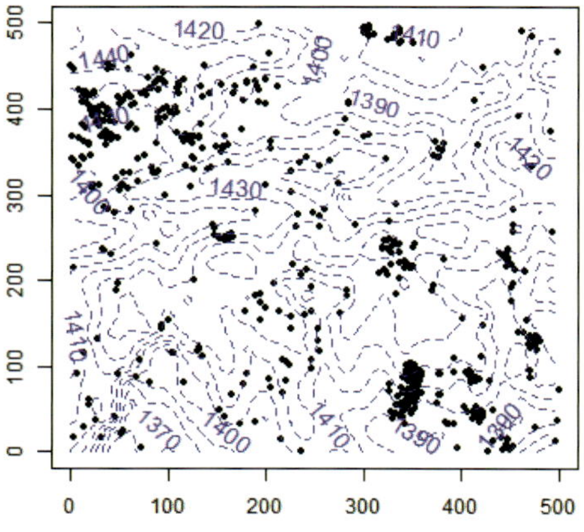

个体分布图 Distribution of individuals

径级分布表 DBH class

胸径区间 (Diameter class) (cm)	个体数 (No. of individuals in the plot)	比例 (Proportion) (%)
1～2.5	100	17.8
2.5～5	119	21.2
5～8	147	26.2
8～11	129	23.0
11～15	49	8.7
15～20	15	2.7
≥20	3	0.5

190 齿缘吊钟花 chǐ yuán diào zhōng huā | Serrulate Enkianthus

Enkianthus serrulatus，吊钟花属
杜鹃花科 | Ericaceae

代码（SpCode）= ENKSER
个体数（Individual number/25 hm^2）= 4613
最大胸径（Max DBH）= 40.23 cm
所在林层（Stratum）= 亚乔木层 (Understory)
重要值排序（Importance value rank）= 5/54

落叶灌木或小乔木。树皮灰褐色，无毛。叶厚纸质，簇生于枝顶，椭圆形或椭圆状长圆形，先端短渐尖或渐尖，基部宽楔形或圆形，叶缘具小齿。伞形花序，顶生，常2～6朵花组成，花冠钟状，白色。蒴果椭圆形，有棱角。花期4～6月，果期7～9月。

Deciduous shrubs or small trees. Bark grayish brown, glabrous. Leaves thickly papery, fascicled on branch apex, blade elliptic or elliptic-oblong, apex shortly acuminate or acuminate, base broadly cuneate or rounded, margin denticulate. Umbels terminal, usually consisting of 2-6 flowers, corolla campanulate, white. Capsules ellipoid, angled. Fl. Apr.–Jun., fr. Jul.–Sept..

树干 Trunk
摄影：卢志军 Photo by: Zhijun Lu

花 Flowers
摄影：徐耀粘 Photo by: Yaozhan Xu

果枝 Fruiting branch
摄影：卢志军 Photo by: Zhijun Lu

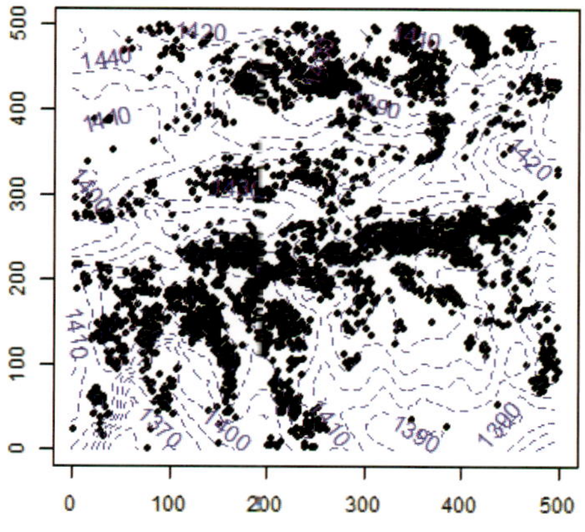

个体分布图 Distribution of individuals

径级分布表 DBH class

胸径区间 (Diameter class) (cm)	个体数 (No. of individuals in the plot)	比例 (Proportion) (%)
1～2.5	499	10.8
2.5～5	1789	38.8
5～8	1339	29.0
8～11	572	12.4
11～15	285	6.2
15～20	98	2.1
≥20	31	0.7

191 灯笼吊钟花 dēng lóng diào zhōng huā | Chinese Enkianthus

Enkianthus chinensis，吊钟花属
杜鹃花科 | Ericaceae

代码（SpCode）= ENKCHI
个体数（Individual number/25 hm^2）= 450
最大胸径（Max DBH）= 31.33 cm
所在林层（Stratum）= 灌木层 (Shrub)
重要值排序（Importance value rank）= 26/128

树干 Trunk
摄影：卢志军 Photo by: Zhijun Lu

落叶灌木或小乔木。树皮灰黑色，无毛。叶纸质，长圆形至椭圆形，先端急尖至钝，基部楔形或宽楔形，叶缘具细齿。花序总状或短总状伞形，花肉红色，下垂。蒴果圆卵形。花期4～5月，果期7～8月。

Deciduous shrubs or small trees. Bark grayish black, glabrous. Leaves papery, blade oblong to elliptic, apex acute to obtuse, base cuneate or broadly cuneate, margin serrulate. Racemes or short racemose-umbels, flowers flesh red, pendulous. Capsules ovoid. Fl. Apr.–May, fr. Jul.–Aug..

花枝 Flowering branch
摄影：刘海波 Photo by: Haibo Liu

小枝和叶背 Branchlet and leaf abaxial surface
摄影：刘海波 Photo by: Haibo Liu

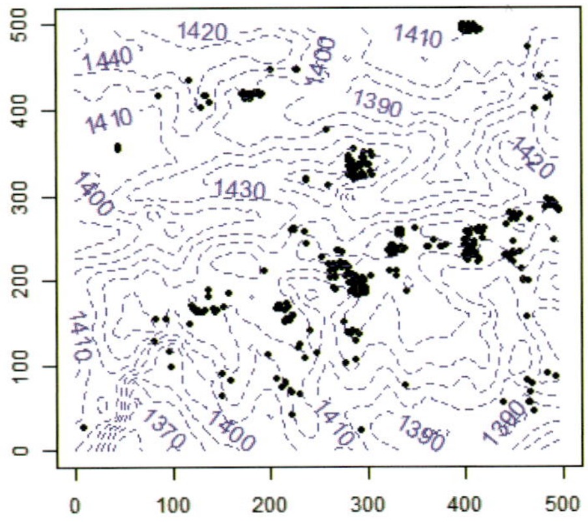

个体分布图 Distribution of individuals

径级分布表 DBH class

胸径区间 (Diameter class) (cm)	个体数 (No. of individuals in the plot)	比例 (Proportion) (%)
1～2	37	8.2
2～3	45	10.0
3～4	54	12.0
4～5	47	10.4
5～7	90	20.0
7～10	79	17.6
≥10	98	21.8

192 小果珍珠花

xiǎo guǒ zhēn zhū huā | Little-fruit Lyonia

Lyonia ovalifolia var. *elliptica*，珍珠花属
杜鹃花科 | Ericaceae

代码（SpCode）= LYOOVA
个体数（Individual number/25 hm^2）= 2269
最大胸径（Max DBH）= 30.10 cm
所在林层（Stratum）= 亚乔木层 (Understory)
重要值排序（Importance value rank）= 7/54

落叶灌木或小乔木。枝幼时有微毛，后无毛。叶纸质或厚纸质，长圆状椭圆形或卵状椭圆形，先端渐尖或短渐尖，基部近心形、圆形或宽楔形。总状花序，腋生，花白色。蒴果球形。花期7～8月，果期9～10月。

Deciduous shrubs or small trees. Young branchlets pubescent, then glabrous. Leaves papery or thickly papery, blade oblong-elliptic or ovate-elliptic, apex acuminate or shortly acuminate, base subcordate, rounded or broadly cuneate. Racemes axillary, flowers white. Capsules globose. Fl. Jul.–Aug., fr. Sept.–Oct..

树干 Trunk
摄影：卢志军 Photo by: Zhijun Lu

果枝 Fruiting branch
摄影：卢志军 Photo by: Zhijun Lu

花枝和叶背 Flowering branch and leaf abaxial surface
摄影：刘海波 Photo by: Haibo Liu

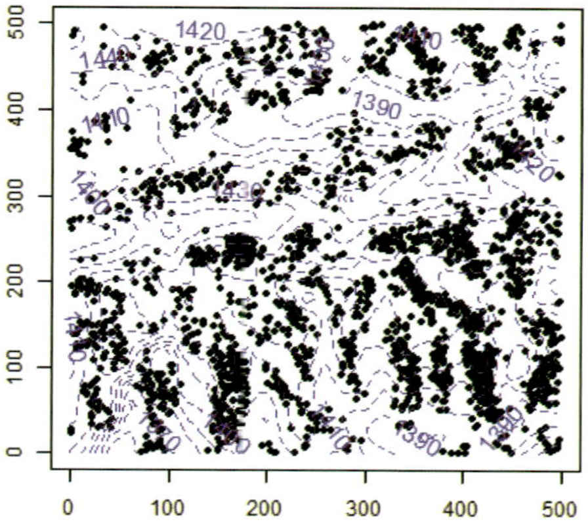

个体分布图 Distribution of individuals

径级分布表 DBH class

胸径区间 (Diameter class) (cm)	个体数 (No. of individuals in the plot)	比例 (Proportion) (%)
1～2.5	57	2.5
2.5～5	445	19.6
5～8	726	32.0
8～11	546	24.1
11～15	294	13.0
15～20	150	6.6
≥20	51	2.2

193 长蕊杜鹃

chǎng ruǐ dù juān | Long-stamen Rhododendron

Rhododendron stamineum，杜鹃花属
杜鹃花科 | Ericaceae

代码（SpCode）= RHOSTA
个体数（Individual number/25 hm^2）= 9523
最大胸径（Max DBH）= 30.12 cm
所在林层（Stratum）= 亚乔木层 (Understory)
重要值排序（Importance value rank）= 1/54

常绿灌木或小乔木。枝条无毛。叶革质，常轮生枝顶，椭圆形至椭圆状卵形，先端渐尖或短渐尖，基部楔形，叶缘稍反卷。伞状花序，侧生于枝顶的叶腋，花白色或少数玫瑰色。蒴果圆柱形，稍弯弓状。花期5～6月，果期7～8月。

树干 Trunk
摄影：卢志军 Photo by: Zhijun Lu

Evergreen shrubs or small trees. Branches glabrous. Leaves leathery, usually verticillate on branch apex, blade elliptic to elliptic-ovate, apex acuminate or shortly acuminate, base cuneate, margin slightly revolute. Umbels, axillary on branch apex, flowers white or rarely roseate. Capsules cylindric, slightly curved. Fl. May–Jun., fr. Jul.–Aug..

花枝 Flowering branches
摄影：刘海波 Photo by: Haibo Liu

叶背 Leaf abaxial surface
摄影：卢志军 Photo by: Zhijun Lu

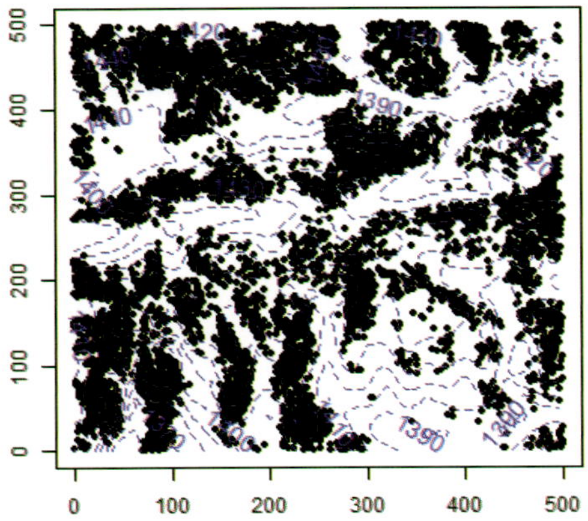

个体分布图 Distribution of individuals

径级分布表 DBH class

胸径区间 (Diameter class) (cm)	个体数 (No. of individuals in the plot)	比例 (Proportion) (%)
1～2.5	222	2.3
2.5～5	1783	18.7
5～8	2810	29.5
8～11	2130	22.4
11～15	1418	14.9
15～20	702	7.4
≥20	458	4.8

194 满山红 măn shān hóng | Maries Azalea

Rhododendron mariesii，杜鹃花属
杜鹃花科 | Ericaceae

代码（SpCode）= RHOMAR
个体数（Individual number/25 hm^2）= 6845
最大胸径（Max DBH）= 26.6 cm
所在林层（Stratum）= 灌木层 (Shrub)
重要值排序（Importance value rank）= 3/128

落叶灌木。枝条轮生，幼枝初被黄色绢毛，后脱落。叶纸质，常轮生枝端，卵状披针形，有时为宽卵形或椭圆形。花簇生枝顶，玫瑰紫色，先叶开放，花萼被黄棕色或灰色伏毛，花冠辐状漏斗形，上方的裂片有红紫色斑点。蒴果圆柱形，有浅沟，密被长柔毛。花期4～5月，果期6～7月。

Deciduous shrubs. Branches verticillate; soft yellowish-brown hairy when young, then glabrescent. Leaves papery, usually verticillate on branch apex, blade ovate-lanceolate, sometimes broadly ovate or elliptic. Flowers fascicled on branch apex, rose-purple, appearing before leaves, calyx yellow-brown or gray appressed haired, corolla funnelform, with purple-red flecks on upper lobes. Capsules cylindric, shallowly ridged, densely villous. Fl. Apr.–May, fr. Jun.–Jul..

树干 Trunk
摄影：卢志军 Photo by: Zhijun Lu

花 Flowers
摄影：徐耀粘 Photo by: Yaozhan Xu

小枝和叶背 Branchlet and leaf abaxial surface
摄影：卢志军 Photo by: Zhijun Lu

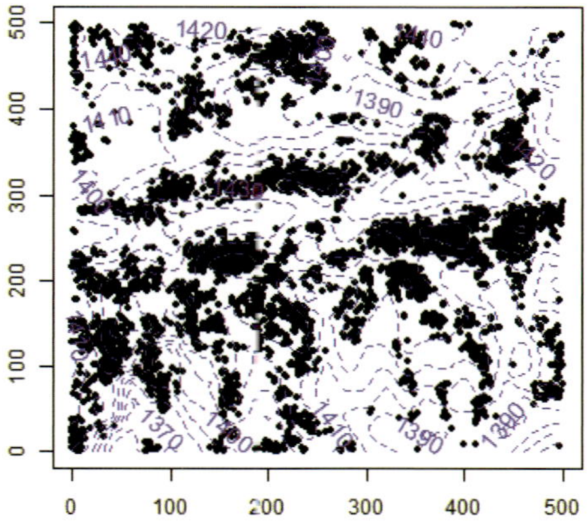

个体分布图 Distribution of individuals

径级分布表 DBH class

胸径区间 (Diameter class) (cm)	个体数 (No. of individuals in the plot)	比例 (Proportion) (%)
1～2	1145	16.7
2～3	1500	21.9
3～4	1492	21.8
4～5	985	14.4
5～7	1026	15.0
7～10	520	7.6
⩾10	177	2.6

195 云锦杜鹃 yún jǐn dù juān | Fortune Rhododendron

Rhododendron fortunei，杜鹃花属

杜鹃花科 | Ericaceae

代码（SpCode）= RHOFOR

个体数（Individual number/25 hm^2）= 2709

最大胸径（Max DBH）= 42.93 cm

所在林层（Stratum）= 亚乔木层 (Understory)

重要值排序（Importance value rank）= 6/54

常绿灌木。树皮灰色，片状脱落。幼枝绿色，近无毛，有腺体。叶革质，簇生枝顶，长圆形至长圆状椭圆形，先端钝尖，基部圆形、宽楔形或心形。花粉红色或白色带粉红色。蒴果长圆形，表面粗糙，有残存腺体。花期5月。

Evergreen shrubs. Bark gray, peeling in flakes. Young branchlets green, sub glabrescent, glandular. Leaves leathery, fascicled on branch apex, blade oblong to oblong-elliptic, apex obtuse, base rounded, broadly cuneate or cordate. Flowers pink or pinkish white. Capsules oblong, surface rough, glandulars remained. Fl. May.

树干 Trunk

摄影：卢志军 Photo by: Zhijun Lu

花 Flowers

摄影：刘海波 Photo by: Haibo Liu

果枝 Fruiting branch

摄影：鲍大川 Photo by: Dachuan Bao

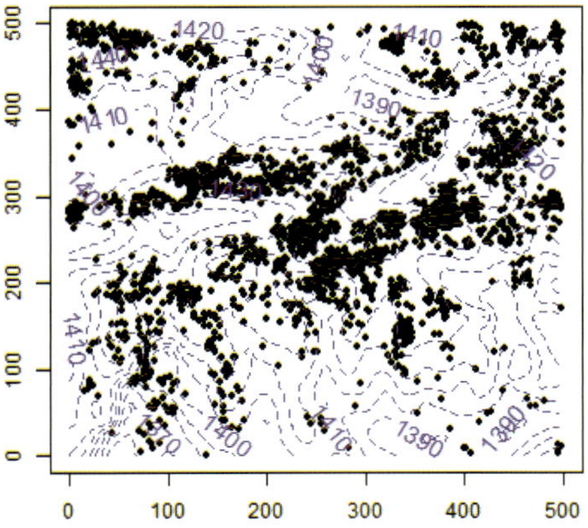

个体分布图 Distribution of individuals

径级分布表 DBH class

胸径区间 (Diameter class) (cm)	个体数 (No. of individuals in the plot)	比例 (Proportion) (%)
1～2.5	234	8.6
2.5～5	529	19.5
5～8	556	20.5
8～11	597	22.0
11～15	479	17.7
15～20	219	8.1
≥20	95	3.5

196 杜鹃

dù juān | Indian Azalea

Rhododendron simsii，杜鹃花属
杜鹃花科 | Ericaceae

代码（SpCode）= RHOSIM
个体数（Individual number/25 hm^2）= 616
最大胸径（Max DBH）= 17.39 cm
所在林层（Stratum）= 灌木层 (Shrub)
重要值排序（Importance value rank）= 31/128

落叶灌木。枝条密被亮棕色或褐色的扁平糙伏毛。叶纸质，卵形、椭圆状卵形或倒卵形，顶端短渐尖，基部楔形或宽楔形，叶缘稍反卷，具细齿。花玫瑰红色至淡红色或深红色，上部裂片带深红斑点。蒴果卵圆形，密被糙伏毛。盛花期4～5月，果期6～8月。

Deciduous shrubs. Branches densely shiny brown or brown appressed flat setose. Leaves papery, blade ovate, elliptic-ovate or obovate, apex shortly acuminate, base cuneate or broadly cuneate, margin slightly revolute, finely toothed. Corolla rose, light to dark red, with dark red flecks on upper lobes. Capsules ovoid, densely strigose. Fl. Apr.–May, fr. Jun.–Aug..

树干 Trunk
摄影：卢志军 Photo by: Zhijun Lu

花 Flowers
摄影：徐耀粘 Photo by: Yaozhan Xu

果枝 Fruiting branch
摄影：卢志军 Photo by: Zhijun Lu

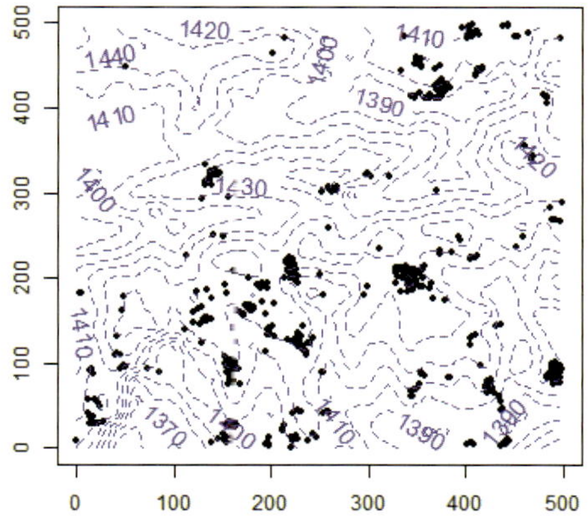

个体分布图 Distribution of individuals

径级分布表 DBH class

胸径区间 (Diameter class) (cm)	个体数 (No. of individuals in the plot)	比例 (Proportion) (%)
1～2	109	17.7
2～3	128	20.8
3～4	123	20.0
4～5	88	14.3
5～7	95	15.4
7～10	56	9.1
≥10	17	2.8

197 四川杜鹃

sì chuān dù juān | Szechwan Rhododendron

Rhododendron sutchuenense，杜鹃花属
杜鹃花科 I Ericaceae

代码（SpCode）= RHOSUT
个体数（Individual number/25 hm^2）= 7
最大胸径（Max DBH）= 18.94 cm
所在林层（Stratum）= 亚乔木层 (Understory)
重要值排序（Importance value rank）= 48/54

树干 Trunk
摄影：卢志军 Photo by: Zhijun Lu

常绿灌木或小乔木。幼枝粗壮，初被灰白色茸毛，后无毛。叶厚革质，簇生枝顶，长圆状倒披针形或长圆状卵形，顶端钝形或圆形，基部楔形。顶生总状花序，花大，粉红带蔷薇色。蒴果木质，长圆形。花期3～4月，果期5～6月。

Evergreen shrubs or small trees. Young branches stout, initially pubescent white villus, then glabrous. Leaves thickly leathery, fascicled on branch apex, blade oblong-oblanceolate or oblong-ovate, apex obtuse or rounded, base cuneate. Raceme on apex, flowers large, roseate pink. Capsules woody, oblong. Fl. Mar.–Apr., fr. May–Jun..

花 Flowers
摄影：徐耀粘 Photo by: Yaozhan Xu

果枝 Fruiting branch
摄影：卢志军 Photo by: Zhijun Lu

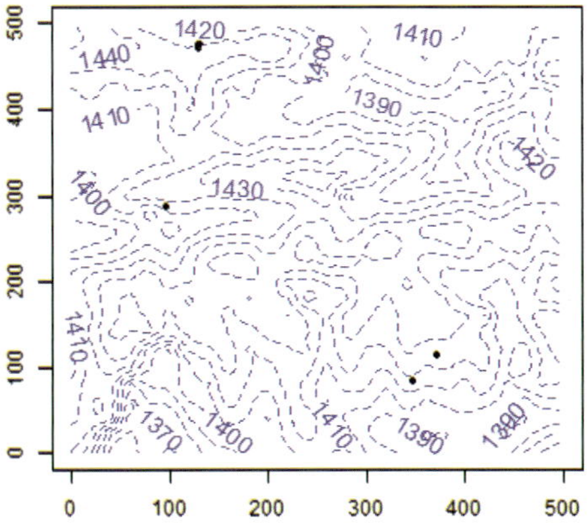

个体分布图 Distribution of individuals

径级分布表 DBH class

胸径区间 (Diameter class) (cm)	个体数 (No. of individuals in the plot)	比例 (Proportion) (%)
1～2.5	1	14.3
2.5～5	2	28.6
5～8	2	28.6
8～11	0	0.0
11～15	1	14.3
15～20	1	14.3
≥20	0	0.0

198 耳叶杜鹃 ěr yè dù juān | Auriculate Rhododendron

Rhododendron auriculatum，杜鹃花属
杜鹃花科 | Ericaceae

代码（SpCode）= RHOAUR
个体数（Individual number/25 hm^2）= 4
最大胸径（Max DBH）= 10.31 cm
所在林层（Stratum）= 亚乔木层 (Understory)
重要值排序（Importance value rank）= 52/54

常绿灌木或小乔木。树皮灰色。幼枝密被长腺毛。叶纸质或薄革质，长圆形或椭圆状披针形，顶端钝尖，基部圆形或耳状，不对称，叶缘骨质，具腺毛。花大，白色或浅红色，芳香。蒴果长圆形，有肋棱和短腺毛。花期7～8月，果期9～10。

Evergreen shrubs or small trees. Bark grayish. Young shoots densely long glandular-setulose. Leaves papery or thinly leathery, blade oblong or elliptic-lanceolate, apex obtuse, cuspidate, base rounded or auriculate, unequal, margin cartilaginous, glandular-ciliate. Flowers large, white or light red, fragrant. Capsules terete, ribbed and shortly glandular haired. Fl. Jul.–Aug., fr. Sept.–Oct..

树干 Trunk
摄影：卢志军 Photo by: Zhijun Lu

叶片 Leaves
摄影：江明喜 Photo by: Mingxi Jiang

叶背 Leaf abaxial surface
摄影：卢志军 Photo by: Zhijun Lu

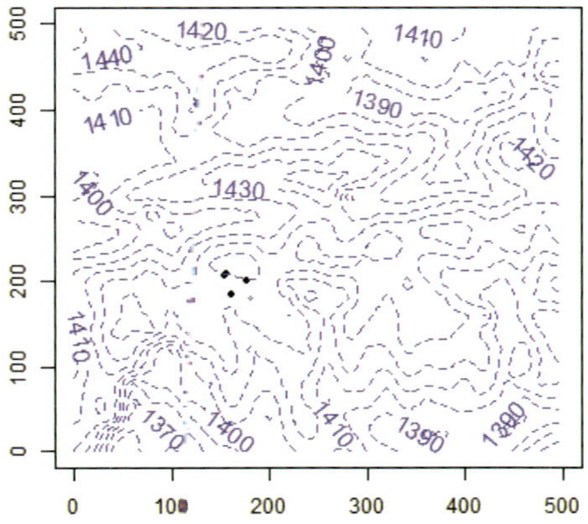

个体分布图 Distribution of individuals

径级分布表 DBH class

胸径区间 (Diameter class) (cm)	个体数 (No. of individuals in the plot)	比例 (Proportion) (%)
1～2.5	1	25.0
2.5～5	2	50.0
5～8	0	0.0
8～11	1	25.0
11～15	0	0.0
15～20	0	0.0
⩾20	0	0.0

199 腺萼马银花 xiàn è mǎ yín huā | Bach Rhododendron

Rhododendron bachii，杜鹃花属
杜鹃花科 | Ericaceae

代码（SpCode）= RHOBAC
个体数（Individual number/25 hm^2）= 1
最大胸径（Max DBH）= 7.48 cm
所在林层（Stratum）= 亚乔木层 (Understory)
重要值排序（Importance value rank）= 54/54

树干 Trunk
摄影：卢志军 Photo by: Zhijun Lu

常绿灌木。幼枝有柔毛和腺毛。叶散生，薄革质，幼时红棕色，卵状椭圆形或宽椭圆形，基部宽楔形或近圆形，叶缘疏具波状圆齿。花紫色或带紫色，花冠上部三个裂片带深红色斑点，单花侧生于枝上部的叶腋。蒴果宽卵形，有腺毛。花期4～5月，果期6～7月。

Evergreen shrubs. Young shoots pubescent and sparsely glandular-setose. Leaves sparsifolious, thinly leathery, reddish brown when young, blade ovate-elliptic or broadly elliptic, base broadly cuneate or subrounded, margin coarsely undulate-crenate. Flowers purple or purplish, with dark red flecks on upper three lobes, single lateral flowers axillary on upper branches. Capsules broadly ovoid, glandular-hairy. Fl. Apr.–May, fr. Jun.–Jul..

花枝 Flowering branches
摄影：江明喜 Photo by: Mingxi Jiang

小枝和叶背 Branchlet and leaf abaxial surface
摄影：卢志军 Photo by: Zhijun Lu

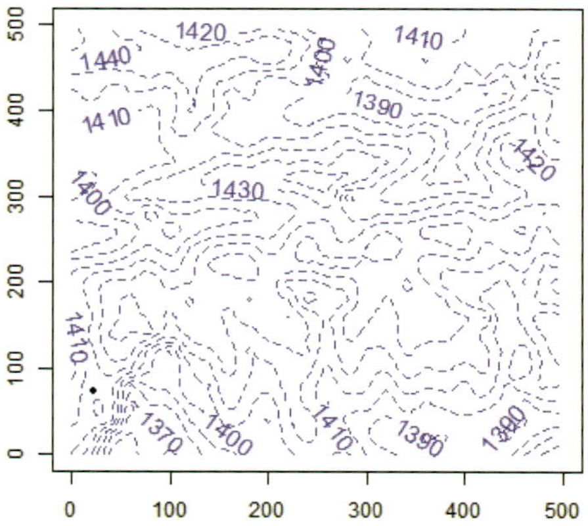

个体分布图 Distribution of individuals

径级分布表 DBH class

胸径区间 (Diameter class) (cm)	个体数 (No. of individuals in the plot)	比例 (Proportion) (%)
1～2.5	0	0.0
2.5～5	0	0.0
5～8	1	1.0
8～11	0	0.0
11～15	0	0.0
15～20	0	0.0
≥20	0	0.0

200 江南越桔 jiāng nán yuè jú | Mandarin Blueberry

Vaccinium mandarinorum，越桔属
杜鹃花科 | Ericaceae

代码（SpCode）= VACMAN
个体数（Individual number/25 hm^2）= 2967
最大胸径（Max DBH）= 20.15 cm
所在林层（Stratum）= 亚乔木层 (Understory)
重要值排序（Importance value rank）= 10/54

常绿灌木或小乔木。幼枝通常无毛，有时被短柔毛。叶片厚革质，卵形或长圆状披针形，先端渐尖，基部楔形至钝圆，边缘有细锯齿。总状花序腋生，花冠白色。浆果，熟时紫黑色，无毛。花期4～6月，果期6～10月。

Evergreen shrubs or small trees. Young branchlets usually glabrous, sometimes pubescent. Leaf blade thickly leathery, ovate or oblong-lanceolate, apex acuminate, base cuneate to obtusely rounded, margin dentate. Raceme axillary. Corolla white. Berry, purple-black at maturity, glabrous. Fl. Apr.–Jun., fr. Jun.–Oct..

树干 Trunk
摄影：卢志军 Photo by: Zhijun Lu

花枝 Flowering branches
摄影：刘海波 Photo by: Haibo Liu

小枝和叶背 Branchlet and leaf abaxial surface
摄影：卢志军 Photo by: Zhijun Lu

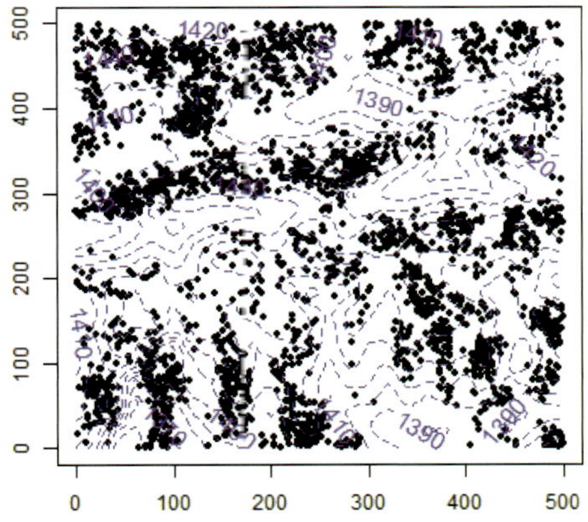

个体分布图 Distribution of individuals

径级分布表 DBH class

胸径区间 (Diameter class) (cm)	个体数 (No. of individuals in the plot)	比例 (Proportion) (%)
1～2.5	440	14.8
2.5～5	1478	49.8
5～8	802	27.0
8～11	190	6.4
11～15	39	1.3
15～20	10	0.3
≥20	8	0.3

201 君迁子

jūn qiān zǐ | Dateplum Persimmon

Diospyros lotus，柿树属
柿树科 | Ebenaceae

代码（SpCode）= DIOLOT
个体数（Individual number/25 hm^2）= 79
最大胸径（Max DBH）= 36.35 cm
所在林层（Stratum）= 灌木层 (Shrub)
重要值排序（Importance value rank）= 38/128

落叶乔木。树皮灰黑色至灰棕色，条裂。小枝被黄灰色短毛。叶片近膜质，椭圆形至长圆形，先端渐尖，基部圆形至宽楔形。花红白色或绿白色。果球形，黄色，常被霜，熟时蓝黑色。花期6月，果期9～11月。

Deciduous trees. Bark grayish black to grayish brown, peeling in strips. Branchlets yellowish gray pubescent. Leaves submembranous, blade elliptic to oblong, apex acuminate, base rounded to broadly cuneate. Flowers white or greenish white. Fruits globose, yellow, usually pruinose, blue-black when mature. Fl. Jun., fr. Sept.–Nov..

树干 Trunk
摄影：卢志军 Photo by: Zhijun Lu

花枝 Flowering branches
摄影：江明喜 Photo by: Mingxi Jiang

小枝和叶背 Branchlet and leaf abaxial surface
摄影：卢志军 Photo by: Zhijun Lu

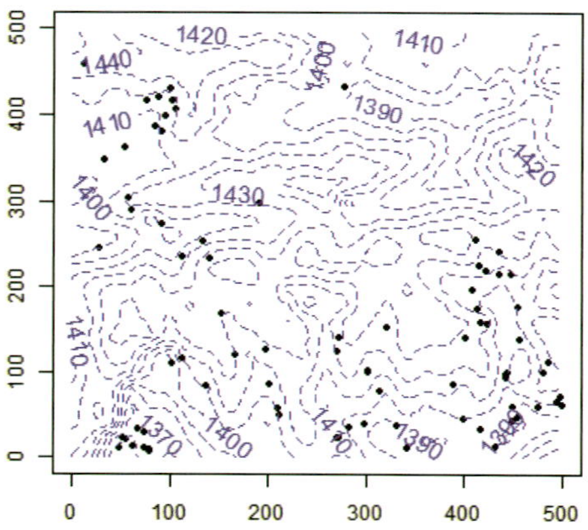

个体分布图 Distribution of individuals

径级分布表 DBH class

胸径区间 (Diameter class) (cm)	个体数 (No. of individuals in the plot)	比例 (Proportion) (%)
1～2	16	20.3
2～3	7	8.9
3～4	4	5.1
4～5	1	1.3
5～7	2	2.5
7～10	5	6.3
≥10	44	55.7

202 山柿

shān shì | Japanese Persimmon

Diospyros japonica，柿树属
柿树科 | Ebenaceae

代码（SpCode）= DIOJAP
个体数（Individual number/25 hm^2）= 63
最大胸径（Max DBH）= 29.68 cm
所在林层（Stratum）= 亚乔木层 (Understory)
重要值排序（Importance value rank）= 25/54

树干 Trunk
摄影：卢志军 Photo by: Zhijun Lu

落叶乔木。树皮灰黑色或灰褐色；枝深褐色或黑褐色，散生纵裂的唇形小皮孔。叶革质，宽椭圆形、卵形或卵状披针形，先端急尖，基部圆形、截形、浅心形或钝。雌雄异株，雄花聚伞花序，雌花单生或2～3一簇。果球形或扁球形，熟时红色，被白霜。花期4～5月，果期9～10月。

Deciduous trees. Bark grayish black or grayish brown. Branches dark brown or black-brown, lenticels sparse, longitudinal peeling, labiate. Leaves leathery, blade broadly elliptic, ovate or ovate-lanceolate, apex acute, base rounded, truncate, shallowly cordate or obtuse. Dioecious, male flowers in cymes, female flowers solitary or in clusters of 2 or 3. Fruits globose or flat globose, red when mature, pruinose. Fl. Apr.–May, fr. Sept.–Oct..

花枝 Flowering branches
摄影：江明喜 Photo by: Mingxi Jiang

叶背 Leaf abaxial surface
摄影：卢志军 Photo by: Zhijun Lu

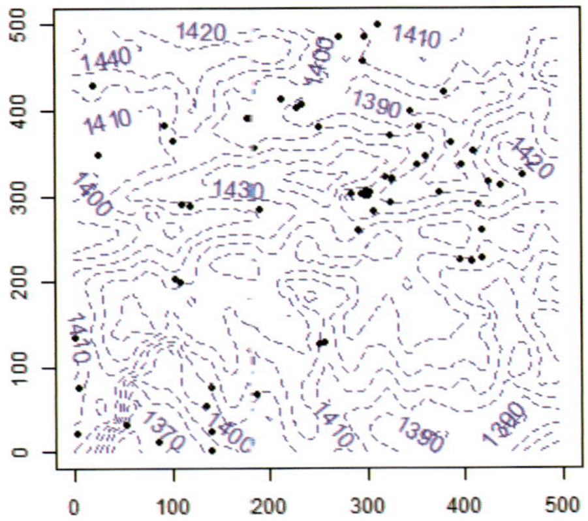

个体分布图 Distribution of individuals

径级分布表 DBH class

胸径区间 (Diameter class) (cm)	个体数 (No. of individuals in the plot)	比例 (Proportion) (%)
1～2.5	4	6.3
2.5～5	1	1.6
5～8	0	0.0
8～11	4	6.3
11～15	9	14.3
15～20	20	31.7
⩾20	25	39.7

203 白辛树

bái xīn shù | Glabrous-leaf Epaulette-tree

Pterostyrax psilophyllus，白辛树属
安息香科 | Styracaceae

代码（SpCode）= PTEPSI
个体数（Individual number/25 hm^2）= 40
最大胸径（Max DBH）= 66.53 cm
所在林层（Stratum）= 乔木层 (Canopy)
重要值排序（Importance value rank）= 21/50

落叶乔木。幼枝疏被星状毛。叶卵圆状椭圆形或倒卵状椭圆形，先端突尖，基部宽楔形，边缘波状。圆锥花序，花白色，萼筒状。核果近纺锤形，具棱而无狭翅，密被黄色长硬毛。花期4～6月，果期7～8月。

树干 Trunk
摄影：卢志军 Photo by: Zhijun Lu

Deciduous trees. Young branchlets stellate pubescent. Leaf blade ovate-elliptic or obovate-elliptic, apex abruptly acute, base broadly cuneate, margin undulate. Panicles, flowers white, calyx tubular. Fruits subfusiform, ribbed, wings absent, densely gray-yellow hirsute. Fl. Apr.–Jun., fr. Jul.–Aug..

果序 Infructescence
摄影：江明喜 Photo by: Mingxi Jiang

果枝和叶背 Fruiting branch and leaf abaxial surface
摄影：刘海波 Photo by: Haibo Liu

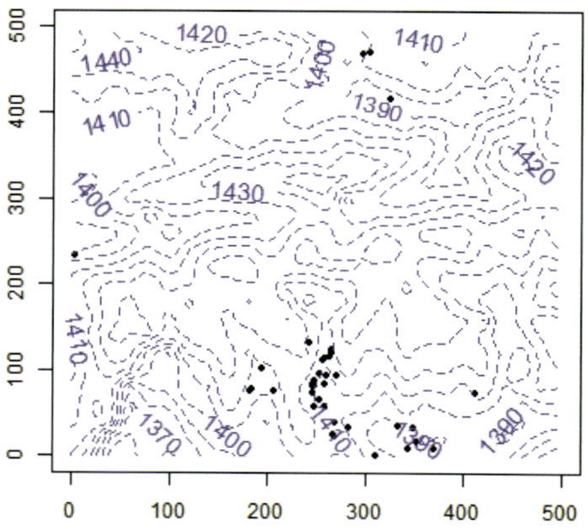

个体分布图 Distribution of individuals

径级分布表 DBH class

胸径区间 (Diameter class) (cm)	个体数 (No. of individuals in the plot)	比例 (Proportion) (%)
1～2.5	15	37.5
2.5～5	1	2.5
5～10	3	7.5
10～25	10	25.0
25～40	3	7.5
40～100	8	20.0
≥100	0	0.0

204 芬芳安息香 fēn fāng ān xī xiāng | Sweet-scented Snow-bell

Styrax odoratissimus，安息香属
安息香科 | Styracaceae

代码（SpCode）= STYODO
个体数（Individual number/25 hm^2）= 35
最大胸径（Max DBH）= 20.05 cm
所在林层（Stratum）= 灌木层 (Shrub)
重要值排序（Importance value rank）= 64/128

树干 Trunk
摄影：卢志军 Photo by: Zhijun Lu

小乔木。树皮灰褐色，不开裂。叶互生，薄革质至纸质，卵形或卵状椭圆形，先端渐尖或急尖，基部宽楔形至圆形，全缘或上部有疏锯齿。总状或圆锥花序，顶生，花白色。果实近球形，顶端骤缩而具弯喙，密被灰黄色茸毛。花期3～4月，果期6～9月。

Small trees. Bark gray-brown, not exfoliating. Leaves alternate, thinly leathery to papery, leaf blade ovate to ovate-elliptic, apex acuminate or acute, base broadly cuneate to rounded, margin entire or remotely serrulate. Inflorescence terminal racemes or panicles. Corolla white. Fruit subglobose, apex with a slightly curved rostrum, densely gray-yellow stellate tomentose. Fl. Mar.–Apr., fr. Jun.–Sept..

花枝 Flowering branches
摄影：卢志军 Photo by: Zhijun Lu

果枝 Fruiting branch
摄影：卢志军 Photo by: Zhijun Lu

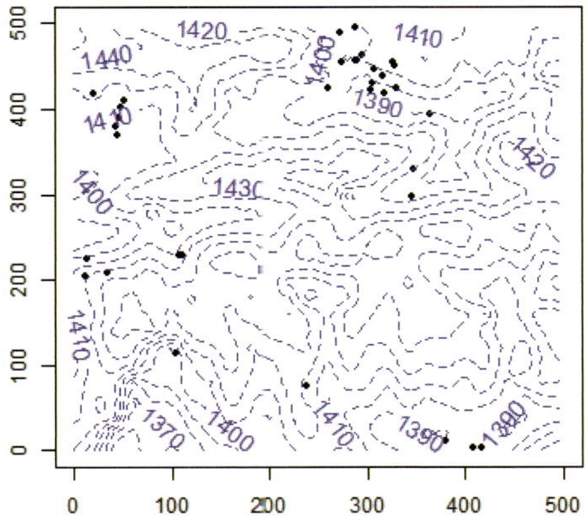

个体分布图 Distribution of individuals

径级分布表 DBH class

胸径区间 (Diameter class) (cm)	个体数 (No. of individuals in the plot)	比例 (Proportion) (%)
1～2	17	48.6
2～3	6	17.1
3～4	0	0.0
4～5	1	2.9
5～7	4	11.4
7～10	2	5.7
≥10	5	14.3

205 薄叶山矾　báo yè shān fán | Thin-leaf Sweet-leaf

Symplocos anomala，山矾属

山矾科 | Symplocaceae

代码（SpCode）= SYMANO

个体数（Individual number/25 hm^2）= 8295

最大胸径（Max DBH）= 20.05 cm

所在林层（Stratum）= 灌木层 (Shrub)

重要值排序（Importance value rank）= 2/128

常绿灌木或小乔木。小枝密生褐色短茸毛。叶披针形或长圆形，先端渐尖，基部楔形，全缘或疏生浅齿牙。总状花序腋生，花冠白色，有香味，裂片长圆形。核果长圆形，被短柔毛，熟后褐色。花期5月，果期7～8月。

Evergreen shrubs or small trees. Young branchlets brown pubescent. Leaf blade lanceolate or oblong, apex acuminate, base cuneate, margin entire or sparsely finely dentate. Racemes axillary, corolla white, fragrant, lobes oblong. Drupes oblong, pubescent, brown when mature. Fl. May, fr. Jul.–Aug..

树干　Trunk

摄影：卢志军　Photo by: Zhijun Lu

花枝　Flowering branches

摄影：路俊盟　Photo by: Junmeng Lu

果枝　Fruiting branches

摄影：鲍大川　Photo by: Dachuan Bao

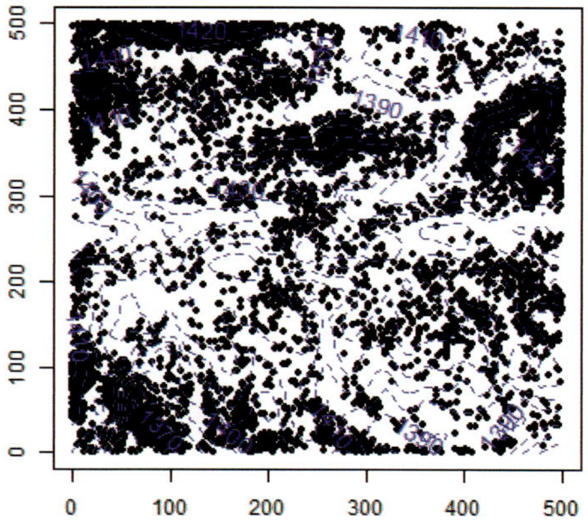

个体分布图 Distribution of individuals

径级分布表 DBH class

胸径区间 (Diameter class) (cm)	个体数 (No. of individuals in the plot)	比例 (Proportion) (%)
1～2	2312	27.9
2～3	1813	21.9
3～4	1434	17.3
4～5	996	12.0
5～7	1129	13.6
7～10	512	6.2
≥10	99	1.2

206 毛山矾

máo shān fán | Groff Sweet-leaf

Symplocos groffii，山矾属
山矾科 | Symplocaceae

代码（SpCode）= SYMGRO
个体数（Individual number/25 hm^2）= 6746
最大胸径（Max DBH）= 21.91 cm
所在林层（Stratum）= 亚乔木层 (Understory)
重要值排序（Importance value rank）= 4/54

树干 Trunk
摄影：卢志军 Photo by: Zhijun Lu

常绿小乔木或灌木。嫩枝和叶柄被展开的灰褐色长硬毛。叶纸质，椭圆形、卵形或倒卵状椭圆形，先端渐尖，基部阔楔形或圆，全缘后弯至反卷或稍具锐尖。穗状花序，苞片三角状阔卵形。核果长圆状椭圆形，被柔毛，顶端宿萼裂片直立。花期4月，果期6~7月。

Evergreen shrubs or small trees. Young branchlets and petioles grayish brown hirsute. Leaf blade papery, elliptic, ovate or obovate-elliptic, apex acuminate, base broadly cuneate to rounded, margin recurved to revolute and entire or remotely sharply dentate. Spikes, bracts triangular broadly ovate. Drupes oblong-elliptic, pubescent, apex with persistent erect calyx lobes. Fl. Apr., fr. Jun.–Jul..

花枝 Flowering branch
摄影：张奎汉 Photo by: Kuihan Zhang

果枝 Fruiting branch
摄影：刘海波 Photo by: Haibo Liu

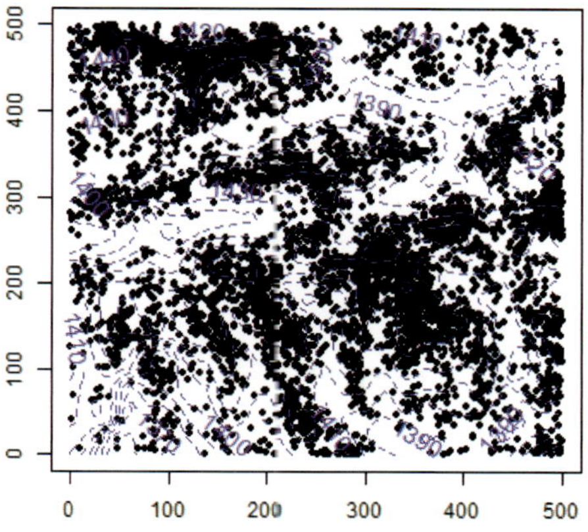

个体分布图 Distribution of individuals

径级分布表 DBH class

胸径区间 (Diameter class) (cm)	个体数 (No. of individuals in the plot)	比例 (Proportion) (%)
1~2.5	2909	43.1
2.5~5	2495	37.0
5~8	814	12.1
8~11	359	5.3
11~15	128	1.9
15~20	33	0.5
≥20	8	0.1

207 铁山矾

tiě shān fán | Iron Sweet-leaf

Symplocos pseudobarberina，山矾属
山矾科 | Symplocaceae

代码（SpCode）= SYMPSE
个体数（Individual number/25 hm^2）= 5604
最大胸径（Max DBH）= 20.57 cm
所在林层（Stratum）= 灌木层 (Shrub)
重要值排序（Importance value rank）= 3/128

常绿乔木。幼枝黄绿色，无毛；老枝深灰色。叶纸质，卵形或卵状椭圆形，先端渐尖或尾状渐尖，基部楔形或稍圆，全缘或具稀疏的浅波状齿。总状花序基部常分枝，无毛，花冠白色。核果绿色或黄色，长圆状卵形，顶端宿萼裂片直立。花期10～11月，果期翌年5月。

Evergreen trees. Young branchlets yellowish green, glabrous; old branchlets dark gray. Leaf blade papery, ovate or ovate-elliptic, apex acuminate or caudate acuminate, base cuneate or slightly rounded, margin entire or sparsely shallowly undulate. Racemes usually branched from base, glabrous, corolla white. Drupes green or yellow, oblong-ovoid, apex with persistent erect calyx lobes. Fl. Oct.–Nov., fr. May of following yr..

树干 Trunk
摄影：卢志军 Photo by: Zhijun Lu

花枝 Flowering branch
摄影：刘海波 Photo by: Haibo Liu

果枝 Fruiting branch
摄影：刘海波 Photo by: Haibo Liu

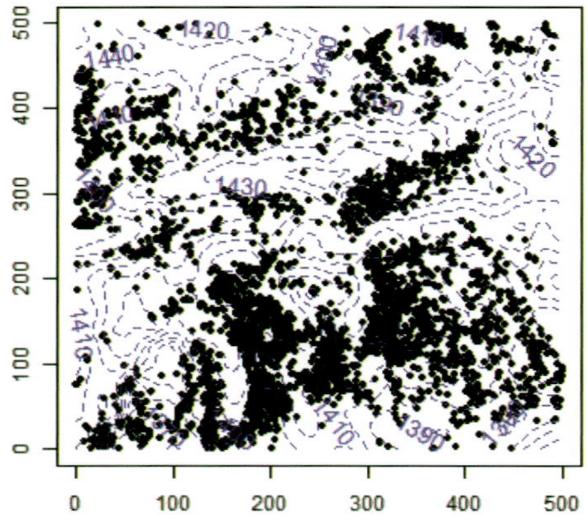

个体分布图 Distribution of individuals

径级分布表 DBH class

胸径区间 (Diameter class) (cm)	个体数 (No. of individuals in the plot)	比例 (Proportion) (%)
1～2	2391	42.7
2～3	1228	21.9
3～4	722	12.9
4～5	439	7.8
5～7	415	7.4
7～10	252	4.5
≥10	157	2.8

208 山矾

shān fán | Sweet-leaf

Symplocos sumuntia，山矾属
山矾科 | Symplocaceae

代码（SpCode）= SYMSUM
个体数（Individual number/25 hm^2）= 4218
最大胸径（Max DBH）= 10.82 cm
所在林层（Stratum）= 灌木层 (Shrub)
重要值排序（Importance value rank）= 9/128

常绿灌木或小乔木。小枝褐色，无毛。叶薄革质，卵状披针形或椭圆形，先端尾状渐尖，基部楔形，边缘有细锯齿。总状花序，花冠白色。核果坛状，黄绿色。花期4～5月，果期7～8月。

Evergreen shrubs or small trees. Young branchlets brown, glabrous. Leaf blade thinly leathery, ovate-lanceolate or elliptic, apex caudate acuminate, base cuneate, margin finely serrate. Racemes, corolla white. Drupes ampulliform, yellowish green. Fl. Apr.–May, fr. Jul.–Aug..

树干 Trunk
摄影：卢志军 Photo by: Zhijun Lu

花枝 Flowering branch
摄影：江明喜 Photo by: Mingxi Jiang

果枝 Fruiting branch
摄影：卢志军 Photo by: Zhijun Lu

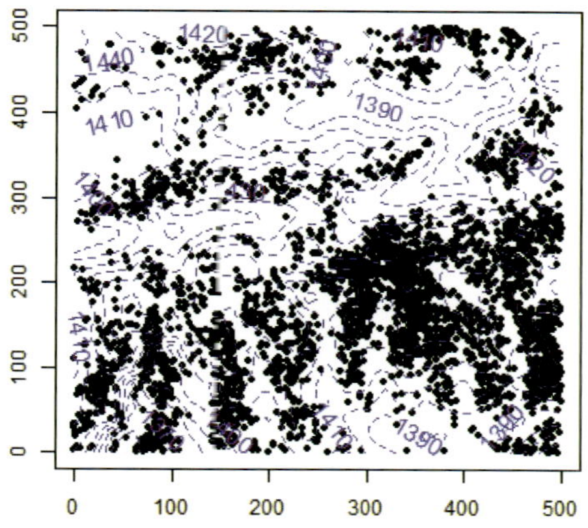

个体分布图 Distribution of individuals

径级分布表 DBH class

胸径区间 (Diameter class) (cm)	个体数 (No. of individuals in the plot)	比例 (Proportion) (%)
1～2	1948	46.2
2～3	1195	28.3
3～4	562	13.3
4～5	240	5.7
5～7	199	4.7
7～10	58	1.4
≥10	16	0.4

209 羊舌树

yáng shé shù | Large-leaf Sweet-leaf

Symplocos glauca，山矾属

山矾科 | Symplocaceae

代码（SpCode）= SYMMAC

个体数（Individual number/25 hm^2）= 3121

最大胸径（Max DBH）= 27.87 cm

所在林层（Stratum）= 亚乔木层 (Understory)

重要值排序（Importance value rank）= 9/54

常绿乔木。幼枝和叶柄被有褐色皱曲茸毛；老枝无毛，黑褐色。叶革质，椭圆形、卵状椭圆形，先端渐尖或急尖，基部圆钝或阔楔形，全缘或上部具细尖齿。团伞花序腋生或生于具叶痕老枝上，花冠白色或浅黄色。核果绿色或浅褐色，圆柱形。花期6～7月。

Evergreen trees. Young branchlets and petioles pubescent with brown crispate hairs; old branchlets glabrous, black-brown. Leaf blade leathery, elliptic, ovate-elliptic, apex acuminate or acute, base rounded and obtuse or broadly cuneate, margin entire or apically finely serrate. Glomerule axillary or on old branches with leaf scars, corolla white or light yellow. Drupes green or light brown, cylindric. Fl. Jun.–Jul..

树干 Trunk

摄影：卢志军 Photo by: Zhijun Lu

花枝 Flowering branch

摄影：刘海波 Photo by: Haibo Liu

果枝 Fruiting branch

摄影：徐文秀 Photo by: Wenxiu Xu

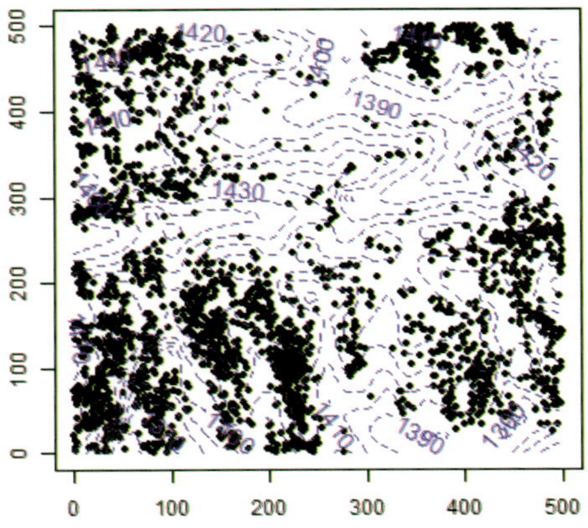

个体分布图 Distribution of individuals

径级分布表 DBH class

胸径区间 (Diameter class) (cm)	个体数 (No. of individuals in the plot)	比例 (Proportion) (%)
1～2.5	1035	33.2
2.5～5	1061	34.0
5～8	491	15.7
8～11	322	10.3
11～15	158	5.1
15～20	41	1.3
≥20	13	0.4

210 老鼠矢 lǎo shǔ shǐ | Star-shape Sweet-leaf

Symplocos stellaris，山矾属
山矾科 I Symplocaceae

代码（SpCode）= SYMSTE
个体数（Individual number/25 hm²）= 1599
最大胸径（Max DBH）= 12.38 cm
所在林层（Stratum）= 灌木层 (Shrub)
重要值排序（Importance value rank）= 19/128

常绿灌木或小乔木。幼枝和芽被红褐色长柔毛。叶厚革质，披针状长圆形至长圆状椭圆形，先端钝或短渐尖，基部阔楔形，全缘。团伞花序或穗状花序，花无梗，花冠裂片长圆状椭圆形。核果长圆形，宿存萼裂片直立。花期9月，果期翌年5月。

Evergreen shrubs or small trees. Young branchlets and buds reddish brown tomentellous. Leaf blade thickly leathery, lanceolate-oblong to oblong-elliptic, apex obtuse or shortly acuminate, base broadly cuneate, margin entire. Glomerule or spikes, peduncle absent, corolla lobes oblong elliptic. Drupes oblong, erect calyx lobes persistent. Fl. Sept., fr. May of following yr..

树干 Trunk
摄影：卢志军 Photo by: Zhijun Lu

小枝和叶背 Branchlet and leaf abaxial surface
摄影：刘海波 Photo by: Haibo Liu

果枝 Fruiting branch
摄影：刘海波 Photo by: Haibo Liu

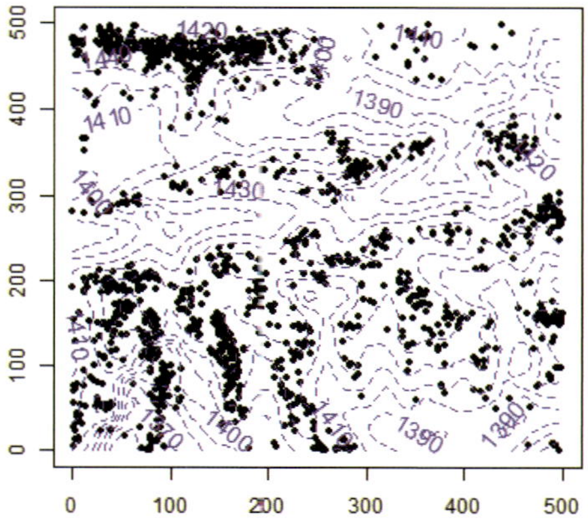

个体分布图 Distribution of individuals

径级分布表 DBH class

胸径区间 (Diameter class) (cm)	个体数 (No. of individuals in the plot)	比例 (Proportion) (%)
1～2	566	35.4
2～3	364	22.8
3～4	242	15.1
4～5	167	10.4
5～7	164	10.3
7～10	69	4.3
≥10	27	1.7

211 白檀

bái tán | Sapphire-berry Sweet-leaf

Symplocos paniculata，山矾属

山矾科 | Symplocaceae

代码（SpCode）= SYMPAN

个体数（Individual number/25 hm^2）= 1566

最大胸径（Max DBH）= 14.10 cm

所在林层（Stratum）= 灌木层 (Shrub)

重要值排序（Importance value rank）= 23/128

落叶灌木或小乔木。幼枝、叶下面，花序及果疏被柔毛。叶椭圆形、倒卵形至倒卵状椭圆形，先端急尖或渐尖，基部楔形至圆形，边缘有细尖锯齿。花白色，芳香，花瓣长圆形。核果斜卵状球形，熟时蓝色，宿存萼裂片直立。花期5～6月，果期7～8月。

Deciduous shrubs or small trees. Young branchlets, leaves abaxially, inflorescences and fruits sparsely pubescent. Leaf blade elliptic, obovate to obovate-elliptic, apex acute or acuminate, base cuneate to rounded, margin finely serrate. Flowers white, fragrant, petals oblong. Drupes oblique ovoid, blue when mature, apex with persistent erect calyx lobes. Fl. May–Jun., fr. Jul.–Aug..

树干 Trunk

摄影：卢志军 Photo by: Zhijun Lu

花枝 Flowering branch

摄影：路俊盟 Photo by: Junmeng Lu

果枝 Fruiting branch

摄影：卢志军 Photo by: Zhijun Lu

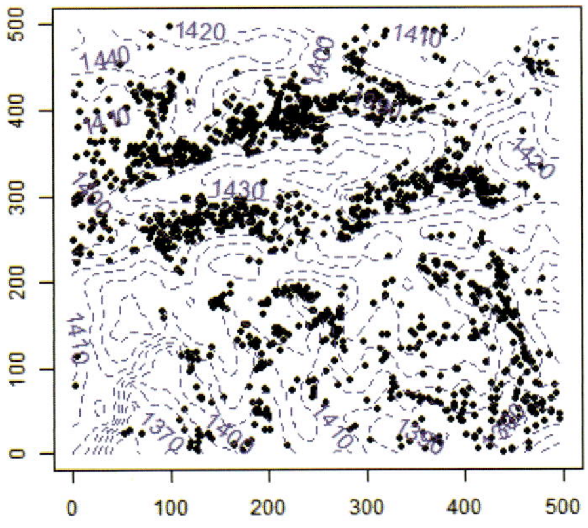

个体分布图 Distribution of individuals

径级分布表 DBH class

胸径区间 (Diameter class) (cm)	个体数 (No. of individuals in the plot)	比例 (Proportion) (%)
1～2	779	49.7
2～3	363	23.2
3～4	181	11.6
4～5	102	6.5
5～7	87	5.6
7～10	39	2.5
≥10	15	1.0

212 光亮山矾 guāng liàng shān fán | Shining Sweet-leaf

Symplocos lucida，山矾属
山矾科 | Symplocaceae

代码（SpCode）= SYMLUC
个体数（Individual number/25 hm^2）= 1315
最大胸径（Max DBH）= 23.82 cm
所在林层（Stratum）= 灌木层 (Shrub)
重要值排序（Importance value rank）= 15/128

常绿灌木或小乔木。小枝黄绿色，无毛，有棱。顶芽棱镖状。叶革质，长圆形或椭圆形，先端短尖，基部楔形，边缘上部具波状浅齿牙。穗状花序腋生，密集，花冠淡黄色，较花萼稍长。核果卵圆形或长圆形，棕色，宿存萼裂片直立，核无棱。花期4～5月，果期7～8月。

Evergreen shrubs or small trees. Branchlets yellowish green, glabrous, ribbed. Terminal buds spearheaded. Leaf blade leathery, oblong to elliptic, apex shortly acuminate, base cuneate, margin apically shallowly undulate. Spikes axillary, dense, corolla light yellow, longer than calyx. Drupes ovoid or oblong, brown, erect calyx lobes persistent, stone ribs absent. Fl. Apr.–May, fr. Jul.–Aug..

树干 Trunk
摄影：卢志军 Photo by: Zhijun Lu

花枝 Flowering branch
摄影：徐耀粘 Photo by: Yaozhan Xu

果枝 Fruiting branches
摄影：刘海波 Photo by: Haibo Liu

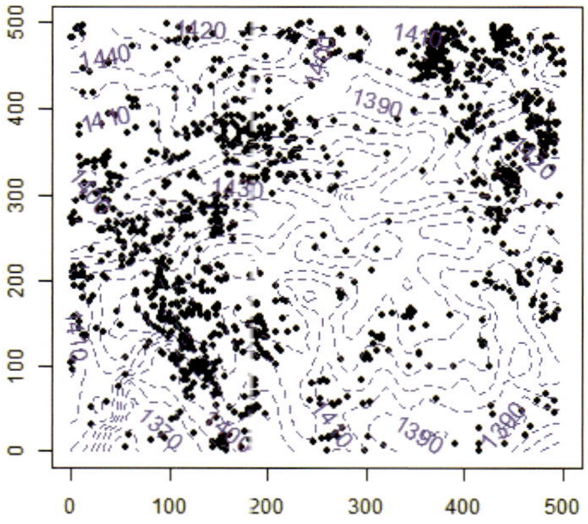

个体分布图 Distribution of individuals

径级分布表 DBH class

胸径区间 (Diameter class) (cm)	个体数 (No. of individuals in the plot)	比例 (Proportion) (%)
1～2	275	20.9
2～3	221	16.8
3～4	167	12.7
4～5	129	9.8
5～7	187	14.2
7～10	178	13.5
≥10	158	12.0

213 苦枥木

kǔ lì mù | Retuse Ash

Fraxinus insularis，白蜡属

木犀科 | Oleaceae

代码（SpCode）= FRAINS

个体数（Individual number/25 hm^2）= 28

最大胸径（Max DBH）= 33.48 cm

所在林层（Stratum）= 乔木层 (Canopy)

重要值排序（Importance value rank）= 22/50

落叶乔木。复叶，小叶3～5，小叶薄革质，卵形至卵状披针形，先端渐尖，基部圆形或窄狭，边缘有锯齿或全缘。圆锥花序大型，多花，白色。翅果线形，顶端微凹。花期5～6月，果期8～9。

Deciduous trees. Leaves compound, leaflets 3-5, blade leathery, ovate to ovate-lanceolate, apex acuminate, base rounded or narrow, margin serrate or entire. Panicles, large, flowers numerous and white. Samaras linear, apex emarginate. Fl. May–Jun., fr. Aug.–Sept..

树干 Trunk

摄影：卢志军 Photo by: Zhijun Lu

果枝 Fruiting branches

摄影：鲍大川 Photo by: Dachuan Bao

复叶背面 Compound leaves abaxial surface

摄影：卢志军 Photo by: Zhijun Lu

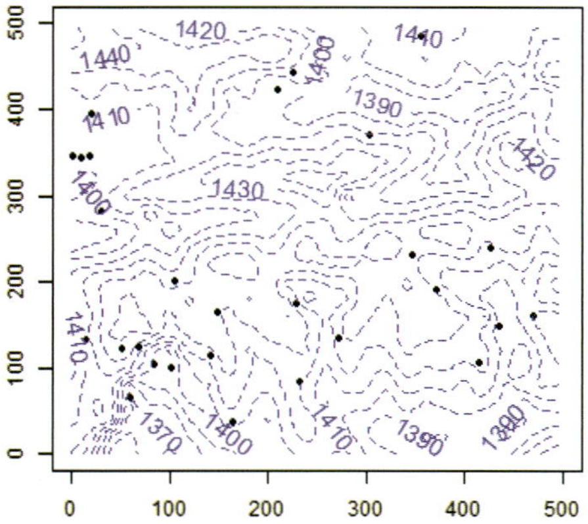

个体分布图 Distribution of individuals

径级分布表 DBH class

胸径区间 (Diameter class) (cm)	个体数 (No. of individuals in the plot)	比例 (Proportion) (%)
1～2.5	10	35.7
2.5～5	4	14.3
5～10	4	14.3
10～25	8	28.6
25～40	2	7.1
40～100	0	0.0
⩾100	0	0.0

214 蜡子树 là zǐ shù | White-flower Privet

Ligustrum leucanthum，女贞属
木犀科 | Oleaceae

代码（SpCode）= LIGLEU
个体数（Individual number/25 hm^2）= 11
最大胸径（Max DBH）= 14.95 cm
所在林层（Stratum）= 灌木层 (Shrub)
重要值排序（Importance value rank）= 87/128

落叶灌木。枝条被长硬毛。叶厚纸质，椭圆形、椭圆状长圆形、卵状长圆形或长圆状披针形，先端锐尖或渐尖，基部楔形或宽楔形。圆锥花序卵形或圆柱形。核果椭圆形，蓝黑色。花期5～6月，果期10月。

Deciduous shrubs. Branchlets hirsute. Leaf blade thickly papery, elliptic, elliptic-oblong, ovate-oblong or oblong-lanceolate, apex acute or acuminate, base cuneate or broadly cuneate. Panicles ovoid or cylindric. Drupes ellipoid, blue-black. Fl. May–Jun., fr. Oct..

树干 Trunk
摄影：卢志军 Photo by: Zhijun Lu

花枝 Flowering branches
摄影：卢志军 Photo by: Zhijun Lu

果枝 Fruiting branches
摄影：卢志军 Photo by: Zhijun Lu

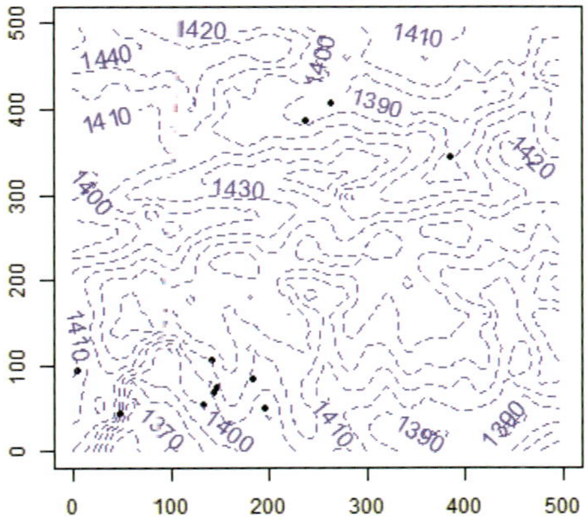

个体分布图 Distribution of individuals

径级分布表 DBH class

胸径区间 (Diameter class) (cm)	个体数 (No. of individuals in the plot)	比例 (Proportion) (%)
1～2	4	36.4
2～3	3	27.3
3～4	0	0.0
4～5	3	27.3
5～7	0	0.0
7～10	0	0.0
⩾10	1	9.1

215 小蜡

xiǎo là | Chinese Privet

Ligustrum sinense，女贞属

木犀科 | Oleaceae

代码（SpCode）= LIGSIN

个体数（Individual number/25 hm²）= 8

最大胸径（Max DBH）= 12.88 cm

所在林层（Stratum）= 灌木层 (Shrub)

重要值排序（Importance value rank）= 93/128

落叶灌木或小乔木。枝条密生短柔毛。叶薄革质，椭圆形至椭圆状长圆形，先端急尖或钝，基部圆形或宽楔形。圆锥花序，花白色。核果近圆形。花期7月，果期9～10月。

Deciduous shrubs or small trees. Branchlets pubescent. Leaf blade thinly leathery, elliptic to elliptic-oblong, apex acute or obtuse, base rounded or broadly cuneate. Panicles, flowers white. Drupes subglobose. Fl. Jul., fr. Sept.–Oct..

树干 Trunk

摄影：卢志军 Photo by: Zhijun Lu

花枝 Flowering branches

摄影：刘海波 Photo by: Haibo Liu

花枝和叶背 Flowering branch and leaf abaxial surface

摄影：刘海波 Photo by: Haibo Liu

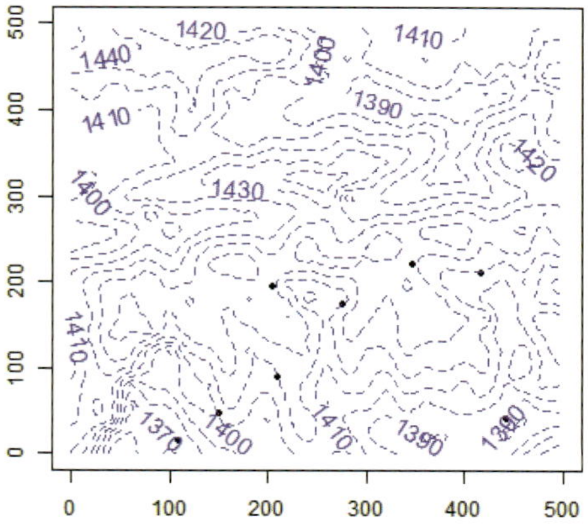

个体分布图 Distribution of individuals

径级分布表 DBH class

胸径区间 (Diameter class) (cm)	个体数 (No. of individuals in the plot)	比例 (Proportion) (%)
1～2	3	37.5
2～3	1	12.5
3～4	1	12.5
4～5	0	0.0
5～7	2	25.0
7～10	0	0.0
⩾10	1	12.5

216 木犀 mù xī | Sweet Osmanthus

Osmanthus fragrans，木犀属
木犀科 | Oleaceae

代码（SpCode）= OSMFRA
个体数（Individual number/25 hm^2）= 140
最大胸径（Max DBH）= 14.58 cm
所在林层（Stratum）= 灌木层 (Shrub)
重要值排序（Importance value rank）= 47/128

常绿乔木或灌木。叶对生，革质，椭圆形至椭圆状披针形，先端急尖或渐尖，基部楔形，全缘或上半部疏生细锯齿。花序簇生于叶腋，花冠4裂，白色或黄色，极芳香。核果椭圆形。花期5～6月，果期9～10月。

Evergreen trees or shrubs. Leaves opposite, leathery, elliptic to elliptic-lanceolate, apex acute or acuminate, base cuneate, margin entire or apically sparsely serrulate. Cymes fascicled in leaf axils, corolla 4-lobed, white or yellow, extremely fragrant. Drupes ellipoid. Fl. May–Jun., fr. Sept.–Oct..

树干 Trunk
摄影：卢志军 Photo by: Zhijun Lu

小枝和叶片 Branchlet and leaves
摄影：卢志军 Photo by: Zhijun Lu

小枝和叶背 Branchlet and leaf abaxial surface
摄影：卢志军 Photo by: Zhijun Lu

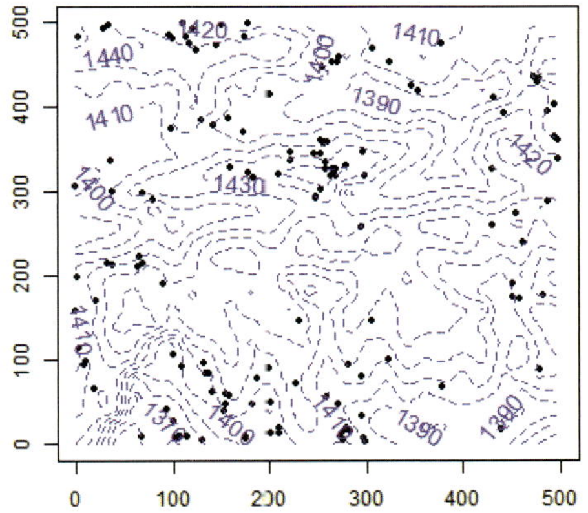

个体分布图 Distribution of individuals

径级分布表 DBH class

胸径区间 (Diameter class) (cm)	个体数 (No. of individuals in the plot)	比例 (Proportion) (%)
1～2	36	25.7
2～3	21	15.0
3～4	26	18.6
4～5	15	10.7
5～7	25	17.9
7～10	12	8.6
≥10	5	3.6

217 香果树

xiāng guǒ shù | Chinese Privet

Emmenopterys henryi，香果树属

茜草科 | Rubiaceae

代码（SpCode）= EMMHEN

个体数（Individual number/25 hm^2）= 9

最大胸径（Max DBH）= 49.53 cm

所在林层（Stratum）= 乔木层 (Canopy)

重要值排序（Importance value rank）= 32/50

落叶乔木。小枝黄褐色，无毛，有皮孔。叶对生，革质，宽椭圆形至宽卵形，先端急尖或骤渐尖，基部阔楔形，全缘。聚伞花序在顶端排列成大型的圆锥花序，花大，淡黄白色，花冠漏斗状。蒴果椭圆形，木质，有纵棱，成熟时红色。种子细小，有宽翅。花期7～8月，果期9～10月。

树干 Trunk

摄影：卢志军 Photo by: Zhijun Lu

Deciduous trees. Branchlets yellow-brown, glabrous, lenticellate. Leaves opposite, leathery, broadly elliptic to broadly ovate, apex acute or abruptly acuminate, base broadly cuneate, margin entire. Panicles consisting of cymes on branch apex, flowers large, light yellow-white, corolla funnelform. Capsules ellipoid, woody, longitudinally ribbed, red when mature. Seeds tiny, broadly winged. Fl. Jul.–Aug., fr. Sept.–Oct..

花枝 Flowering branches

摄影：江明喜 Photo by: Mingxi Jiang

小枝和叶背 Branchlet and leaf abaxial surface

摄影：卢志军 Photo by: Zhijun Lu

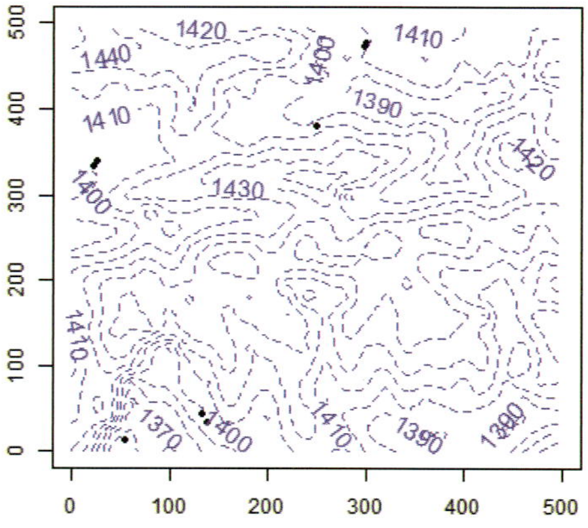

个体分布图 Distribution of individuals

径级分布表 DBH class

胸径区间 (Diameter class) (cm)	个体数 (No. of individuals in the plot)	比例 (Proportion) (%)
1～2.5	0	0.0
2.5～5	0	0.0
5～10	1	11.1
10～25	5	55.6
25～40	2	22.2
40～100	1	11.1
≥100	0	0.0

218 老鸦糊　　lǎo yā hú | Girald Beautyberry

Callicarpa giraldii，紫珠属
马鞭草科 | Verbenaceae

代码（SpCode）= CALGIR
个体数（Individual number/25 hm^2）= 232
最大胸径（Max DBH）= 22.60 cm
所在林层（Stratum）= 灌木层 (Shrub)
重要值排序（Importance value rank）= 44/128

落叶灌木。小枝和花序被黄褐色星状柔毛。叶椭圆形或卵状椭圆形，先端渐尖，基部楔形。聚伞花序，花萼杯状或钟状，萼齿钝三角形，花冠紫红色。果紫色，球形。花期5～7月，果期8～10月。

Deciduous shrubs. Branchlets and inflorescences pubescent with yellow-brown stellate hairs. Leaf blade elliptic or ovate-elliptic, apex acuminate, base cuneate. Cymes, calyx cupulate or campanulate, calyx tooth obtusely triangular, corolla purple-red. Fruit purple, globose. Fl. May–Jul., fr. Aug.–Oct..

树干　Trunk
摄影：卢志军　Photo by: Zhijun Lu

花枝　Flowering branch
摄影：刘海波　Photo by: Haibo Liu

小枝和叶背　Branchlet and leaf abaxial surface
摄影：卢志军　Photo by: Zhijun Lu

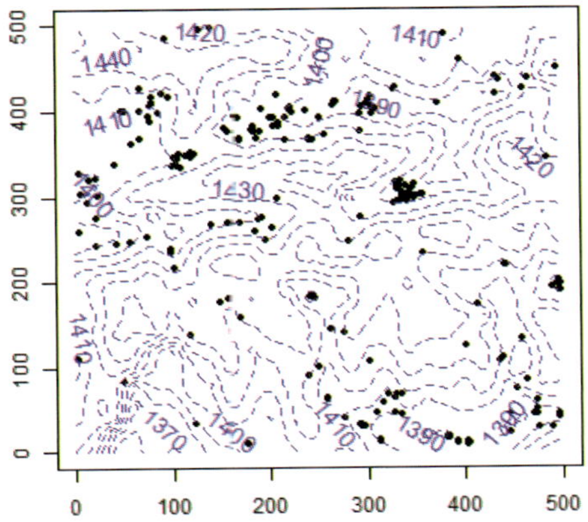

个体分布图 Distribution of individuals

径级分布表 DBH class

胸径区间 (Diameter class) (cm)	个体数 (No. of individuals in the plot)	比例 (Proportion) (%)
1～2	84	36.2
2～3	65	28.0
3～4	34	14.7
4～5	23	9.9
5～7	17	7.3
7～10	8	3.4
≥10	1	0.4

219 海州常山 hǎi zhōu cháng shān | Harlequin Glorybower

Clerodendrum trichotomum，大青属
马鞭草科 | Verbenaceae

代码（SpCode）= CLETRI
个体数（Individual number/25 hm^2）= 27
最大胸径（Max DBH）= 7.32 cm
所在林层（Stratum）= 灌木层 (Shrub)
重要值排序（Importance value rank）= 76/128

树干 Trunk
摄影：卢志军 Photo by: Zhijun Lu

灌木或小乔木，有臭味。嫩枝有黄褐色短柔毛。叶宽卵形或卵状椭圆形，先端急尖，基部宽楔形或截形，全缘或有波状齿。伞房状聚伞花序顶生或腋生，疏松而开展，花萼紫红色，花冠白色或粉红色。核果近球形，熟时蓝紫色。花期6～8月，果期9～11月。

Shrubs or small trees, smelly. Young branchlets pubescent with yellow-brown short tender hairs. Leaf blade broadly ovate or ovate-elliptic, apex acute, base broadly cuneate or truncate, margin entire or undulate. Corymbose cymes terminal or axillary, sparse and spreading, calyx purple-red, corolla white or pink. Drupes subglobose, blue-purple when mature. Fl. Jun.–Aug., fr. Sept.–Nov..

小枝和叶背 Branchlet and leaf abaxial surface
摄影：卢志军 Photo by: Zhijun Lu

果实和种子 Fruits and seeds
摄影：卢志军 Photo by: Zhijun Lu

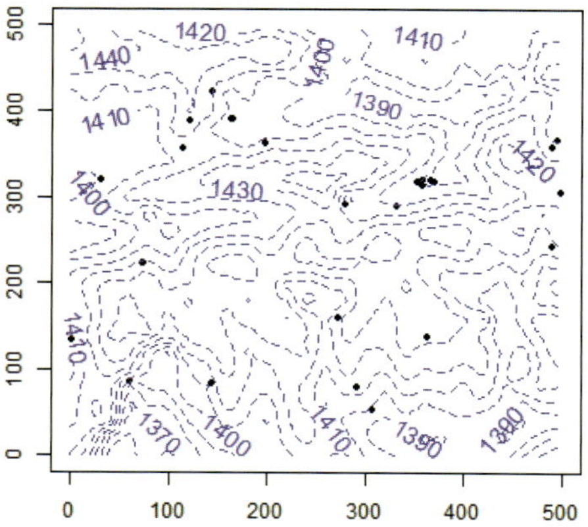

个体分布图 Distribution of individuals

径级分布表 DBH class

胸径区间 (Diameter class) (cm)	个体数 (No. of individuals in the plot)	比例 (Proportion) (%)
1～2	12	44.4
2～3	7	25.9
3～4	1	3.7
4～5	1	3.7
5～7	3	11.1
7～10	3	11.1
≥10	0	0.0

220 毛泡桐 máo pào tóng | Royal Paulownia

Paulownia tomentosa，泡桐属
玄参科 | Scrophulariaceae

代码（SpCode）= PAUTOM
个体数（Individual number/25 hm^2）= 2
最大胸径（Max DBH）= 51.50 cm
所在林层（Stratum）= 乔木层 (Canopy)
重要值排序（Importance value rank）= 43/50

落叶乔木。树皮灰褐色，枝上皮孔明显。叶卵状心形，先端锐尖全缘或波状浅裂。聚伞圆锥花序顶生，花冠紫色，漏斗状钟形至狭钟形。蒴果多，卵圆形。花期5月，果期6～8月。

树干 Trunk
摄影：卢志军 Photo by: Zhijun Lu

Deciduous trees. Bark grayish brown, lenticels conspicuous. Leaf blade ovate-cordate, apex acute, margin entire or shallowly lobed. Thyrses terminal, corolla purple, funnelform-campanulate to shallowly campanulate. Capsules numerous, ovoid. Fl. May, fr. Jun.–Aug..

果枝 Fruiting branch
摄影：卢志军 Photo by: Zhijun Lu

小枝和叶背 Branchlet and leaf abaxial surface
摄影：卢志军 Photo by: Zhijun Lu

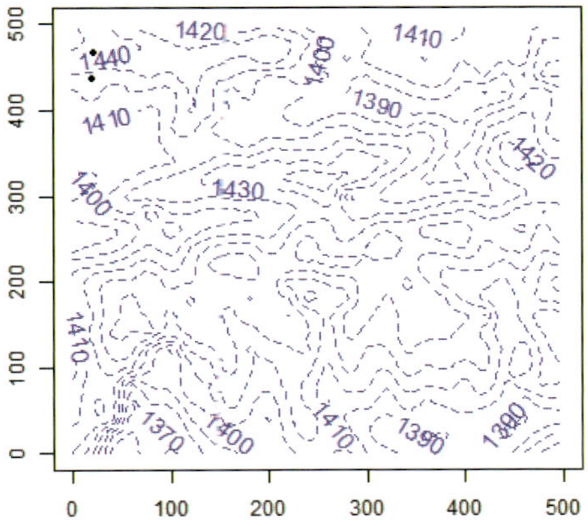

个体分布图 Distribution of individuals

径级分布表 DBH class

胸径区间 (Diameter class) (cm)	个体数 (No. of individuals in the plot)	比例 (Proportion) (%)
1～2.5	1	50.0
2.5～5	0	0.0
5～10	0	0.0
10～25	0	0.0
25～40	0	0.0
40～100	1	50.0
≥100	0	0.0

221 云南双盾木 yún nán shuāng dùn mù | Yunnan Dipelta

Dipelta yunnanensis，双盾木属
忍冬科 I Caprifoliaceae

代码（SpCode）= DIPYUN
个体数（Individual number/25 hm^2）= 7
最大胸径（Max DBH）= 14.26 cm
所在林层（Stratum）= 灌木层 (Shrub)
重要值排序（Importance value rank）= 82/128

落叶灌木。树皮片状剥落。幼枝被短柔毛。叶椭圆形至宽披针形，先端渐尖，基部宽楔形至近圆形，全缘，少数稍具齿。伞房状聚伞花序生于短枝顶部叶腋，花冠白色至粉红色，略带紫色，钟形。核果卵形，被柔毛，具宿存增大的小苞片。花期5～6月，果期5～10月。

Deciduous shrubs. Bark peeling in sheets. Young branchlets pubescent. Leaf blade elliptic to broadly lanceolate, apex acuminate, base broadly cuneate to subrounded, margin entire or rarely remotely dentate. Corymbose cyme axillary on short branch apex, corolla white to pink, purplish, campanulate. Drupes ovoid, pubescent, bracts persistent and enlarged. Fl. May–Jun., fr. May–Oct..

树干 Trunk
摄影：卢志军 Photo by: Zhijun Lu

果枝 Fruiting branch
摄影：卢志军 Photo by: Zhijun Lu

叶片 Leaves
摄影：卢志军 Photo by: Zhijun Lu

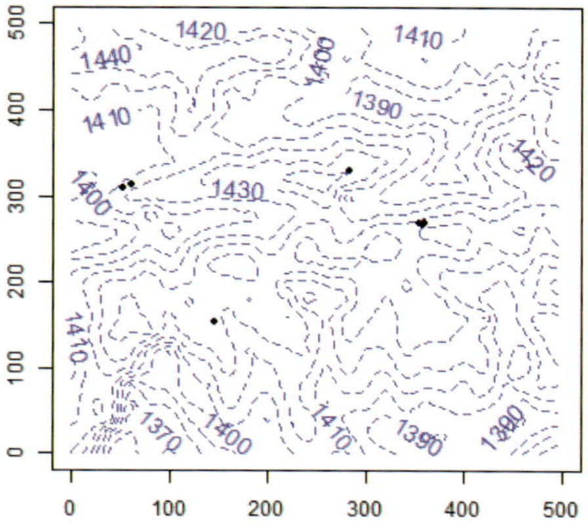

个体分布图 Distribution of individuals

径级分布表 DBH class

胸径区间 (Diameter class) (cm)	个体数 (No. of individuals in the plot)	比例 (Proportion) (%)
1～2	1	14.3
2～3	1	14.3
3～4	1	14.3
4～5	0	0.0
5～7	0	0.0
7～10	2	28.6
≥10	2	28.6

222 蕊被忍冬

ruǐ bèi rěn dōng | Bilabiate Honey-suckle

Lonicera gynochlamydea，忍冬属
忍冬科 | Caprifoliaceae

代码（SpCode）= LONGYN
个体数（Individual number/25 hm^2）= 60
最大胸径（Max DBH）= 14.59 cm
所在林层（Stratum）= 灌木层 (Shrub)
重要值排序（Importance value rank）= 63/128

落叶灌木。幼枝常带紫色，后变灰色，无毛。叶卵状披针形至线状披针形，先端长渐尖，基部圆形至楔形，叶缘具短硬毛。花序腋生，花冠白色，有红斑。浆果灰紫色至白色；种子黑色。花期5月，果期7～9月。

Deciduous shrubs. Young branchlets purplish, then gray, glabrous. Leaf blade ovate-lanceolate to linear lanceolate, apex long acuminate, base rounded to cuneate, margin with short stiff hairs. Inflorescences axillary, corolla white, with red flecks. Berries grayish purple to white. Seeds black. Fl. May, fr. Jul.–Sept..

树干 Trunk
摄影：卢志军 Photo by: Zhijun Lu

花枝 Flowering branches
摄影：刘海波 Photo by: Haibo Liu

小枝和叶片 Branchlets and leaves
摄影：卢志军 Photo by: Zhijun Lu

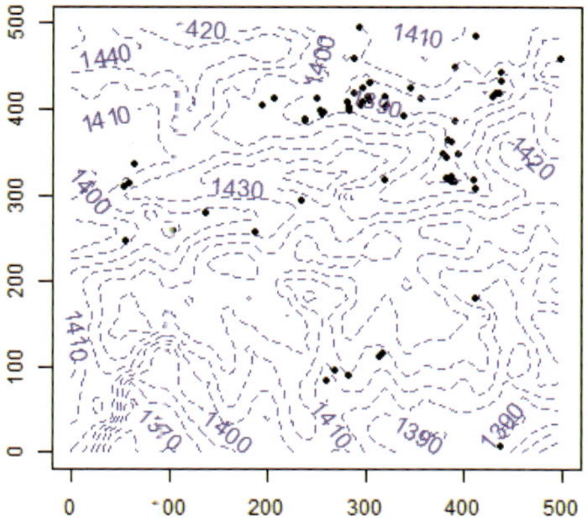

个体分布图 Distribution of individuals

径级分布表 DBH class

胸径区间 (Diameter class) (cm)	个体数 (No. of individuals in the plot)	比例 (Proportion) (%)
1～2	20	33.3
2～3	24	40.0
3～4	7	11.7
4～5	4	6.7
5～7	2	3.3
7～10	1	1.7
≥10	2	3.3

223 宜昌荚蒾

yí chāng jiá mí | Matsuda Viburnum

Viburnum erosum，荚蒾属
忍冬科 | Caprifoliaceae

代码（SpCode）= VIBERO
个体数（Individual number/25 hm^2）= 2318
最大胸径（Max DBH）= 9.02 cm
所在林层（Stratum）= 灌木层 (Shrub)
重要值排序（Importance value rank）= 18/128

落叶灌木。树皮浅棕色当年小枝连同芽、叶柄和花序均密被簇状短毛和长柔毛，二年生小枝带灰紫褐色，无毛。叶纸质，形状变化很大，卵状披针形、卵状矩圆形、狭卵形、椭圆形和倒卵形。复伞形聚伞花序，花冠白色，辐状。果实红色，宽卵圆形。花期4~5月，果熟期8~10月。

Deciduous shrubs. Bark light brown. Branchlets of current year, bud, petiole and inflorescence densely stellate-pubescent and mixed with long hairs; branchlets of previous year gray-brownish, glabrous. Leaves papery, blade greatly variated, ovate-lanceolate, ovate-elliptic, narrowly ovate, elliptic and ovate. Compound umbellate cymes, corolla white, radial. Fruits red, broadly ovoid. Fl. Apr.–May, fr. Aug.–Oct..

树干 Trunk
摄影：卢志军 Photo by: Zhijun Lu

花枝 Flowering branches
摄影：谷志容 Photo by: Zhirong Gu

果枝 Fruiting branches
摄影：江明喜 Photo by: Mingxi Jiang

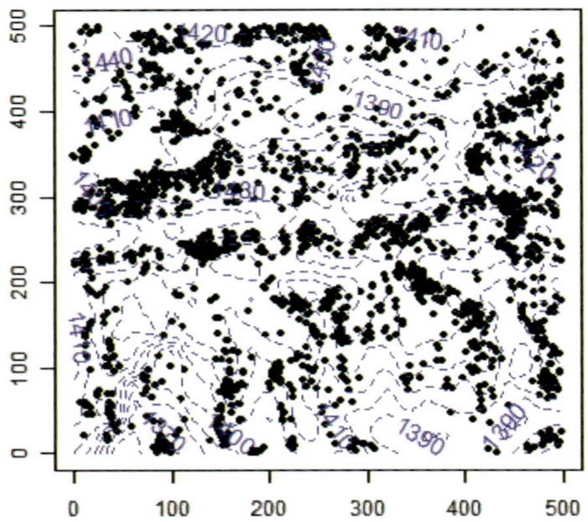

个体分布图 Distribution of individuals

径级分布表 DBH class

胸径区间 (Diameter class) (cm)	个体数 (No. of individuals in the plot)	比例 (Proportion) (%)
1~2	1192	51.4
2~3	666	28.7
3~4	265	11.4
4~5	117	5.0
5~7	64	2.8
7~10	14	0.6
≥10	0	0.0

224 桦叶荚蒾

huà yè jiá mí | Birch-leaf Viburnum

Viburnum betulifolium，荚蒾属
忍冬科 | Caprifoliaceae

代码（SpCode）= VIBBET
个体数（Individual number/25 hm^2）= 1262
最大胸径（Max DBH）= 27.39 cm
所在林层（Stratum）= 灌木层 (Shrub)
重要值排序（Importance value rank）= 27/128

落叶灌木或小乔木。树皮淡棕色。小枝紫色或深紫褐色或黑褐色。叶宽卵形至菱状卵形或宽倒卵形，少数椭圆状长圆形，先端急尖或短渐尖，基部宽楔形，少截形，边缘基部上半部有浅波状牙齿。复伞形聚伞花序顶生或生于1对叶的侧生短枝上，花冠白色，辐状，裂片圆卵形。核果红色。花期5～7月，果期6～11月。

Deciduous shrubs or small trees, deciduous. Bark light brown. Branchlets purple, dark brown or black-brown. Leaf blade broadly ovate to rhombic-ovate, or broadly obovate, rarely elliptic-oblong, apex acute or shortly acuminate, base cuneate, rarely truncate, margin base apically shallowly undulate. Compound umbel-like cymes terminal or on lateral short branches with 2 leaves, corolla white, radial, lobes ovate. Drupes red. Fl. May–Jul., fr. Jun.–Nov..

树干 Trunk
摄影：卢志军 Photo by: Zhijun Lu

花枝 Flowering branch
摄影：刘海波 Photo by: Haibo Liu

果枝 Fruiting branches
摄影：刘海波 Photo by: Haibo Liu

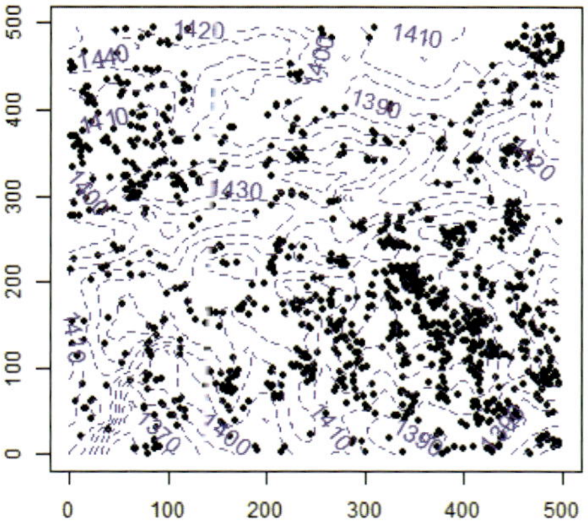

个体分布图 Distribution of individuals

径级分布表 DBH class

胸径区间 (Diameter class) (cm)	个体数 (No. of individuals in the plot)	比例 (Proportion) (%)
1～2	506	40.1
2～3	342	27.1
3～4	205	16.2
4～5	100	7.9
5～7	83	6.6
7～10	19	1.5
≥10	7	0.6

225 茶荚蒾 chá jiá mí | Tea Viburnum

Viburnum setigerum，荚蒾属
忍冬科 | Caprifoliaceae

代码（SpCode）= VIBSET
个体数（Individual number/25 hm^2）= 1021
最大胸径（Max DBH）= 15.02 cm
所在林层（Stratum）= 灌木层 (Shrub)
重要值排序（Importance value rank）= 29/128

落叶灌木或小乔木。幼枝无毛；二年生枝灰色、灰褐色或紫褐色。叶卵状长圆形至卵状披针形，先端渐尖或长渐尖，基部圆形，边缘在基部以上疏生尖锯齿。复伞形聚伞花序，花冠白色，辐状。核果红色，果序弯垂。花期3～4月，果期5～11月。

Deciduous shrubs or small trees. Young branchlets glabrous, 2-yr branchlets gray, grayish brown or purplish brown. Leaf blade ovate-oblong to ovate-lanceolate, apex acuminate or long acuminate, base rounded, margin sparsely serrate since base. Compound umbel-like cymes, corolla white, radial. Drupes red, pendulous. Fl. Mar.–Apr., fr. May–Nov..

树干 Trunk
摄影：卢志军 Photo by: Zhijun Lu

花枝 Flowering branches
摄影：江明喜 Photo by: Mingxi Jiang

果枝 Fruiting branch
摄影：刘海波 Photo by: Haibo Liu

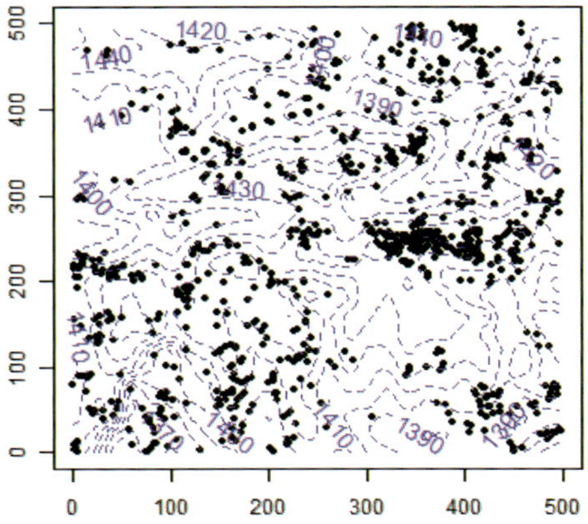

个体分布图 Distribution of individuals

径级分布表 DBH class

胸径区间 (Diameter class) (cm)	个体数 (No. of individuals in the plot)	比例 (Proportion) (%)
1～2	478	46.8
2～3	251	24.6
3～4	164	16.1
4～5	70	6.9
5～7	38	3.7
7～10	16	1.6
≥10	4	0.4

226 合轴荚蒾

hé zhóu jiá mí | Sympodial Viburnum

Viburnum sympodiale，荚蒾属
忍冬科 | Caprifoliaceae

代码（SpCode）= VIBSYM
个体数（Individual number/25 hm^2）= 402
最大胸径（Max DBH）= 13.8 cm
所在林层（Stratum）= 灌木层 (Shrub)
重要值排序（Importance value rank）= 36/128

落叶灌木或小乔木。当年生枝被黄褐色鳞片状或糠状簇状毛。叶纸质，卵形至椭圆状卵形或圆状卵形，先端渐尖或急尖，基部圆形，很少浅心形，边缘具不规则牙齿状尖锯齿。聚伞花序，周围有大型、白色的不孕花，花冠白色或带微红。果实红色，熟时紫黑色，卵圆形。花期4~5月，果期8~9月。

Deciduous shrubs or small trees, deciduous. Branchlets of current year yellow-brownish lepidote or furfuraceous stellate-pubescent. Leaves papery, blade ovate to elliptic-ovate or orbicular-ovate, apex acuminate or acute, base rounded, rarely shallowly cordate, margin irregularly serrate. Cyme with large white sterile flowers, corolla white or reddish. Fruit red, purple-black at maturity, ovoid. Fl. Apr.–May, fr. Aug.–Sept..

树干 Trunk
摄影：卢志军 Photo by: Zhijun Lu

果枝 Fruiting branch
摄影：卢志军 Photo by: Zhijun Lu

小枝和叶背 Branchlet and leaf abaxial surface
摄影：卢志军 Photo by: Zhijun Lu

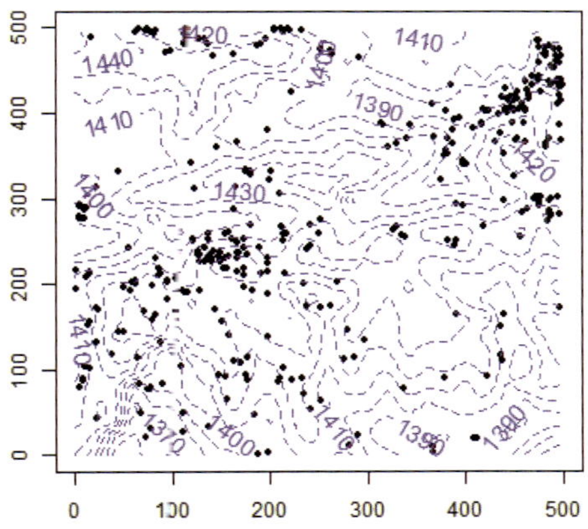

个体分布图 Distribution of individuals

径级分布表 DBH class

胸径区间 (Diameter class) (cm)	个体数 (No. of individuals in the plot)	比例 (Proportion) (%)
1~2	97	24.1
2~3	79	19.7
3~4	71	17.7
4~5	50	12.4
5~7	58	14.4
7~10	36	9.0
≥10	11	2.7

227 蝴蝶戏珠花 hú dié xì zhū huā | Tomentose Japanese Snow-bell

Viburnum plicatum var. *tomentosum*，荚蒾属
忍冬科 | Caprifoliaceae

代码（SpCode）= VIBPLI
个体数（Individual number/25 hm^2）= 140
最大胸径（Max DBH）= 18.1 cm
所在林层（Stratum）= 灌木层 (Shrub)
重要值排序（Importance value rank）= 50/128

落叶灌木。幼枝浅黄褐色；二年生枝灰褐色或灰黑色。叶较狭，长圆状卵形，有时椭圆状倒卵形或宽卵形，先端有时渐尖，基部宽楔形。花序周围有4～6朵大型的黄白色不孕花，中央的可孕花花冠白色至乳白色；雄蕊高出花冠。核果先红色，后变黑色。花期4～6月，果期5～9月。

Deciduous shrubs. Young branchlets light yellow-brown; 2-yr branchlets gray-brown or gray-black. Leaf blade narrow, oblong-ovate, sometimes elliptic-obovate or broadly ovate, apex sometimes acuminate, base broadly cuneate. Inflorescence surrounded by 4-6 yellow unfertile flowers; central fertile flower corolla white to milk white, stamen higher than corolla. Drupes red then black. Fl. Apr.–Jun., fr. May–Sept..

树干 Trunk
摄影：卢志军 Photo by: Zhijun Lu

花序 Inflorescence
摄影：徐耀粘 Photo by: Yaozhan Xu

果枝 Fruiting branches
摄影：江明喜 Photo by: Mingxi Jiang

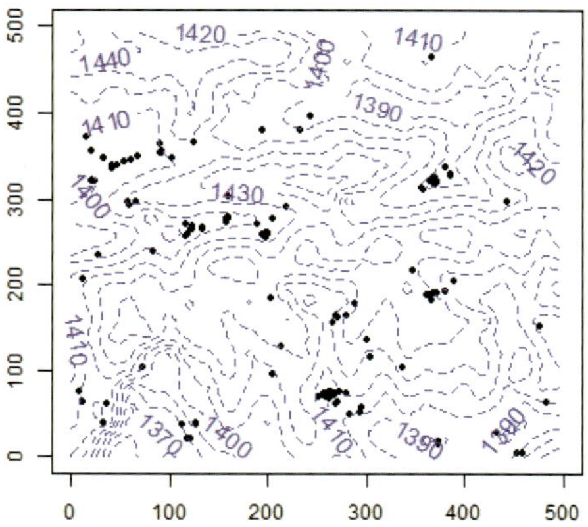

个体分布图 Distribution of individuals

径级分布表 DBH class

胸径区间 (Diameter class) (cm)	个体数 (No. of individuals in the plot)	比例 (Proportion) (%)
1～2	53	37.9
2～3	26	18.6
3～4	20	14.3
4～5	15	10.7
5～7	17	12.1
7～10	7	5.0
≥10	2	1.4

228 荚蒾 jiá mí | Linden Viburnum

Viburnum dilatatum，荚蒾属
忍冬科 | Caprifoliaceae

代码（SpCode）= VIBDIL
个体数（Individual number/25 hm^2）= 81
最大胸径（Max DBH）= 7.25 cm
所在林层（Stratum）= 灌木层 (Shrub)
重要值排序（Importance value rank）= 60/128

落叶灌木。幼枝、叶柄和花序均密被黄褐色或黄绿色小刚毛状粗毛及簇状短毛，二年生枝暗褐色。叶宽倒卵形、倒卵形或宽卵形，先端急尖，基部圆形至宽楔形，边缘基部上有粗锯齿，齿端突尖。复伞形聚伞花序稠密，生于有1对叶的短枝顶端；花冠白色，辐状。核果红色。花期4～6月，果期5～10月。

Deciduous shrubs. Young branchlets, petiole and inflorescence densely yellow-brown or yellow-green bristlelike hairy and stellate-pubescent, 2-yr branchlets dark brown. Leaf blade broadly obovate, obovate or broadly ovate, apex acute, base rounded to broadly cuneate, margin base coarsely serrate, with acute apex. Compound umbellate cymes dense, terminal on apex of short branch with one pair of leaves, corolla white, radial. Drupes red. Fl. Apr.–Jun., fr. May–Oct..

树干 Trunk
摄影：卢志军 Photo by: Zhijun Lu

花枝 Flowering branches
摄影：刘海波 Photo by: Haibo Liu

果枝 Fruiting branch
摄影：卢志军 Photo by: Zhijun Lu

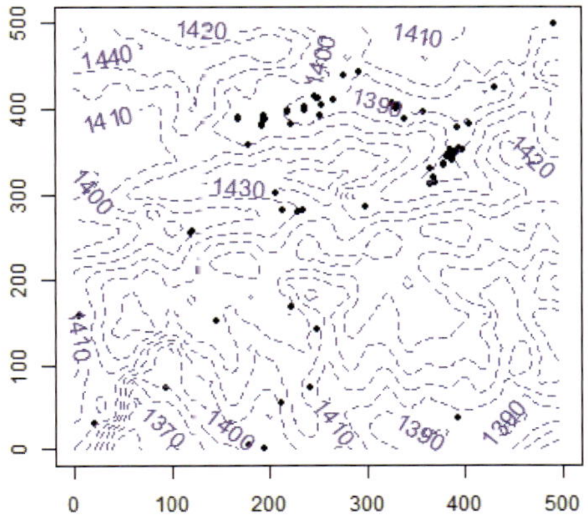

个体分布图 Distribution of individuals

径级分布表 DBH class

胸径区间 (Diameter class) (cm)	个体数 (No. of individuals in the plot)	比例 (Proportion) (%)
1～2	33	40.7
2～3	28	34.6
3～4	9	11.1
4～5	7	8.6
5～7	3	3.7
7～10	1	1.2
≥10	0	0.0

229 巴东荚蒾 bā dōng jiá mí | Henry Viburnum

Viburnum henryi，荚蒾属
忍冬科 | Caprifoliaceae

代码（SpCode）= VIBHEN
个体数（Individual number/25 hm^2）= 75
最大胸径（Max DBH）= 14.60 cm
所在林层（Stratum）= 灌木层 (Shrub)
重要值排序（Importance value rank）= 45/128

常绿灌木或小乔木。当年生枝带紫色，二年生枝灰褐色，无毛。叶倒卵状长圆形至长圆形或狭长圆形，先端渐尖，基部楔形至圆形，边缘在基部以上有浅锐锯齿，齿端常有硬凸头。聚伞花序圆锥状，顶生；花冠白色，有芳香，辐状。核果先红色，后变黑色。花期5～7月，果期6～10月。

Evergreen shrubs or small trees. Branchlets of current yr purplish, 2-yr branchlets gray-brown, glabrous. Leaf blade obovate-oblong to oblong or narrowly oblong, apex acuminate, base cuneate to rounded, margin shallowly serrate since base, apex usually stiffly tipped. Cymes panicle-like, terminal, corolla white, fragrant, radial. Drupes red then black. Fl. May–Jul., fr. Jun.–Oct..

树干 Trunk
摄影：卢志军 Photo by: Zhijun Lu

花枝 Flowering branches
摄影：刘海波 Photo by: Haibo Liu

果枝 Fruiting branches
摄影：徐耀粘 Photo by: Yaozhan Xu

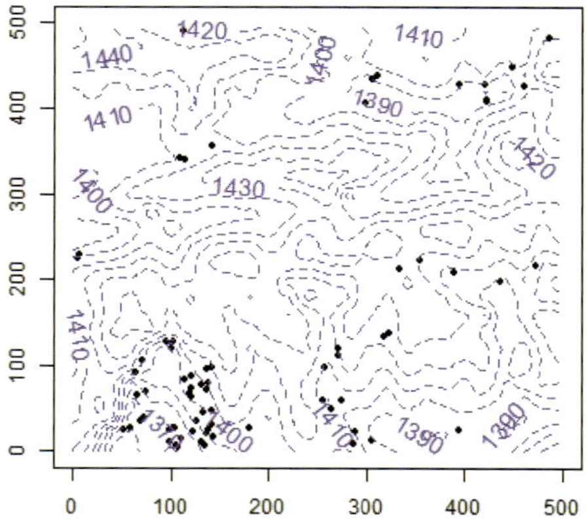

个体分布图 Distribution of individuals

径级分布表 DBH class

胸径区间 (Diameter class) (cm)	个体数 (No. of individuals in the plot)	比例 (Proportion) (%)
1～2	9	12.0
2～3	4	5.3
3～4	5	6.7
4～5	7	9.3
5～7	9	12.0
7～10	18	24.0
≥10	23	30.7

230 球核荚蒾 qiú hé jiá mí | Globose-pyrene Viburnum

Viburnum propinquum，荚蒾属
忍冬科 | Caprifoliaceae

代码（SpCode）= VIBPRO
个体数（Individual number/25 hm^2）= 11
最大胸径（Max DBH）= 6.31 cm
所在林层（Stratum）= 灌木层 (Shrub)
重要值排序（Importance value rank）= 95/128

常绿灌木。幼枝红褐色，二年生枝灰色，无毛。幼叶常紫色，后变革质，卵形至卵状披针形或椭圆形至椭圆状长圆形，先端渐尖，基部近圆形或宽楔形至楔形，两侧稍不对称，边缘通常疏生浅锯齿。聚伞花序无毛，花冠绿白色或浅黄色，辐状，有芳香。核果蓝黑色，有光泽。花期4～6月，果期6～10月。

Evergreen shrubs. Young branchlets red-brown, 2-yr branchlets gray, glabrous. Young leaves usually purple, then leathery, blade ovate to ovate-lanceolate or elliptic to elliptic-oblong, apex acuminate, base subrounded or broadly cuneate to cuneate, slightly asymmetric, margin usually sparsely shallowly serrate. Cymes glabrous, corolla greenish white or light yellow, radial, fragrant. Drupes blue-black, shiny. Fl. Apr.–Jun., fr. Jun.–Oct..

树干 Trunk
摄影：卢志军 Photo by: Zhijun Lu

叶片 Leaves
摄影：江明喜 Photo by: Mingxi Jiang

叶背 Leaf abaxial surface
摄影：卢志军 Photo by: Zhijun Lu

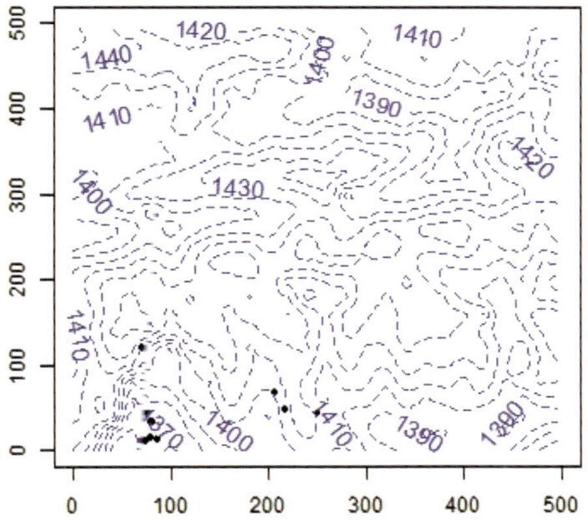

个体分布图 Distribution of individuals

径级分布表 DBH class

胸径区间 (Diameter class) (cm)	个体数 (No. of individuals in the plot)	比例 (Proportion) (%)
1～2	3	27.3
2～3	1	9.1
3～4	5	45.5
4～5	1	9.1
5～7	1	9.1
7～10	0	0.0
≥10	0	0.0

231 伞房荚蒾 sǎn fáng jiá mí | Corymb-flower Viburnum

Viburnum corymbiflorum，荚蒾属
忍冬科 | Caprifoliaceae

代码（SpCode）= VIBCOR
个体数（Individual number/25 hm^2）= 2
最大胸径（Max DBH）= 2.10 cm
所在林层（Stratum）= 灌木层 (Shrub)
重要值排序（Importance value rank）= 123/128

落叶灌木或小乔木。叶纸质，很少亚革质，矩圆形至矩圆状披针形，先端急尖，基部圆形至宽楔形，边缘除基部外疏生外弯的尖锯齿。伞房状圆锥花序，花冠白色。果实红色，椭圆形。花期4月，果期6～7月。

Deciduous shrubs or small trees. Leaves papery, very rarely subleathery, oblong or oblong-lanceolate, apex acute, base rounded to broadly cuneate, margin remotely serrate except at base. Corymb paniculate, corolla white. Fruit red, ellipsoid. Fl. Apr. fr. Jun.–Jul..

树干 Trunk
摄影：卢志军 Photo by: Zhijun Lu

叶片 Leaves
摄影：卢志军 Photo by: Zhijun Lu

叶背 Leaf abaxial surface
摄影：卢志军 Photo by: Zhijun Lu

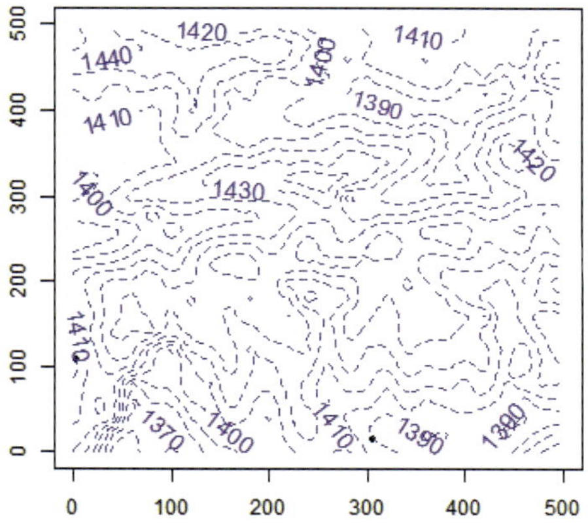

个体分布图 Distribution of individuals

径级分布表 DBH class

胸径区间 (Diameter class) (cm)	个体数 (No. of individuals in the plot)	比例 (Proportion) (%)
1～2	1	50.0
2～3	1	50.0
3～4	0	0.0
4～5	0	0.0
5～7	0	0.0
7～10	0	0.0
≥10	0	0.0

232 半边月

bàn biān yuè | Japanese Weigela

Weigela japonica，锦带花属
忍冬科 | Caprifoliaceae

代码（SpCode）= WEIJAP
个体数（Individual number/25 hm^2）= 600
最大胸径（Max DBH）= 22.11 cm
所在林层（Stratum）= 灌木层 (Shrub)
重要值排序（Importance value rank）= 22/128

落叶灌木。小枝光滑，稍呈四方形，有2列短柔毛。叶长椭圆形、矩圆形至倒卵状椭圆形，顶端渐尖，基部楔形至圆形，边缘具细锯齿。聚伞花序生于侧生短枝的叶腋，花冠初时白色，后变红色。果实光滑。花期4～5月，果期8～9月。

Deciduous shrubs. Branchlets glabrous, slightly square, 2-row pubescent. Leaf blade long elliptic, short rounded to obovate-elliptic, apex acuminate, base cuneate to rounded, margin serrulate. Cymes axillary on lateral short branch, corolla initially white, then red. Fruits smooth. Fl. Apr.–May, fr. Aug.–Sept..

树干　Trunk
摄影：卢志军　Photo by: Zhijun Lu

花枝　Flowering branch
摄影：刘海波　Photo by: Haibo Liu

花枝和叶背　Flowering branch and leaf abaxial surface
摄影：卢志军　Photo by: Zhijun Lu

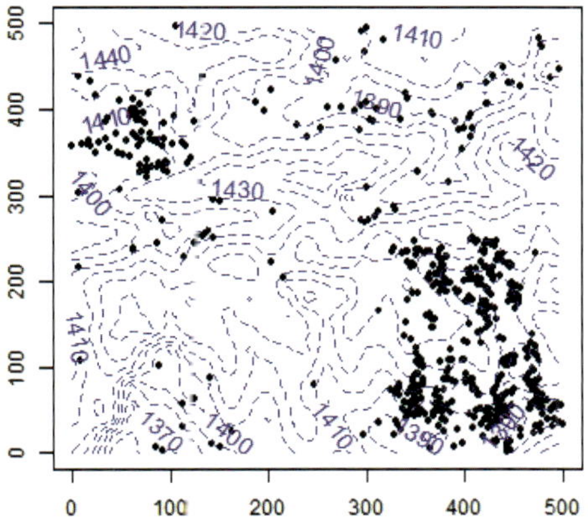

个体分布图 Distribution of individuals

径级分布表 DBH class

胸径区间 (Diameter class) (cm)	个体数 (No. of individuals in the plot)	比例 (Proportion) (%)
1～2	60	10.0
2～3	36	6.0
3～4	29	4.8
4～5	42	7.0
5～7	110	18.3
7～10	186	31.0
⩾10	137	22.8

附录I 植物中文名索引
Appendix I Chinese Species Name Index

附录II 植物拉丁名索引
Appendix II Scientific Species Name Index